Whole Earth
Field Guide

Whole Earth
Field Guide

edited by **Caroline Maniaque-Benton**
with **Meredith Gaglio**

The MIT Press
Cambridge, Massachusetts
London, England

This book is supported by a grant from the Graham Foundation for Advanced Studies in the Fine Arts and a grant from IPRAUS/École nationale supérieure d'architecture Paris-Belleville.

This book was set in Windsor, Helvetica Neue, and Warnock Pro by the MIT Press.

Library of Congress Cataloging-in-Publication Data

Names: Maniaque-Benton, Caroline, editor. | Gaglio, Meredith, editor.
Title: Whole Earth field guide / [compiled by] Caroline Maniaque-Benton with
    Meredith Gaglio.
Description: Cambridge, MA : The MIT Press, [2016] | Anthology of 80 texts
    selected from the Last whole Earth catalog (1971). | Includes
    bibliographical references and index.
Identifiers: LCCN 2016007200 | ISBN 9780262529280 (pbk. : alk. paper)
Subjects: LCSH: Green technology--Catalogs. | Manufactures--Catalogs. |
    Handicraft--Equipment and supplies--Catalogs. | Appropriate
    technology--Catalogs.
Classification: LCC TS199 .W485 2016 | DDC 628--dc23 LC record available at
http://lccn.loc.gov/2016007200

10 9 8 7 6 5

Contents

Preface

There is a large and growing literature on the *Whole Earth Catalog*, which was first published in California in 1968. Most of these studies have sought to situate the catalog in the context of the West Coast's counterculture or emerging digital culture. However, the *Whole Earth Catalog* extended far beyond the alternative sphere. Between 1968 and 1972, almost two million copies were sold (Random House distributed the 1971 edition), and an entire generation turned to the *Catalog* for advice and stimulation. This is because the books recommended for, or by, the readers included all kinds of practical information, ranging from where to buy seeds, tools, and clothing, to reflections on the universe, including philosophical, scientific, and geographical texts. In fact, countercultural publications are a distinct minority among the books that the *Catalog* recommended to its readers. The aim of this anthology is to introduce the reader to the intellectual world to which the *Catalog* opened a door.

The Last Whole Earth Catalog (*LWEC*) was published in 1971 and is considered the definitive edition. This is the version that won the National Book Award in the Contemporary Affairs category and the version we have used for the selection of texts and for the appendix that lists the books in the catalog. We decided to follow the structure of the *LWEC*, which meant organizing our selections according to the catalog's nine sections. We followed the order of their appearance in the catalog, always referring to the catalog's page number.

In selecting eighty from the almost one thousand items in the *LWEC*, we have tried to give an overall impression of the diversity of texts cited. We have included some well-known books alongside many ephemeral publications published before 1971.

The texts have been edited to eliminate references that cannot be explained within the space available and, where necessary, to preserve the flow of the argument. In the headnotes, citations of reviewers' comments accompanying the text come from the page of the *LWEC* for the item cited.

The introduction to this anthology does not explicitly address the wider political and cultural context of the *Catalog*, since this has been well covered in the existing literature. However, it is important to stress that the emergence of the *Catalog* in the context of American counterculture cannot be understood without consideration of the legacies of beat culture, the civil rights movement, the 1960s student protests at Berkeley and the free speech movement, the vehement anti–Vietnam War movement, the beginning of the women's liberation movement in 1968, and the increasing awareness of damage being done to the environment. By the time the *Whole Earth Catalog* was founded in 1968, a generation was turning against the values of what they called "the military-industrial complex," and many of the tools and skills that the *Catalog* presents can be seen as ways for individuals to find personal solutions to what was viewed increasingly as an oppressive system. The scope of the introduction is more modest, however: to uncover the origins of the *Whole Earth Catalog* as a publishing enterprise.

Acknowledgments

Giving birth to this book was quite an adventure. It began, in collaboration with Simon Sadler, as a collection of primary sources relating to countercultural architectural themes of the 1960s and 1970s. When Sadler decided to focus on the architectural issues, I decided instead to focus on the knowledge base of the *Whole Earth Catalog*.

My foremost thanks go to the copyright holders who have permitted us to reproduce and edit the texts included here. A heartfelt thanks also goes out to Stewart Brand, who gave permission to use *WEC* material. I am particularly grateful to Robert G. Trujillo and the Green Library Collections and University Archives at Stanford University for making the study of the *WEC* archives a pleasure.

The research laboratory IPRAUS/UMR Architecture, Urbanisme, Société: savoirs, enseignement, recherche—CNRS 3329 supported the project, providing much-needed funds to travel in the United States. This allowed me to work at the New York Public Library, Stanford University, and UC Berkeley. I would like to thank Estelle Thibault for her care in managing this process. A fellowship at the Center for Advanced Study in the Visual Arts at the National Gallery of Art in Washington also enabled me to work intensively at the Library of Congress. A grant from the Graham Foundation for Advanced Studies in the Fine Arts provided much-needed funds for the publication.

I am also grateful to Meredith Gaglio, who brought the project a fresh look that was well informed by her expertise on the subject. She made a number of the selections and intelligently contributed some of the section headings and headnotes. She also worked on the long and difficult process of copyright permissions. Many thanks also go to Jay Baldwin, Pedro Bandeira, Greg Castillo, Lionel Devlieger, Jean-Philippe Garric, André Guillerme, Jeremy McGowan, Cristiana Mazzoni, Joaquim Moreno, Joan Ockman, Antoine Picon, Sim Van der Ryn, Yannis Tsiomis, Simon Sadler, Lukasz Stanek, Marc Treib, Rebecca Williamson, and Jean-Louis Violeau for their helpful insights.

I am additionally grateful to Roger Conover, who believed in this project and was patient enough to see it through to completion. Justin Kehoe was also helpful in keeping things going. Wilma Wols provided much support for the introduction, which was translated by Tricia Meehan. The book owes a special debt to Tim Benton, who, in addition to his unflinching support, assisted in collecting information from the Stanford archives, as well as in compiling the list of books cited in the *WEC* (see the appendix).

Abbreviations

WEC: *Whole Earth Catalog*

LWEC: *The Last Whole Earth Catalog* (1971)

L(U)WEC: *The Last (Updated) Whole Earth Catalog* (1974)

SBP M1237: Stewart Brand Papers M1237, Special Collections, Stanford
University Libraries, Stanford, CA.

WEAP M1045: *Whole Earth Catalog* Archive, Special Collections, Stanford
University Libraries, Stanford, CA.

Note: All illustrations for the introduction come from the Stanford
University Library Archives, Special Collections SBP M1237 and WEAP
M1045, held in Stanford, CA, unless indicated below the image.

Introduction

The aim of this introduction is to shed light on material aspects of *The Last Whole Earth Catalog* (1971)—its modes of production and the behind-the-scenes debates—and to better understand the intentions of its protagonists.

After tracing the origins of the catalog and the ideas of its founder, Stewart Brand, I examine the production process, discussing some of the actions taken by Brand and his team. My research is based on the archives of the *Whole Earth Catalog and CoEvolution Quarterly* given by Stewart Brand to Stanford University.[1] One of the clarifications emerging from this research is that, far from being an improvised publication, the catalog was meticulously prepared and at the forefront of the latest editorial and typographic technologies. Even if the contents and language of the publication resemble in some ways that of so-called little magazines, the catalog's success came from its broad appeal. I finish with a brief review of the reception of the catalog, its relation to its readers, and some more recent publications.

The Last Whole Earth Catalog

In 1972, after receiving the National Book Award for Contemporary Affairs for the 447 pages of *The Last Whole Earth Catalog* (*LWEC*),[2] Stewart Brand thanked the jury and declared: "As you know, The *WEC* is less a book than it is a public event. And if the award encourages still more self-initiated, amateur, use-based, non–New York publishing, good deal."[3]

He also added that he would directly reinvest the prize money to cover the transportation costs for participating in the United Nations Conference on the Human Environment in Stockholm, where he proposed to organize a Life Forum on the margins of the official events. He announced that the $1,000 grant (equivalent to $5,755 in 2016) would finance the participation of "one Hopi Indian mad at what uninvited mega-technology does to native cultures, one scientist mad at what mega-warfare does to life in all its forms in Southeast Asia, one thoughtful poet, and one cheerful, skillful member of the Hog Farm commune."[4] He continued: "They and the rest of us are going to Stockholm to continue the action we had a piece of in the catalog: helping raise planetary consciousness, helping raise ecological good faith and good technique."[5]

Promoting environmental awareness by identifying a series of tools—books, techniques, know-how—and transforming this knowledge into action were the great challenges of the catalog. Edited at Menlo Park, near Palo Alto, where Stanford University is located, the catalog took the form of a magazine, slightly larger than tabloid format, 27.5 cm by 38.5 cm, illustrated in black and white. The seven divisions of its first sixty-four pages included "Understanding Whole Systems," "Shelter and Land Use," "Industry and Craft," "Communications," "Nomadics," and "Learning." The later versions had nine sections, with "Shelter and Land Use" and "Industry and Craft" being split into four categories. Each section was then divided into themes, often represented on a single spread. A reader or one of the *WEC* reviewers introduced each text or group of objects.

The Last Whole Earth Catalog

access to tools

$5

Updated August 1972

*Evening.
Thanks again.*

Figure 0.1
Cover of *The Last Whole Earth Catalog*. Stewart Brand, ed., *The Last Whole Earth Catalog* (Menlo Park, CA: Portola Institute, 1971). Courtesy of Stewart Brand and the Department of Special Collections and University Archives, Stanford University Libraries.

Brand published the first issue of the catalog in 1968; two more in 1969; one in 1970; and the *LWEC* in June 1971. In addition, he published four *Supplements* in 1969, four more in 1970, and two in 1971. Beginning in 1974, a new publication in magazine format, the *CoEvolution Quarterly* (1974–1984), offered feature articles following an editorial lead and was more comparable to a scholarly journal.[6] A new version of the *LWEC* reappeared in October 1974 under the title of *The Last (Updated) Whole Earth Catalog* and *The WEC Epilog*.[7]

The publication of 1971 is recognizable by its cover representing "the first American photograph of the 'Whole Earth,'" from November 1967, taken by NASA during the *Apollo 4* mission. The back cover is similar to the cover of the first autumn 1968 catalog.

From a six-page mimeographed newsletter presenting nearly 120 items for sale in the spring of 1968 to the award-winning version of 1971, the *Whole Earth Catalog* was a team endeavor. Under the leadership of Brand, about a dozen people managed this editorial enterprise, from fabricating the catalog to distributing books that the readers could order directly from the Whole Earth Truck Store.

Readers could use the *LWEC* in different ways. One could order the suggested books either from the publishers or from the Whole Earth Truck Store,[8] or one could consult the catalog for the mere pleasure of leafing through the extracts and illustrations. The rich and abundant information it contained satisfied several generations of readers, happy to browse its pages without resorting to a more in-depth consultation of the works.[9]

The commentary in 1972 by two journalists for the North American magazine *Contemporary Affairs* summarizes the *LWEC* well, comparing the catalog to Thoreau's *Walden* (1854): a philosophical and optimistic exploration of individualism (text 14). It also testifies to the material and intellectual interests of the 1960s:

This Space age WALDEN affirms the ability of man, the individual, to survive in a world of increasingly dangerous technology. Not only does this work provide access to practical tools of free-form-education, but also more pervasively, it explores an optimistic philosophy of individualism. In form and content the *LWEC* mirrors the contemporary era, as does no other work. It is a field guide for young and old to the archeology of the contemporary. Were this the only artifact of American culture 1972 left remaining 100 years from now, our survivors would be able to put it all back together.[10]

Stewart Brand's Background

To understand the origins and intentions of the *WEC*, we need to look in more depth at its founder. Born in 1938 in Rockford, Illinois, a city that specialized in making machine tools, Brand was the son of an advertising copywriter who was a radio enthusiast. His mother was fascinated by space exploration, as

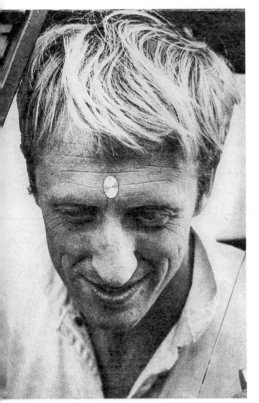

Figure 0.2
Portrait of Stewart Brand, ca. 1966. In Lawrence Dietz, "Middle Management Takes a Trip," *New York/World Journal Tribune*, January 29, 1967, 18. M1237, box 109. Courtesy of Stewart Brand and the Department of Special Collections and University Archives, Stanford University Libraries.

well as technology that opened up the potential of communicating with the universe. The media historian Fred Turner points out that these interests are astonishing for a middle-class midwestern family.[11]

Brand's diary and notebooks demonstrate the role that writing played for him as a means of reflection since his teenage years. The introspective character of these documents reveals a young man of nineteen with a strong inclination for writing, as well as hints of irony and self-mockery. On January 29, 1957, Brand noted in his diary a "capacity" for work that he found surprising:

For Stew Brand, a notoriously lazy, procrastinate, and ineffectual individual, Jan. 29, '57 marks a departure. I sweated like a steam pipe today. … The day is extraordinary for two reasons: 1) I accomplished everything I set out to do in the day, even homework … 2) there was at no one place a breathing spell of more than 5 minutes; I rushed for 15 hours. The day went fast, for I never stopped to contemplate it. Yet I am not nervous, hurried, or fatigued. Perhaps there is something in this infamous American way of life. If I were trying to forget or recover from something—say an unfortunate love affair—this is the ideal manner, for during the entirety of the day I neither reflected nor thought. I merely reacted.[12]

Here we see the first signs of an entrepreneur and organizer. As a teenager, Brand attended Phillips Exeter Academy, a renowned private boarding school in New Hampshire, fifty-three miles north of Boston.[13] The school is well known for having developed an interactive teaching pedagogy, which took place around an oval table.[14] The teacher and his group of students worked together, exchanging ideas and information, similar to the Socratic method used in some universities. A teaching slogan at Exeter was "Ideas must lead to action."

From 1958 to 1960, Brand studied biology at Stanford University. He was passionate about the cybernetic biology of Paul Ehrlich, a young professor who investigated "the complexity of relationships in the functions of human beings, the processes according to which human beings receive information, translate it according to their own energy, to transmit it to their principal organs."[15] Ehrlich later won acclaim for his book *The Population Bomb*, published in 1968 (text 11).[16] After his graduation, in 1961, Brand trained as an officer in the army reserve.

The Army: A Leadership Model

Historians today are skeptical about the influence of Brand's army training on his personality, seeing it as a paradox in relationship to his countercultural ambitions.[17] Brand, however, acknowledges the importance of this general training, explaining that it taught him the skills of organization and leadership.[18] It certainly allowed him to become aware of organizational theories: "Man's long experience in warfare has taught military leaders that the following four basic principles of military organization must be observed in all operations, tactical or non tactical, in peace as well as in war: unity in command; chain of command; functional specialization; and functional grouping."[19]

We now know that "the functioning of post-war American science, as military as it might be, is not organized on a vertical, hierarchical and rationalizing model."[20] New collaborative practices multiplied, but Brand underscored the importance of leadership: "[The army] gave me the thing that's almost impossible to get anywhere else, which is leadership training. It turns out leadership is a skill....I was taught that it's no good giving an order if you don't explain why the thing you want to have happen should happen."[21]

The army, in fact, gave him time to reflect, as it was a meditational interlude away from the preoccupations of daily life: "Both Jerry Brown and I believe in doing time....He did it in the seminary. Ken Kesey, my friend, did it in jail. I did it in the Army as an infantry officer."[22]

A group leader, quick to react, Brand knew how to mobilize the skills required to be an officer: mastering the art of leadership and knowing how to bring people together.[23] Sure of his intelligence, his ability, and his critical capacities, he had a somewhat curious vision of himself: "I was taught that there are four kinds of officers—stupid and industrious, stupid and lazy, smart and industrious and smart and lazy. And the worst of these is stupid and industrious and the best of these is smart and lazy. And I've always tried to live up to that."[24]

The army experience may have also influenced Brand's attitude toward a certain kind of elite, capable of transforming society. In a text written in 1975, Brand explained that the elite corresponds not necessarily to those who earn a good living or have studied at the best universities, but rather to those who are capable of thinking holistically:

One of the driving forces of finely textured life style is elites—in what I claim is a good sense. An elite is a number of people defining for each other by their actions what good is. It is the way humans use each other to surpass themselves. In your case it is the people you keep an eye on and who keep an eye on you, and whose esteem means more than money. Besides their values to the members, elites are a potent force for excellence in the community at large. The more the better that any ill effects from their strengths will offset each other.[25]

Writing Material

Jumping from an airplane is a unique experience, almost surreal—one that Brand used as material and inspiration for writing. Three typescript drafts of a text about his first jump are conserved in his archives and reveal a taste for transmitting the lived experience, as well as his efforts to refine his writing. He recalled: "I made my first jump as an Army lieutenant in the Airborne School at Fort Benning, Georgia. I have a record of what happened because I wrote it down directly afterward, but all I really remember is the strangeness felt in the plane, the terrible anticipation, and, of all things, the gaiety."[26]

He remarked on this gaiety in another draft:

Jumping out of an airplane is simply unreal. And the anticipation was due largely to the four weeks of intensive training that preceded being able to sit there in the web seat of a C-123....But the gaiety...I am a somber person. I since recall myself whispering to the jumper next to me, "Did you know your chute has no static line? Haw haw haw." I smiled climbing into the plane, smiled when we crowded into our seats, when the motors started, when the long taxiing began, when we took off....I smiled going out the door.[27]

The euphoria he felt, the bliss and heightened awareness he conveyed in describing these phenomena, evoke a psychedelic experience.[28] A remark made during the Alloy conference (text 31) emphasizes this aspect: "You can get flashes all kinds of ways. I got a flash once when my parachute didn't open."[29]

Another fundamental experience had to do with changes in perception, which he identified when noting the difference between what one sees on the level of the street and what one sees from the sky: "Hideous as New Jersey is on the ground, from the air it looks well. You see clouds on their own level, the variety of coast toward the east, and, below, an expanse of deep green crisscrossed with manmade lines. Recognizable things like blimp hangars are pitifully small."[30] Brand's fascination with the view from the air was reflected in some of the choices of books for the catalog (text 3).

It also may have influenced Brand's choices in page layout, as he often evaluated the catalog by taking a bird's-eye view and looking at it from above. He adjusted some of the double-page spreads—especially in the "Understanding Whole Systems" section—as if they were landscapes seen from a plane, where the ratio transforms perception.

Another influence on Brand was the technique of debriefing. As he noted: "There's an after-action report, debriefing tradition in the military that actually I haven't seen much of in corporations."[31] He would take up this technique for the postevent evaluation widely used in making the *WEC*.[32] It became a common practice in corporate group management in the 1960s and 1970s.[33]

Brand also became army photographer at Fort Benning in Georgia,[34] then at Fort Dix in New Jersey, and briefly at the Pentagon. We'll look more closely at the importance of photography to Brand's work later.

Two Affiliations: USCO and Merry Pranksters

In New York, in 1962, Brand became involved with Steve Durkee and Gerd Stern's USCO group, with whom he organized psychedelic multimedia happenings. USCO, a contraction of "Company of Us," was a collective of artists, poets, and engineers, based, from 1966, in an abandoned church near New York.[35] The performances of the USCO group would lead to two major modifications in artistic events in the tradition of John Cage, Robert Rauschenberg, and Allan Kaprow. As Fred Turner wrote, "they not only aspired that their public become more aware of the environment, but they also incited them to personally perceive themselves as members of a mystical community."[36] However, USCO used everyday materials, as well as new electronic communication tools, to facilitate this understanding.[37]

It was not long before Brand moved on to San Francisco: "I was looking for San Francisco beauty, San Francisco people, San Francisco happiness— the bohemian style.... Therefore, resolved—to go posh. To frequent the theaters, music halls, galleries, and homes not as an interloper taking all he can learn, but as a learning participant."[38] Brand then studied design at the San Francisco Art Institute and photography at San Francisco State College.

Brand organized the Trips Festival from January 21 to 23, 1966, at San Francisco's Longshoremen's Hall. The event marked the encounter between the psychedelic and multimedia scenes and, in particular, allowed Brand to realize that organizing events could contribute to "changing the world" and that it really wasn't so difficult after all.[39] The flyers, designed by Wes Wilson, claimed: "The Trip—or electronic performance—is a new medium of communication and entertainment." With the Merry Pranksters, Brand participated in San Francisco's Summer of Love, the 1967 youth gathering celebrating drugs, mystical forces, and strobe technology. Like many intellectuals of his generation, Brand studied the works of Norbert Wiener, Marshall McLuhan, and Buckminster Fuller and began to imagine a synthesis between cybernetic theory and countercultural politics. He was realizing that communication could be a means by which to shape the structure of minds.

Figure 0.3
"A Sky of Parachutes" (photographer unknown). Headquarters and Service Co., *The Airborne Classbook*, graduation of February 10, 1961. Courtesy of Stewart Brand and the Department of Special Collections and University Archives, Stanford University Libraries.

Stewart Brand and Photography

If today Stewart Brand is viewed mainly as a writer,[40] in 1962 he saw himself as a photographer. In 1963, he began a new research venture into Native American societies. Inspired both by the transcendentalism of Emerson and by Ansel Adams's photographs of the American West, Brand went to live among the Indians of Warm Springs, a reservation in Oregon, and began to photograph them: "Wild horses were difficult but quite a learning experience. The pictures are good, but are available at present only to the Indians."[41] He was especially fascinated by the attitudes of the American Indians toward territory and nature.[42]

He deepened his knowledge of the American Indians by exploring the archival collections of the Museum of Natural History, as well as films and sound recordings at the Library of Congress. He also consulted the Wittick Collection at the New Mexico History Museum in Santa Fe.[43] This collection was built around the photographs of George Ben Wittick, who began to photograph New Mexico's landscapes and inhabitants while he was documenting the construction of the railway.

Brand drew on this intensive research to propose a multimedia exhibition, America Needs Indians,[44] to the National Museum of Natural History, using slideshows, three films, and two sound recordings. He also organized dance performances. Charles Perry commented that "the effect was an inundation of Indian faces, images of Nature, Indian music, everything that America had been before the Europeans came."[45] These multimedia projections were meant to create a thrilling spectacle for the public. The experience gave Brand important lessons in organization, as well as the ability to convince and engage an audience.[46] Once again, he tested his ability to bring together a group and put in place the principles of military organization that he had learned: "Very early, men discovered that they could accomplish more through their joint cooperative efforts than they could by working separately."[47]

Only a few of Brand's photos are available in his archives. But the archives do include a framed photo that he gave his parents; it is the one that he used for the first cover of the *Difficult but Possible Supplement to the Whole Earth Catalog*, of September 1969 (see the image on the next page). The photograph depicts a lunar halo over the ocean. The only illuminated part, the zigzag road, is characteristic of the winding coastal highway bordering the Pacific Ocean, twenty miles or so north of San Francisco.

Presumably this particular photo was important. What was he trying to create? In a notebook dated summer 1963, he specified the role of photography for him: "I want to reveal, use photographs for revelation."[48] He simultaneously sought strength and simplicity. He was aware that his goal was to awaken the viewer: "In my mind at present is, whatever draws powerful response is right, anything to wake the viewer up, get him reacting, involved....I see I'm working toward strength, toward the effort to really stand the viewer's hair up with an image that doesn't mess around. That goes straight."[49]

In his notebook, he commented on reading images: "All these notions about reading photographs are left over from photojournalism....Once you've woken the guy up, and let him see something, he'll come back the more readily the next time, and will esteem you as friend and teacher."[50] He was extremely interested in images; the works of Marshall McLuhan analyzing the persuasive power of photographs had not escaped him.

It was Brand's particularly skillful use of symbols that allowed radical ideas to reach a much wider audience. The cover of the March 1970 *WEC* supplement, edited by Gurney Norman with Diana Shugart, showed a multiethnic

Figure 0.4 (opposite)
Stewart Brand and Zach Stewart, detail of the scenographic project for America Needs Indians, an exhibition at the National Museum of Natural History, New York, 1963. The drawing is a version for the Museum of Anthropology, University of California, Berkeley, May 18, 1963. M1237, box 109. Courtesy of Stewart Brand and the Department of Special Collections and University Archives, Stanford University Libraries.

group of young people playing volleyball. The ball was substituted by his trademark image of the globe, a potent metaphor for the idea that saving the planet depends on all of us.

His attention to image comes through, as we will see later, in his choices of recommended books and simple and powerful images. In 1971 Brand discussed this in relationship to the book *Geology Illustrated*:

As a means of communicating geological concepts, the pictures are fully important as the words that accompany them. On most pages the photographs represent the facts, the words supply the interpretation. Many of the illustrations will, therefore, repay a little of the kind of attention that would be accorded the real feature in the field. In keeping with this, almost no identifying marks have been placed on the photographs and very few on the drawings. The text (which almost invariably concerns an illustration on the same or a facing page) serves as an expanded legend.[51]

In his book *From Counterculture to Cyberculture*, Fred Turner addressed Brand's recent career, placing him in the context of various groups and engagements.[52] Turner was particularly concerned with situating Brand's innovations in the developing history of digital culture in California. Other authors have focused on different aspects of Brand's subsequent career.[53]

MARCH
Whole Earth Catalog $1

THE WORLD GAME

"I travel around the world a great deal, and everywhere I hear
humanity saying, 'We are not against any other human beings;
we feel the world ought to work properly.' Everywhere they
say it's our politicians that get us into trouble. This is the
majority viewpoint all around the earth today."
—R. Buckminster Fuller
See page 30

Figure 0.7
Cover of "The World Game" supplement.
Stewart Brand, Ann Helmuth, Joe
Bronner, Tom Duckworth, Lois Brand,
and Hal Hershey, eds., "The World Game"
supplement, *Whole Earth Catalog*
(Menlo Park, CA: Portola Institute, March
1970). Courtesy of Stewart Brand and the
Department of Special Collections and
University Archives, Stanford University
Libraries.

Preparing for the World of Publishing

Although Brand had acquired some publication skills through organizing various events in San Francisco, by the late 1960s he looked for more professional
advice.[54] In 1968 he met a number of important personalities in the field of
publishing. He did not seek out small-scale publishers but rather approached
the most highly recognized people in the field.

L. W. "Bill" Lane, co-owner and publisher of *Sunset* magazine,[55] whose
headquarters were also at Menlo Park,[56] initiated Brand into his working
method: plan the articles three to four months in advance, set up a table distributing the tasks, and empower each contributor. Lane also suggested that
Brand invite the readers to participate; *Sunset*, for example, had readers contribute recipes. We will see that Brand took good note of all of this, and the
supplements to the *Whole Earth Catalog* would be in a way a testimony to
this interactive effort.

Lane also advised Brand to take his own photographs, probably to master
the visual content as much as the text; it would also reduce costs and delays.
Lane recommended using the top-quality Hasselblad camera, the same model
used to photograph the first moon landing on July 20, 1969.

Lane added: "Must keep your mag evolving and swinging—or you're dying.
Take a risk every month—Surprise your reader, but stay in character. Introduce
change slowly, play by ear."[57] Lane suggested that it would be equally good

Sunset

Offices in Menlo, L.A. Seattle, & Honolulu.

Article planning begins 3–4 months ahead

Set up & works on a map (dummy)

Each writer largely responsible for his own story. Several issues going at a time.
Everyone secondarily a travel writer. Do buy addition some free-lance material.

Food much reader-edited — contribute recipes, etc.
100% staff photographed

Gimmick on photographing fruit — Throw in a deep-freeze. Take out, & it thaws thru a lovely stage — "devine dewy-fresh".

Shoot with a Hasselblad.

Garden gives & biggest differences in & 3 issues.

"Everything we right about we give some action that people can do on it." And give them accurate guidance.

Call Tibor.

WHOLE EARTH TRUCK STORE AND CATALOG

PORTOLA INSTITUTE

1115 MERRILL STREET • MENLO PARK, CALIFORNIA 94025 • PHONE 415 323-5155

MESSAGE   REPLY

TO

DATE   22 Nov 68

RECEIVED
DEC -3 1968
R. BUCKMINSTER FULLER

Dear Mr. Fuller

Thank you for helping inspire this Catalog into existence.

Are there tools you would suggest for listing?

Are there categories of tool you would add to expand the usefullness of the Catalog?

Stewart Brand

DATE   December 8, 1968

Thanks very much for sending me the "Whole Earth Catalogue." And, congratulations! It is truly magnificent.

I am sending you, under separate cover, Volume 5 of the Design Science Decade. I would like to call your attention particularly to my letter to Dr. Constantin Doxiadis published therein. You may find other items of interest in this volume too.

Faithfully yours,

R. Buckminster Fuller

BY   SIGNED

Figure 0.8 (top)
Detail of Steward Brand's notes from his meeting with L. W. "Bill" Lane, co-owner and publisher of *Sunset* magazine, 1968. "Gimmick on photographing fruit—Throw in the deep-freezer." "Shoot with a Hasselblad." "Everything we write about we give some action that people can do on it. And give them accurate substance." M1237, box 26. Courtesy of Stewart Brand and the Department of Special Collections and University Archives, Stanford University Libraries.

Figure 0.9 (bottom)
Stewart Brand, letter to Buckminster Fuller, November 22, 1968, and Fuller's response. M1237, box 6. Courtesy of Stewart Brand and the Department of Special Collections and University Archives, Stanford University Libraries.

to establish a well-targeted and thoughtful questionnaire: "Questionnaires should not be vague or be sent to inexpert people (dilutes your reply)."[58] Again, Brand followed this advice; he used a questionnaire to ask selected contributors about their reading habits.[59]

The smallest details of the conversation with Lane at the *Sunset* headquarters were hastily jotted down in a yellow-paged notebook: "'Get full scoop and caption any picture'; 'Take an outstanding caption and hold the reader'; 'Comment on every item and interest in the picture'; 'Short as possible, but rich.'"[60] Here again Brand would follow these tips meticulously. The strength of the *WEC* lay precisely in capturing the readers' imagination by one or two insightful phrases, sometimes offset, thus provoking surprise. One thing is certain, affirmed his interlocutor: "A big effort always spills over into other articles."[61]

Brand likewise met with William Flynn, managing editor at the San Francisco magazine *Newsweek*, in early 1968. Flynn advised Brand on another editorial strategy: "Use all the quotations you can. You should have shorthand for note taking. Interviewing—some techniques. Be silent—look."[62] Still other advice included "Must have pictures with all stories. ... Get the best conversations on the truck or plane (and keep pad and pencil out of sight)."[63]

Other tips included: know your subject well before the interview and use your networks of friends to obtain good articles or subjects for articles; adopt a provocative attitude that will allow things to move, which is to say, disturb preconceptions. To be taken seriously by readers, he received a last bit of advice: "Always check what other people say. Maintain your objectivity."[64]

Brand completed his study of the publishing world by reading and taking notes from a book published in 1964 by Theodore Bernard Peterson, *Magazines in the Twentieth Century*, based on Peterson's doctoral dissertation. The book gave an inventory of numerous magazines, describing their characteristics, target readership, and number of copies distributed. It included a number of periodicals specializing in political information, such as *Newsweek* and *Time*; magazines in the world of business, such as *Fortune* and *Business Week*; and magazines that catered specifically to women, such as *Ladies' Home Journal* and *Good Housekeeping.* There were also "Shelter & Home mags" like *Better Homes and Gardens*, with 4.5 million readers, as well as *House and Garden* and *House Beautiful*, the last two being "more sophisticated." With Peterson's book, Brand could survey the magazine press to target more strategically the audience of his own intended publication.

Figure 0.10
"Trips Festival." Photo by Susan Elting. Warren Hinckle, "A Social History of the Hippies Trips Festival," *Ramparts Magazine*, March 1967, 8. M1237, box 109. Courtesy of Stewart Brand and the Department of Special Collections and University Archives, Stanford University Libraries.

During a meeting with Lane at the *Sunset* headquarters, some leaders of large-circulation periodicals were identified for Brand. They included Edward Weeks, former editor in chief of the *Atlantic* (Boston); Russell Lynes, former managing editor of *Harper's Magazine* (New York);[65] and certain key people at the *New York Times Magazine*. Brand decided to use proven techniques of mass communication, without resorting to the modest techniques of underground publications such as "wood-block printing, potato stamping or stenciling, [and] mimeography, offset lithography and multicolor, silk-screen printing."[66] In fact, as the architectural historian Beatriz Colomina has pointed out, some of the large-circulation professional journals modified their page layout and typography to conform to the libertarian spirit of the underground magazines. She cites in particular the example of *Casabella*, directed by Alessandro Mendini from 1970 to 1976, or *Architectural Design* under the leadership of art editor Peter Murray.[67]

Six years later, once the *Whole Earth Catalog* adventure was up and running, these contacts would prove fruitful. In 1974, Tony Jones, associate editor of *Harper's Magazine*, invited Brand to edit a sixteen-page section called *WRAPAROUND* with a print run of 37,000 copies. An editorial statement in *Harper's* explained the thinking behind the invitation: "The section relies for its vitality on the willingness of its readers to share their experiences with one another. It is this sense of a family tie between the two enterprises that brought Brand and *WRAPAROUND* together."[68] Jones wrote a letter

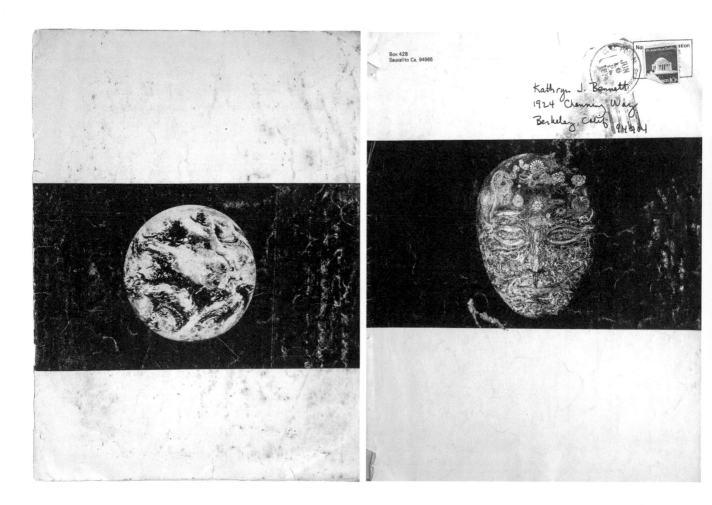

demonstrating *Harper's* interest in Brand's undertaking and indicating the reciprocal advantages that each party could draw from such a collaboration, and he invited him to New York to discuss the opportunity:

I would hate to feel that proceeding with this project as a normal article, with the usual relationship between author and magazine, was locking us out of a fuller kind of collaboration. It seems to me we are both involved with very similar kinds of editorial process where the principle of reader feedback and participation structures our thinking. Given that commonality of interest—and our different capacities, experiences, and tools—there seems to be a great deal of room to explore a partnership that would work to everyone's advantage. Certainly there's the possibility of creating an unconventional linkage between different forms of publication, since it's not hard to conceive a *WRAPAROUND* and the *Epilog* [the *Whole Earth Epilog* contained 320 pages] feeding into each other. Obviously we will have to become better acquainted with each other before we can give form to such airy possibilities. But I think *WRAPAROUND* demonstrates *Harper's* seriousness about moving in new directions, and our sense of common enterprise might lead us toward some interesting experiments.[69]

Once *Harper's* published the two eight-section *WRAPAROUND*s guest edited by Brand, he assessed the experience: "We're grateful to *Harper's* for the considerable freedom we've been given in a careful magazine. Guest editorship is always a kind of trespass, hopefully worth it. In *Catalog* tradition we should mention some numbers—*Harper's* paid Point approximately $4,200 [$21,500 in 2016] for preparation and production of the 'Wraparound' in April 1974; the print run was around 350,000 copies."[70]

If this was good business, as it was without a doubt for *Harper's Magazine*, the big advantage for Brand was its distribution throughout the entire United States, allowing him to reach a new audience, especially on the East Coast. An additional 38,000 copies of the section were printed for his benefit, which he distributed to the *WEC* subscribers, as well as to his contacts, helping him build up his professional network.

For the additional copies of *WRAPAROUND*, sent to *WEC* readers as a promotional document, image again played a major role. The sixteen-page *WRAPAROUND* was promoted with a flyer divided into thirds, with two images contained in the center band, presented against a black background; the image on the front cover was of a mask representing a human face, and the one on the back was a photograph of the earth. This established the human–nature comparison through a homology between the human face and the globe, which were the same size.

The Founding of the *Whole Earth Catalog*

In the columns of the *LWEC*, Brand recounted in 1971 that the idea for the catalog came to him in March 1968 while he was flying back from his father's funeral. Reading *Spaceship Earth* by Barbara Ward[71] and looking out at the magnificent landscape through the airplane window, he wondered if he could somehow make life easier for his friends who had decided to live a communal life away from cities.

Brand repeated this story about the founding of the *WEC* on various occasions. In fact, he had already recalled the anecdote in a letter to a potential sponsor in 1969, adding: "One thing that might be of interest to Foundations and other supporters of things worthy; If I had to go through the pure hell of

Figures 0.11 and 0.12
Front and back covers of flyer to promote *WRAPAROUND*, *Harper's Magazine*, April 1974. M1045, box 35. Courtesy of Stewart Brand and the Department of Special Collections and University Archives, Stanford University Libraries.

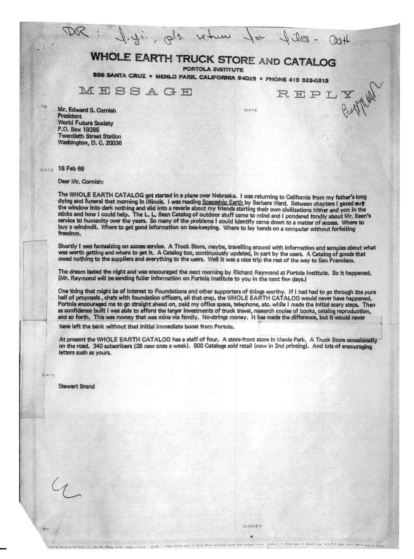

Figure 0.13
Stewart Brand, letter to Edward S. Cornish, president of the World Future Society, February 18, 1969. M1237, box 6, fol. 5. Courtesy of Stewart Brand and the Department of Special Collections and University Archives, Stanford University Libraries.

proposals, chats with foundation officers, all that crap, the *WEC* would never have happened."[72]

Brand's idea of a catalog of tools was prompted by mail-order catalogs like L. L. Bean, which specialized notably in outdoor sporting goods, and others, such as the long-running Sears and Roebuck mail-order catalog.[73] It is in volumes like these, "comparable to telephone directories, that it is necessary to look for...the sources of the *Whole Earth Catalog*."[74]

Inspired by Buckminster Fuller, who had said that an idea remains in your head only ten minutes or so, Brand quickly jotted down his thoughts one day and on the next morning hastened to meet Dick Raymond. Raymond, the head of the Portola Institute, a nonprofit organization to which Brand contributed, was dedicated to the development of knowledge outside university institutions and was encouraging any initiative of this sort.[75] Facilitating the distribution of printed works through an itinerant shop, the Truck Store—a 1962 red Dodge pickup truck that Brand and his wife Lois would use to visit the communities created far from cities—seemed to him to be a good idea. As for Brand, he was ready to invest several thousand dollars that his family had bequeathed to him to hire staff.

Money

In 1971 Brand was not afraid to come clean about the financial details of the *WEC*, including his own personal investment in it. At the end of the *LWEC*, he published a frank statement about the financial sources of the undertaking:

You may or may not think that capitalism is nice, and I don't know if it's nice. But we should both know that the *Whole Earth Catalog* is made of it. Capital was invested by my parents and parents' parents' parents in such activities as iron-mining in Minnesota and Eastman Kodak. They paid nicely enough, and by family attentiveness-to-business and flat-out parental generosity, I wound up with a bundle of money without having done a lick of work for it. Stock had been bought in my name; my parents handled it but it was mine to work with: it's a good system, like giving your kid a tough horse to ride when he's young. By the time I was 29 the stock came to over $100,000. I had ignored it all through my twenties, living in $20 apartments and not travelling much, occasionally wage-earning in photography, design, Army. The idea for the *Catalog* hit me plenty hard, but I think I could never have raised money for it. … So I invested, comrade. I took the profits from old investments and put them into a new one. … Why am I saying all this? Because many who applaud the *Catalog* and wholeheartedly use it, have no applause for the uses of money, of ego, of structure (read uprightness), of completion, of business as usual. All the things, plus others, which make the *Catalog*, and make the selective applauders into partial liars.[76]

Brand admitted that his entry into business had involved along a steep learning curve:

One of the main things that drove me into business was ignorance. Liberally educated young man, I hadn't the faintest idea how the world worked. Bargaining, distribution, markup, profit, bankruptcy, lease, invoice, fiscal year, inventory—it was all mystery to me and usually depicted as sordid.[77]

Production

As already mentioned, Brand's initial idea in March 1968 was to use the mobile Truck Store to make available books, tools, camping equipment, and construction plans for houses. Readers could either order books from the Menlo Park store or turn to the publishers or retail outlets, whose addresses were supplied. After the summer of 1968, Brand spent most of his time making the catalog and its list of books and items. Clearly this was a major undertaking, and Brand was not alone in the venture.[78]

Others helped significantly. His wife Lois Jennings, trained as a mathematician, played an important part in the general organization, as he pointed out: "From the beginning the pretty little Indian girl Lois, who still has to show her ID to bartenders, was the hard core of the business. She applied her math background to our bookkeeping, and her sharp tongue to our laziness and forgotten promises. She had the administrative qualities you look for in a good First Sergeant. In my experience every working organization has one overworked underpaid woman in the middle of things carrying most of the load."[79] The couple was supported by a dozen or so people and sometimes more, as can be seen in the team photographs taken in front of the display window of the Menlo Park store. And, of course, Brand had the financial and emotional support of his parents.

TOOLS

SWISS ARMY KNIVES
MANY MODELS
Starting at $7.50

SMO-SEAL

WOKS
— WITH TOPS —
Large $16.50
Small $14.00

$19.15
SHADE EXTRA
ALADDIN LAMP

Earthchild pack:
$22.00

Diet for a Small Planet
Frances M. Lappé

THE Natural Way to Draw
NICOLAIDES

M.G. KAINS
Five Acres and Independence

A Key to the Psilocybin Mushroom

The Indian

THE WOK
a chinese cook book

MOTHER EARTH'S Hassle-Free Indoor Plant Book
by Lynn & Joel Rapp

A New Method of MULCH GARDENING
How to Have a Green Thumb Without an Aching Back
by Ruth Stout

SEVEN ARROWS
HYEMEYOHSTS STORM

HOW WORK TOOLS WO

COOKBOOK FOR THE NEW AGE
EARTH WATER FIRE AIR

THE EX-URBANITE'S COMPLETE EASY-DOES-IT
FIRST-TIME FARMER'S
A Useful Book

THE WELL BODY BOOK

Figures 0.14 and 0.15
"Tools," left and right pages of spread, *Whole Earth Truck Store*, July 1968. First six-page mimeographed newsletter presenting 120 articles for sale. M1045, box 37, 6-flat. Courtesy of Stewart Brand and the Department of Special Collections and University Archives, Stanford University Libraries.

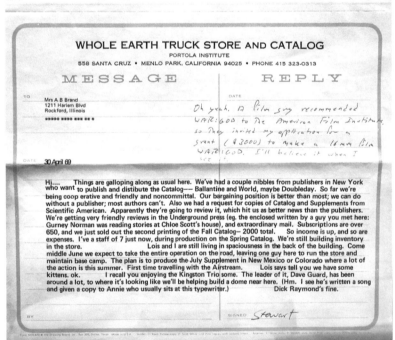

Running the *Whole Earth Catalog*

Brand explained the production process in a seven-page essay in the 1971 *LWEC*.[80] He distinguished three steps: gathering information, evaluating it, and giving it an appropriate form. This amounted to selecting the books and other "tools," reviewing them in a short text, and deciding how to present them in photographic or textual forms. In the beginning, Brand confessed that he had imagined that the undertaking would be like a sort of laboratory, "with nifty projects everywhere, earnest folk climbing around on new dome designs, solar generators, manure converts; comparing various sound systems, horse breeds, teaching methods."[81] To get readers involved, the catalog offered a cash compensation of ten dollars for every review written and several dollars for every suggestion published. On the first page of the first catalog published in 1968, the *WEC* printed the selection criteria for the books and objects to be reviewed: "An item is listed in the catalog if it is deemed: useful as a tool, relevant to independent education, high quality or low cost, easily available by mail."[82]

First and foremost, the catalog was the result of a methodical organization of information. Brand sought out work on bibliometrics and scientific data. He selected the 1968 book *Data Study* to be included in the *WEC*. A reference guide on the science of organizing books and concepts, the text highlighted the complexity of ordering books in a library.

In the same way, one can conceive of the *Whole Earth Catalog*—a sort of atlas of knowledge—as a library with each section corresponding to a shelf. The works are classified by proximity of interest. It is, moreover, not uncommon that the works selected for the *WEC* are in dialogue with one another, rather like elements of an Exquisite Corpse stretching over several pages. Brand took pleasure in organizing the pages and creating a sort of conversation between the works. The most convincing example of this technique is the juxtaposition of photographic representations with their contrasting scales from macroscopic to microscopic views, in the "Understanding Whole Systems" section.

The editors' main sources of information were publishers' catalogs, bibliographies of good books, their own bookcases, and friends' bookcases, as well as the bookshelves of the Stanford University and Menlo Park libraries:[83] "The literature for us consisted of *Publishers' Weekly*, *Forthcoming Books in Print* (both from R. R. Bowker, source of all the basic

Figure 0.18
Page layout for "Organic Gardening," in the "Land Use" section, *LWEC* (1971), 51. Courtesy of Stewart Brand and the Department of Special Collections and University Archives, Stanford University Libraries.

19

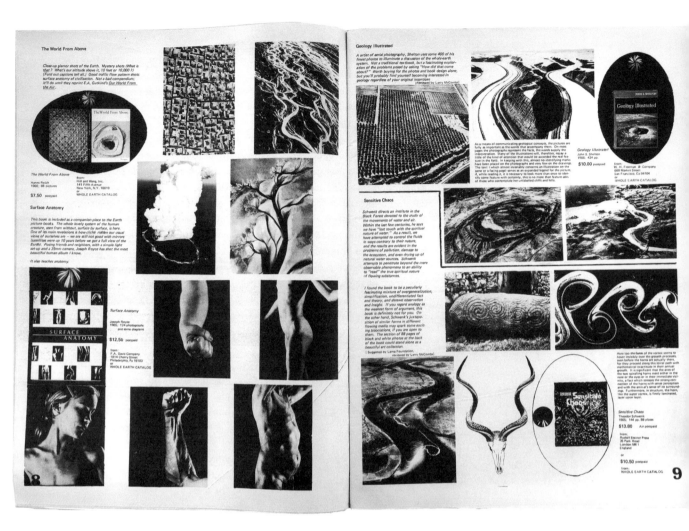

Figure 0.19 (top)

"The World from Above" and "Geology Illustrated," double spread in the "Understanding Whole Systems" section, *Whole Earth Catalog*, fall 1968, 8–9. Marc Treib Collection. Courtesy of Stewart Brand and the Department of Special Collections and University Archives, Stanford University Libraries.

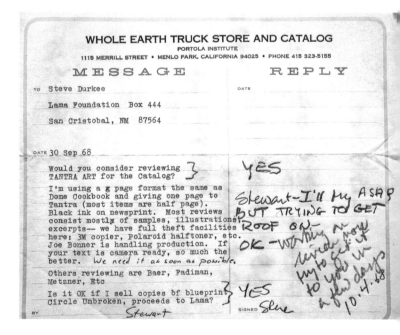

cataloging information of books in the U.S. I consider R. R. Bowker a major pillar of Western Civilization; they labor endlessly, invaluably, without bias and of course unheralded)."[84]

Once again, we find Brand far from amateurism in his efforts to gather information from the most official institutions; R. R. Bowker is an American company furnishing bibliographic information on published books to publishers, booksellers, libraries, and individuals. Brand also relied on *Scientific American* and *Popular Science*, as well as on the *WEC*'s sister publications, notably *Mother Earth News*, *Big Rock Candy Mountain* (text 73), *Canadian Whole Earth Almanac*, and *Natural Life Styles*.[85]

Brand recounted: "I never spent my time reading the good stuff—whose quality was usually evident in 2–3 minutes. I spent the yellow-brown hours reading the lousy books, digging past their promising facades to the hollow within. Some of these wound up in the stove, where their publishers belonged."[86] The reviews assessed the items on two levels: first as pure information, accurately transmitting the contents of the books,[87] and then as encouragement to readers to examine the whole book.

Brand developed a particular reading technique, trying to approach the point of view of the reader who is not an expert in the subject matter:

Usually I review a book before I read it. These are almost always shorter, pithier, more positive and useful reviews. You're approaching the book from the same perspective as the reader—unfamiliarity—and you're not apt to fall into imitation of the author's style or petty argument with his views, as critics do. So, I

Figure 0.20 (opposite bottom)
Stewart Brand, letter to Steve Durkee, Lama Foundation, September 30, 1968. "Would you consider reviewing TANTRA ART for the Catalog? I am using a page format the same as Dome Cookbook and giving one page to Tantra (most items are half page). Black ink on newsprint. Most reviews consist mostly of samples, illustrations, excepts—we have full theft facilities here: 3M copier, Polaroid halftoner, etc. Joe Bonner is handling production. If your text is camera ready, so much the better." M1237, box 6, fol. 7. Courtesy of Stewart Brand and the Department of Special Collections and University Archives, Stanford University Libraries

Figure 0.21
Victor Papanek, letter to Stewart Brand, December 27, 1970. "We talked about the book being reviewed in *WEC*'s last issue and I said that I thought that the book would not be ready by then. You said that if I could get you galleys by late March, that would be cool. Since then I have flown to NY to see my publishers, Pantheon Books." M1237, box 6, fol. 7. Courtesy of Stewart Brand and the Department of Special Collections and University Archives, Stanford University Libraries.

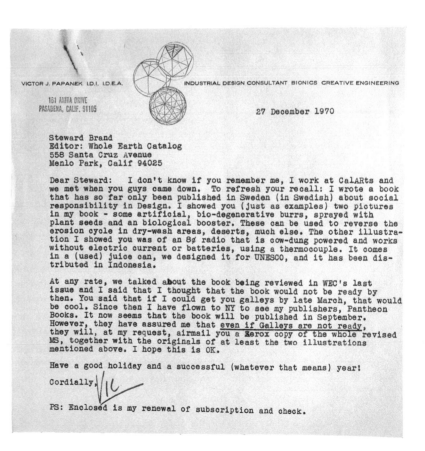

review the book, enthusiastically, on what I know from its title, its subject, the author, my own experience, and a hasty glance at the pages. Then I look a little more deeply to see if the review is fulfilled. If not, I either rewrite the review or discard the book.[88]

Despite this apparently glib statement, Brand was an attentive and critical reader, always attentive of capturing the most interesting quotations, illustrations, and paraphrases. His introductory statements are invariably punchy and to the point. Elaborating further on his selection technique for the works, Brand confided that the quickest way to evaluate a book's authority is to look at its illustrations and academic apparatus, that is, the index and bibliography. For Brand, photography had to be meaningful, and this applied to the captions as well. Readers also greatly appreciated diagrams because they facilitated understanding by providing a simplified representation of complex ideas.[89]

The selection of extracts—in word or image—was particularly important. The extract must sum up the book by palpably transmitting its contents while delivering some important ideas that would be useful regardless of whether or not readers bought the book: "I always attempt to gut the book with the excerpts, extract its central value. Really good books like *On Growth and Form*, or *Stick and Rudder*,[90] or [*The*] *Natural Way to Draw* will not be gutted; practically any line or picture in them can be used."[91] For example, take his introduction to *Nomadic Furniture*, by James Hennessey and Victor Papanek (1972): "If you move your dwelling-stuff more than once every four years, this lightweight book can save you the aggravation of feeling so stupid while wrestling that heavy bed (couch, table, chest, and bookshelf) down that narrow stairway to that overloaded car."[92]

Brand required a particular discipline for the reviews. The commentator not only had to give a quick overview of the document but also had to present a well-informed and enlightened assessment of the subject. Brand stressed this last point by stating: "An ideal review gives the reader a quick idea of what the item is, what it's useful for, how it compares to others like it, and how competent the reviewer is to judge. (This last is why I stopped having unsigned reviews—the reader gradually grows familiar with the weaknesses and strengths of the various reviewers.)"[93]

Participation and Selection

The *WEC* was based on a network of experts—not necessarily professionals but rather enthusiasts who had developed an expertise—who found, evaluated, and shared ideas and skills. The *WEC* network also relied on a wide range of amateurs, from the layperson to the true expert on a subject.[94]

Brand used two methods to establish and then extend his network; he directly approached people who he knew were competent in certain subjects, and then he asked them to review this or that book. But also, beginning with the supplement that followed the first *WEC* issue of autumn 1968, Brand challenged the readers directly and invited them to participate either in suggesting books that they believed essential or by adding to the information already published in the *WEC*. The active participation of "amateur" experts gives a particular flavor to the *WEC*, reflecting highly personal, and sometimes quirky, skills and enthusiasms.[95]

WHOLE EARTH TRUCK STORE AND CATALOG
PORTOLA INSTITUTE
1115 MERRILL STREET • MENLO PARK, CALIFORNIA 94025 • PHONE 415 323-5155

M E S S A G E   R E P L Y

TO   Steve Baer   DATE

Box 442

Corrales, NM 87048

DATE 23 Aug, 68

Steve,

Lots of things developing here. I worked for 2 hours last night on line with a
fantastically sophisticated interactive computer system, and then dreamt most of the
night in computer terms-- adventures among the tree of choices and clarities.
Today I'm disappearing into the mountains for an underground summit conference that
Kesey is calling (probably with subsequences, next step meetings; if you'd make one,
let me know).
And a request to you. Many of the evaluations and descriptions in the Catalog are being
written by people who know a lot about the particular items. I'm going to carry
Daniel"s sun book. Would you care to write it up ? The Catalog pays writers $10
per item and by-lines their reviews. The piece should be about 200 words, addressed
to others of us, evaluating the thing for usefulness and for quality among its kind,
describing it and howiit may be useful. Also pick 3 or 4 rich samples of text and
illustration from the book-- two of these will accompany the review. (I have the
paper edition; if you have it to work from, then page references will do.)
I don't know anyone better qualified; I hope you'll do it.   SIGNED   Stewart

WHOLE EARTH TRUCK STORE AND CATALOG
PORTOLA INSTITUTE
558 SANTA CRUZ AVENUE • MENLO PARK, CALIFORNIA 94025 • PHONE 415 323-0313

M E S S A G E   R E P L Y

TO   Steve Baer   DATE
     1780 Menaul Blvd NE
     PO Box 712
     Albuquerque, NM 87103

DATE 4 Dec 70

I'm reading Pentagon of Power now. The only
thing that slows me on Mumford is his occasionall
pious ought-to-be approach to this or that, which
is ineffective. Fuller does it too. I used to
want the universe to line up with my desires; now
I prefer my desires to line up with the universe.
Understandings too. Ought-to is what the missionaries
did to the Indians. I'm glad Mumford deplores it,
but deplore is no solution, more often aggravation.

Technics & Civ will be in the last Catalog. Any
additional Mumford you'll have to add; you're the
only user I know. How about a review of the new
book?   We had a good time at CIA.
LA was awful I thought. Self-inflicted wounds.
Have you seen Three by Eastlake--- 3 fine novels
of just north of you.   Good infor from Pida on islands, too mother costly..
SIGNED Stewart

Form N3-R72-B The Drawing Board, Inc., Box 505, Dallas, Texas Made in U.S.A.   Sender: 1. Keep Yellow copy. 2. Send White and Pink copies with carbons intact.   Receiver: 1. Write reply. 2. Detach stub, keep Pink copy, return White copy to Sender.

Figure 0.22 (top)
Stewart Brand, letter to Steve Baer, August
23, 1968. M1237, box 6, fol. 5. Courtesy of
Stewart Brand and the Department of Special
Collections and University Archives, Stanford
University Libraries.

Figure 0.23 (bottom)
Stewart Brand, message to Steve Baer
about his reading of Lewis Mumford's
Pentagon of Power, December 4, 1970.
M1237, box 6, fol. 7. Courtesy of Stewart
Brand and the Department of Special
Collections and University Archives, Stanford
University Libraries.

"How to Do a *Whole Earth Catalog*"

In "How to Do a *Whole Earth Catalog*" (1971), Brand gave precise details for the making of the catalog, from the initial idea to the financial resolution, from investments to profits.[96] He named and set out the roles of each team member. He cited the critical role of Dick Raymond and the Portola Institute, who provided the initial funding and workplace. From summer 1968, the collaborators were Sandra Tcherepnin ("Sandra fell in love with the IBM composer"[97]), Joe Bonner ("a young artist"), Annie Helmuth, and Tom Duckworth. Brand also mentioned his agent Don Gerrard and Don Burns, who in 1968 founded Bookpeople, a small book distributor based in Berkeley.[98] The team members were the sole distributors until the March 1971 supplement, when the magazine the *Realist* took charge of half the distribution.

Brand's article addressed all aspects of the undertaking. He detailed the catalog's production methods and recounted the number of subscriptions. He continued by describing the highs and lows of working through all the different versions of the catalog and the supplements. The photographs accompanying the story showed his collaborators in action at their workplaces.

Organization and Management

Brand sought a form of work organization that would simultaneously be maintainable and managed as a self-regulating system, where each person had his or her part to play. He believed that the company must encourage the personal development of each member through the learning process. Making the *WEC* thus required an appetite for self-knowledge and responsibility.[99] In "Texture of Life," Brand stated:

Figure 0.24
The daily volleyball game at the Whole Earth Truck Store, 1970. Photo by Sam Falk. M1237, box 26, fol. 12. Courtesy of Stewart Brand and the Department of Special Collections and University Archives, Stanford University Libraries.

Texture in the work-place, at least in my workplace and the one I ally with, means greater continuity of staff, more familial amenities (in our case daily volley ball and home-cooking for lunch), more individual direction of hours and vacation, more individual responsibility for tasks, less supervision from above (and perhaps more by peers).[100]

Here are the principles of a friendly work environment, where lunch breaks were set aside to establish relationships that went beyond the professional dimension.[101] Brand understood well the necessity of creating an environment where team members would feel invested and thus motivated.

In the 1960s and '70s, this reflection on the place of each person within a business organization was already part of the managerial culture of the big capitalist companies like IBM. It came to be known in the early 1990s as "Learning History," theorized by Peter Senge and developed at an MIT laboratory, the Center for Organizational Learning.[102] During the '90s, friends close to Brand, Stephanie Mills and Art Kleiner among others, who contributed articles to *CoEvolution Quarterly*[103] and were supporters of the *Whole Earth Catalog* adventure,[104] continued to deepen these approaches in their own work. Mills, for example, developed the "Listening Projects" in 1996, which showed how writing a company's history through the eyes of its participants could be a tool for reinforcing community building, and that the process of opening up dialogue with the members of a group allowed each participant's perspective to be captured.[105]

The physical organization of the workplace was also important. The photographer Sam Falk, a former correspondent for the *New York Times* who is known for his portrait of Marlon Brando, photographed the *Whole Earth Catalog* office in 1970, giving us a glimpse of the company's spatial organization.

Figure 0.25
Whole Earth Truck Store, Menlo Park, 1970. Photo by Sam Falk. The kitchen table plays an important role for a moment of conviviality during the making of the *Whole Earth Catalog*. M1237, box 26, fol. 12. Courtesy of Stewart Brand and the Department of Special Collections and University Archives, Stanford University Libraries.

Figure 0.26 (top)
The offices of the Whole Earth Truck Store, Menlo Park, 1970. Photo by Sam Falk. On the back of the photograph, Stewart Brand wrote the names Les Rosen, Peter Ratner, Marry Jo Morra, Jerry Fihn, Austin Jenkins, John Clark, as working in the "mail order department." M1237, box 26, fol. 12. Courtesy of Stewart Brand and the Department of Special Collections and University Archives, Stanford University Libraries.

Figure 0.27 (middle)
Lois Jennings Brand at her desk, 1970. Photo by Sam Falk. M1237, box 26, fol. 12. Courtesy of Stewart Brand and the Department of Special Collections and University Archives, Stanford University Libraries.

Figure 0.28 (bottom)
The store at Menlo Park, 1970. Photo by Sam Falk. M1237, box 26, fol. 12. Courtesy of Stewart Brand and the Department of Special Collections and University Archives, Stanford University Libraries.

The work ethic was an important aspect of the *WEC* company, as indeed it already was for other West Coast businesses at the time. Psychological research, as well as Brand's military training, had shown that one works better if one understands the significance of the task. This did not preclude the exhausting editorial rhythm imposed by producing the different versions of the *WEC*. Cappy McClure, who complained about this from the beginning, left the adventure after only one year, annoyed by the authoritarian tone of Lois Brand. Stewart Brand sent McClure the following message: "Catalog is a product. If we were process, we might be as good as Hog Farm or Libre."[106] By this, he distinguished between a communal lifestyle like that of the Hog Farm or the Libre commune (text 22), where decisions were based on consensus and negotiations, and a business, where decisions had to be centralized. Brand added that it "took two people to replace you," indicating that he recognized the large workload she had carried.[107]

Brand explained his system for respecting deadlines and organizing the contributions of multiple collaborators:

There are two main work governors tacked to the wall—a calendar showing days of production and a page chart. If we have 8 weeks to do 447 pages, then we have to finish a signature of 64 pages every 6 working days, or about 11 pages a day. The signature finished points are marked on the calendar so I know exactly how far behind we are and when we'll have to start working nights to get copy to the printer on time. ... The page chart is big, a couple square inches for each page. On each page I write the basic information for the three layout people. As they finish a pair of pages they mark them off on the chart and look for the next ready pair. From the chart they get the number and name of the pages, the titles of the items (and whether they're new or to be cut out of old flats), plus the appropriate piece of Divine Right, and any headings.[108]

In the catalog production process, Brand recalled that the editor, typist, and photographer, as well as the three people in charge of page layouts, had to work together. The editor needed to be two days ahead of the actual page

Figure 0.29
Stewart Brand working at the light table on the layout, 1970. M1237, box 26, fol. 12. Courtesy of Stewart Brand and the Department of Special Collections and University Archives, Stanford University Libraries.

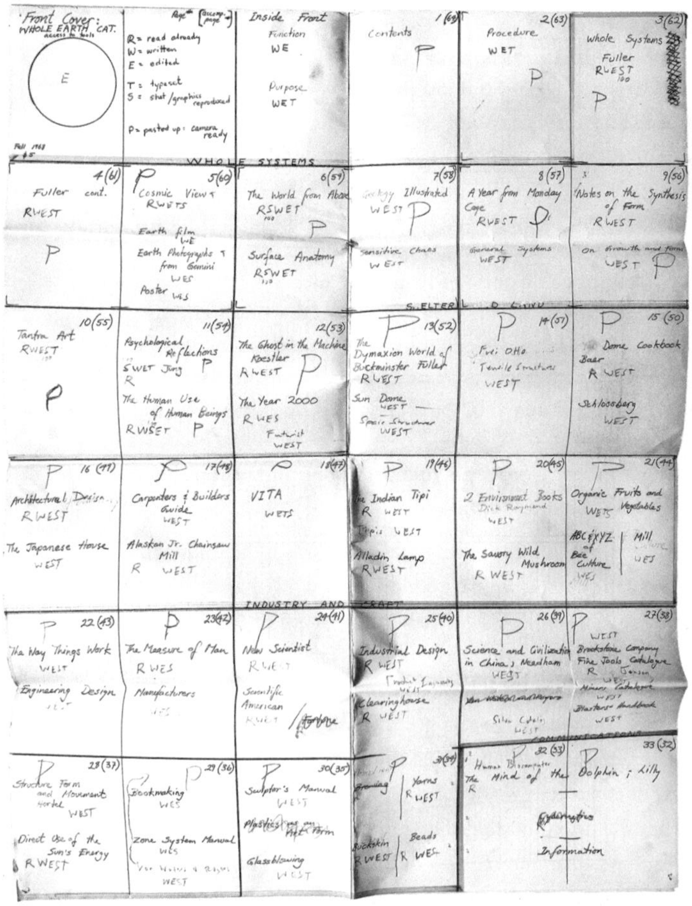

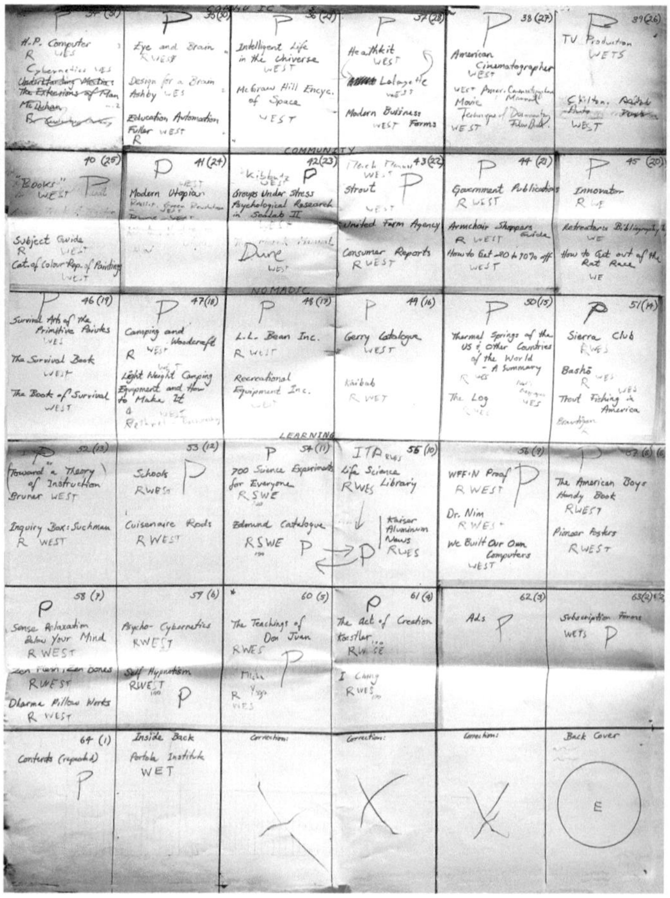

Figure 0.30

The page chart for the first thirty-three pages of the first edition of the *Whole Earth Catalog*, fall 1968. This document mentions three sections: "Whole Systems," "Shelter and Land Use," and "Industry and Craft." The last two sections will be divided into four in the *LWEC*: "Shelter," "Land Use," "Industry," and "Craft." The letters in the document signify as follows: R = Read Already; W = Written; E = Edited; T = Typescript; S = Shot, Graphics. "WEST" means "Written, Edited, Shot, Typescript." M1237, box 6, fol. 7. Courtesy of Stewart Brand and the Department of Special Collections and University Archives, Stanford University Libraries.

Figure 0.31

The last four sections of the page chart are as follows: "Communications," "Community," "Nomadics," and "Learning." M1237, box 6, fol. 7. Courtesy of Stewart Brand and the Department of Special Collections and University Archives, Stanford University Libraries.

layout work to select the elements for each page; the typist and photographer needed to be one day ahead. This close cooperation between the various stakeholders is an operation requiring a back-and-forth among the different tasks, which is characteristic of the field: "When a layout guy has the copy all gathered he calls me over to see what the space situation is and determine what to leave out and what to retype or reshoot so it will fit. After he's finished I'm called over again for any revisions and to try to catch the mistakes while they're easy to correct."[109] The process of correcting the proofs involved the work of two copy editors, who also took care of the indexing, before a series of pages was finalized and sent to the printer. Brand would proofread the proofs, making any last corrections, before sending the catalog to print.

The *WEC* reality was therefore one of a rather tightly controlled system, as it was necessary to have a firm grip on operations for business to run smoothly. This again shows one of the lessons learned from Brand's military training, that of operational planning. In this case, he evaluated production processes in terms of time, comparing a participative type of organization with one of strong central direction.[110] Brand stated that "some publications make all their editorial decisions by continual discussion and consensus. I admire the ones who can make it work. I've gone the faster and possibly more limited route of strong central direction."[111]

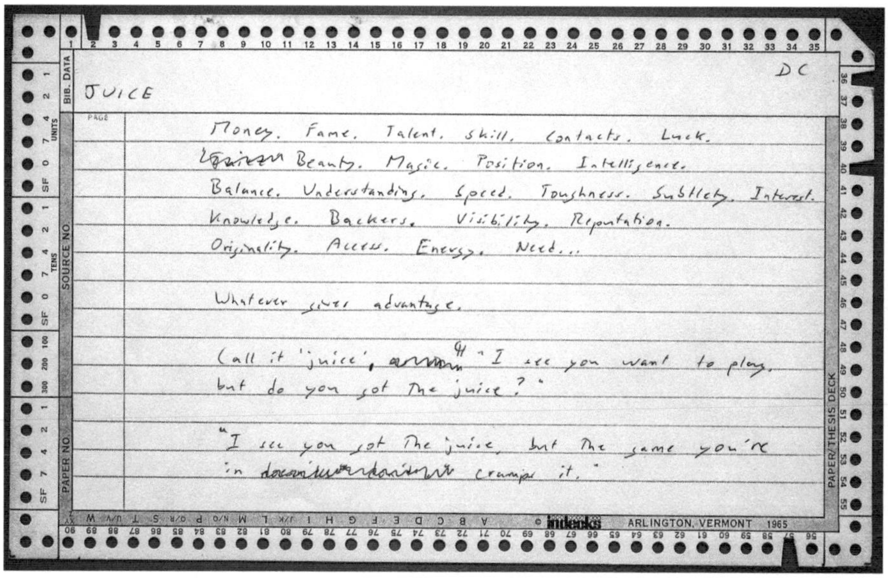

Figure 0.32
Indecks card titled "Juice," used for rough
editing. M1237, box 17, diary 66–69. Courtesy
of Stewart Brand and the Department of
Special Collections and University Archives,
Stanford University Libraries.

He also explained in detail his indexing system: "I use McBee cards, one for
each item, for rough editing."[112] The McBee cards have two rows of holes and
zones corresponding to a summary preclassification. The cards are coded by
punching the holes, so that when a knitting needle is inserted into a particular
hole in a stack of cards and shaken, the coded cards will fall to the floor.[113] (The
Belgian Paul Otlet had already developed a similar hypertext system at the end
of the nineteenth century.[114]) The Indecks punch cards, which Brand also used,
are similar to the McBee system:

What do you have a lot of? Students, subscribers, notes, books, clients, projects?
Once you're past 50 or 100 or whatever, it's tough to keep track, time to exter-
nalize your store and retrieve system. One handy method this side of a high-
rent computer is Indecks. It's funky and functional: cards with a lot of holes in
the edges, a long blunt needle, and a notcher. Run the needle through a hole
in a bunch of cards, lift, and the cards notched in that hole don't rise; they fall
out. So you don't need to keep the cards in order. You can sort them by feature,
number, alphabetically or whatever; just poke, fan, lift and catch. We've used
the McBee cards to manipulate (edit) and keep track of the 3,000 or so items in
this catalog.[115]

Brand then went on to explain the use of the classification cards:

I know from looking at previous Catalogs and the new material approximately
how many pages should be in the, say, Nomadics Section—61 pp. So I take the
stack of McBee cards punch-coded for that section and break them down into
categories—mountain stuff, car-stuff, outdoor suppliers, survival books, etc.
Then those sub-piles are put in some sensible sequence. Then on a big table the
cards are separated further into 61 little page-stacks, by pairs (the reader sees 2
pages at a time, not one).[116]

**The McBee cards can be seen as prefiguring computerized databases. Brand,
moreover, was one of the first countercultural members to recognize the
potential of the personal computer and information technology to link individuals and organizations with a view to societal change.[117]**

Spring 1969
$4

Spring 1970
$3

Figure 0.33
Cover of *Whole Earth Catalog: Access to Tools*, spring 1969. Graphic design by Peter Bailey. Courtesy of Stewart Brand and the Department of Special Collections and University Archives, Stanford University Libraries.

Figure 0.34
Cover of *Whole Earth Catalog*, spring 1970. Graphic design by Peter Bailey. The image is borrowed from the *Andromeda Galaxy* poster, distributed by Edmund Scientific, 100 Edscorp Building, Barrington, NJ 06007. Courtesy of Stewart Brand and the Department of Special Collections and University Archives, Stanford University Libraries.

The *Whole Earth Catalog* Covers

Although *Lunar Orbiter 1* photographed the earth from near the moon on August 23, 1966, NASA barely distributed the photographs.[118] The then twenty-eight-year-old Brand organized a campaign that very year to convince NASA to make the images public. He made badges with the inscription "Why haven't we seen a photograph of the whole Earth yet?" and distributed them on the Berkeley campus, an indication of his willingness to shake up institutions.

NASA disseminated images of the moon shortly after, following the *Apollo 8* lunar orbit in December 1968, during which the astronauts photographed both the earth and the moon. Brand often used the photographic representation of planet Earth for the cover of the *WEC* issues. The terrestrial globe quickly became a part of the iconography of North American environmentalism. These images would contribute to the great appeal of the catalog. Brand used the same image, or two different images taken from the same series, for the front and back covers, then introducing an additional element that was meant to challenge the reader. Thus the image of the Milky Way, used for the *WEC* spring 1970 issue, adds a question mark on one side and an exclamation point on the other. This simple iconographic gesture suggests the idea that the Milky Way is responding to what is in the volume. In the 1971 edition of the *LWEC*, Brand stated that the photographs' impact on the media reinforced the environmentalist movement that had already been gaining momentum by 1969. For the back cover of the *Last Supplement of the Whole Earth Catalog*, the editors—Paul Krassner and Ken Kesey—presented a cutaway globe against a pink background, showing the two Americas, filled with earthworms breaking through the surface.

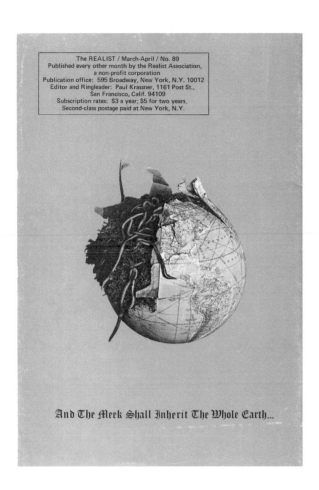

The REALIST / March-April / No. 89
Published every other month by the Realist Association,
a non-profit corporation
Publication office: 595 Broadway, New York, N.Y. 10012
Editor and Ringleader: Paul Krassner, 1161 Post St.,
San Francisco, Calif. 94109
Subscription rates: $3 a year; $5 for two years.
Second-class postage paid at New York, N.Y.

And The Meek Shall Inherit The Whole Earth...

Figure 0.35
Back cover image of a cutaway globe
against a pink background showing the two
Americas, filled with earthworms breaking
through the surface. Paul Krassner and Ken
Kesey, eds., "The Realist Presents: The Last
Supplement to the Whole Earth Catalog,"
Realist 89 (March 1971). Courtesy of Stewart
Brand and the Department of Special
Collections and University Archives, Stanford
University Libraries.

Layout and Typography

In 1971, Brand also clearly articulated the catalog's layout principles: "In descending order of importance, our layout guidelines are: Accuracy; Clarity; Quantity of information; Appearance.... Glamorous white space has no value in a catalog except as occasional eye rest. I figure the reader can close his eyes when he's tired."[119] Not everyone was impressed by the layout. Graphic designer Lorraine Wild attributed the high density to a desire to cram in as many items as possible.[120]

The company, once it took form in 1969, was quick to use the latest developments in office equipment Brand took a lease on the up-to-date IBM Selectric golf ball typewriter. The machine is visible in a Super 8 film shot during the production of the 1970 *Difficult but Possible Supplement* out in the desert. This was not a cheap machine, selling in 1969 for $520 ($3,413 in 2015).[121]

With the golf-ball-type heads, which could quickly be rotated, each Selectric could use more than one font, including italics, scientific notation, and other languages. The reports were typed in Univers Italic,[122] and the book information in Teen. In the $LWEC$, the reader could immediately identify the block of Teen bold text corresponding to Gurney Norman's fictional work *Divine Right's Trip*.[123]

As for the format, Brand's choice was pragmatic:

We use a tabloid-sized page, like the magazines in the Sunday papers. Steve Baer's *Dome Cookbook* [text 23] was what convinced me it's a good format. You have enough space on each page and spread (facing pair of pages) to lay out a

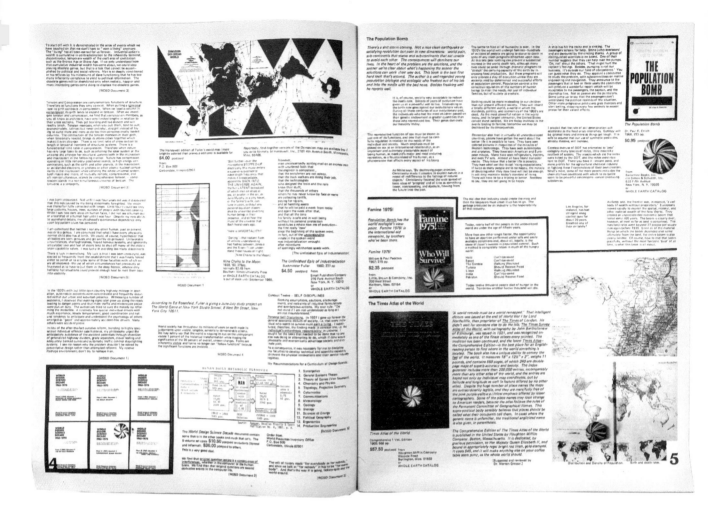

Figure 0.36
Double page devoted to Richard Buckminster Fuller and to Paul Ehrlich's book, *The Population Bomb. Whole Earth Catalog*, spring 1969, 4–5. Marc Treib Collection. Courtesy of Stewart Brand and the Department of Special Collections and University Archives, Stanford University Libraries.

graphic array of information with multiple visual relationships and plenty of freedom for the reader to pick his own path. Also it's an economical size for printing on a web press. The two main disadvantages are that booksellers don't like the display space a tabloid book takes up, and some readers get tired holding the big page up.[124]

In Brand's view, the choice of the tabloid format allowed for flexible deployment of text and image layout. His editorial techniques evolved, however, over time:

At first we laid out our pages just the way a newspaper does (the Menlo-Atherton recorder showed us how). We used flats (blank-line 2-page-sized pieces of paper provided by the printer) on light tables (made of plywood, fluorescent lights, white plastic, and glass). The flats were placed line-side-down on the light tables, so the lines showed through and you can align items on them but they wouldn't show on the final layout. Also like the newspaper we used beeswax to stick the items down with.[125]

He also explained the aesthetic changes introduced by the graphic artists Gordon Ashby and Doyle Phillips:

We went along using light tables and wax until Gordon Ashby and Doyle Phillips revolutionized our procedure with the July 70 Supplement. They showed us that working on a portable drafting table, with a T-Square and rubber

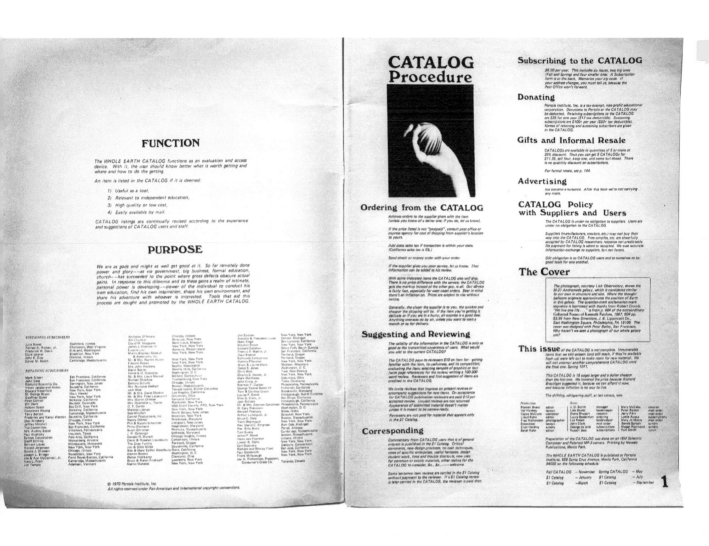

Figure 0.37

Layout before suggestions were made by Gordon Ashby and Doyle Phillips, who worked on the July 1970 supplement. *Difficult but Possible Supplement to the Whole Earth Catalog* (Menlo Park, CA: Portola Institute, July 1969), 4–5. Courtesy of Stewart Brand and the Department of Special Collections and University Archives, Stanford University Libraries.

cement for adhesive was really much easier and gave us better layout. We still had lines on the flats but they were on the top surface in light blue (invisible to the camera).[126]

In 1963, Brand had followed Ashby's class at the San Francisco Art Institute. Ashby had worked with Charles and Ray Eames, notably on the film *Glimpses of the USA* (1959), and had contributed psychedelic videos to the Trips Festival in January 1966.[127] In a recent interview, Ashby "prided himself as a skilled mediator between 'those who conformed to rules and those who broke them,' [and] cultivated inlaw and outlaw personae as an eco-activist."[128] He was in charge of the California Hall of Ecology installation at the Oakland Museum, displayed from 1964 to 1969, combining natural science, history, and art collections within a holistic framework, showing the interconnections between the ecological regions from ocean shoreline to high sierra forest.[129]

In the 1970 *WEC* supplement, Ashby introduced a particular iconography and page layout, devoting several pages to psychedelic and mystical compositions. Greg Castillo has reported that Ashby commented that the choice of the mandalas showed their "unique capacity to express the hippie Zeitgeist."[130]

Although the supplement was indeed very different from earlier *WEC* publications, Castillo claims that it "provided a Rosetta stone with which to decrypt the whole systems ideology encoded within Brand's hippie testament."[131]

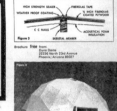

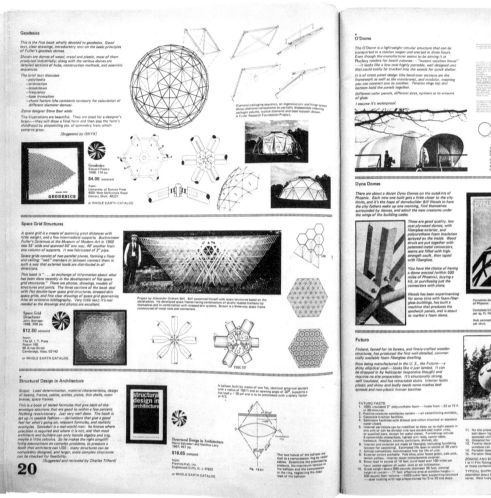

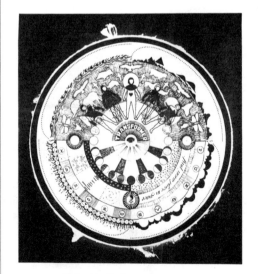

Figure 0.38 (top)
"Shelter and Land Use." In Stewart Brand, ed., *Whole Earth Catalog* (Menlo Park, CA: Portola Institute, 1969), 20–21. Marc Treib Collection. Courtesy of Stewart Brand and the Department of Special Collections and University Archives, Stanford University Libraries.

Figure 0.39 (bottom)
Gordon Ashby, "Transformation," from *$1 Whole Earth Catalog*, July 1970. Greg Castillo Collection. Courtesy of Stewart Brand and the Department of Special Collections and University Archives, Stanford University Libraries.

Figure 0.40
Fred Richardson using the Polaroid MP-3 for making the *Difficult but Possible Supplement to the Whole Earth Catalog*, 1971. Stills from *Saline Valley Production* (Whole Earth Catalog, 1971), filmed by Stewart Brand. http://www.youtube.com/watch?v=OqRdGfDPCDE (accessed August 4, 2014). Courtesy of Stewart Brand and the Department of Special Collections and University Archives, Stanford University Libraries.

The Polaroid Instant Camera

The Polaroid was another device much in vogue in the 1960s. It produced photographic prints directly, without the separate steps of developing the film and then printing on paper. By liberating photographers from the work of developing film in the darkroom, the camera became a cult object, much admired by artists such as Andy Warhol. In a remarkable 10.5-minute publicity film produced in 1972, the designers Charles and Ray Eames analyzed the Polaroid SX-70 and demonstrated how the device could be used to photograph a document from a book.[132]

Brand used several Polaroid cameras. However, even if one could immediately see the results, the Polaroid's image, once produced, existed at a fixed scale; it was not possible to adjust the dimensions for the page layouts. The second drawback was the mediocre quality of the documents, especially for printing. The Polaroid was not the only photographic device used: "The production photography that I do for *WEC* includes all halftones up to 3-1/2" × 4-1/2" and line shots up to page size."[133] The company also had a 11 x 17 inch Visual Graphics Stat King camera. Fred Richardson recounted: "It is a fancy Photostat machine that will enlarge/reduce 2000%–50%. Cost $4,000 [equivalent to $23,000 in 2015]. It can have prints out and dry in 5 minutes or so."[134]

Delegating Control, 1970

Following the advice of Bill Lane, Brand created the unexpected by engaging a guest editor in chief; Brand was somewhat exhausted by three years of intense labor and asked the writer Gurney Norman to edit the 1971 *LWEC*.[135] Norman, born in 1937, studied journalism at the University of Kentucky and received the Wallace Stegner Fellowship in creative writing at Stanford University. Like Brand, Norman joined the army for two years, from 1962 to 1963. He then pursued a career in journalism, but it was not long before he dedicated himself to writing fiction.

Norman negotiated with Brand before taking on the responsibility of editor in chief. He wanted the editing job to accommodate his work as a writer.[136] The novel he was working on, *Divine Right's Trip*, was a story about two hippies, D. R. Davenport and Estelle, who leave California to live in Kentucky, where they plan to farm and raise rabbits.[137]

Norman contributed his novel to the *LWEC*, with passages prominently displayed on each page. A broken and fragmented narrative, a collage or montage novel, the text did not appear as a continuous fabric. Proof of Brand's judgment, the book became a countercultural classic. However, Brand did not follow Norman's proposed management method, and Norman, in the end, played only a minor role in the issue.

On various occasions, Brand delegated the editorship of whole or parts of the catalog. Lloyd Kahn, for example, edited the "Shelter" section of autumn 1969 and spring 1970. Wendell Berry took over a part of the "Land Use" section. Brand also passed on the composition of several supplements to others: Ashby and Doyle Phillips were responsible for the July 1970 supplement, and Paul Krassner and Ken Kesey for *The Last Supplement to the Whole Earth Catalog* of March 1971.

Actions

"Action for creating a better world" could be the *WEC*'s motto. One of the texts cited in the *LWEC* (p. 17) is Hannah Arendt's 1958 book *The Human Condition*. Arendt's description of the value of work was highlighted in the *WEC*: "The task and potential of mortals lie in their ability to produce things—works and deeds and words—which would deserve to be and, at least to a degree, are at home in everlastingness, so that through them mortals could find their place in a cosmos where everything is immortal except themselves."[138]

Stewart Brand was a born organizer who liked to get people together and see what happened. The organization of events played a significant role in *WEC* culture. The Alloy conference, which took place in New Mexico in 1969, was supported by the *WEC* but organized by the engineer Steve Baer; it was a good test of the catalog's strategy (text 31).

A range of creative and diverse minds were invited to an abandoned tile factory in La Luz, New Mexico, "between the Trinity bomb-test site and the Mescalero Apache reservation,"[139] for four days, from March 20 to 23, 1969. The Alloy conference was attended by engineers, inventors, architects, educators, event organizers, and others, including Jay Baldwin, Dean Fleming, Lloyd Kahn, and Brand. To house all these people, a small encampment sprang up consisting of a series of domes and other lightweight structures:

Who were they (who are we?). Persons in their late twenties or early thirties mostly. Havers of family, many of them. Outlaws, dope fiends, and fanatics naturally. Doers, primarily, with a functional, grimy grasp on the world. World thinkers, drop outs from specialization. Hope freaks. ... They camped amid the tumbleweed in weather that baked, rained, greyed, snowed, and blew a fucking dust storm.[140]

The meeting's strengths were likewise its weaknesses: there were many wide-ranging ideas and strong personalities coming together without a clear focus. Nonetheless, Steve Durkee stressed the powerful interconnecting element of the conference: "Try to understand that although we are different we are all trying to do something. People won't stick with it. This is the tough point. This is the hinge. What are we really doing together? Is it connected? Where is it

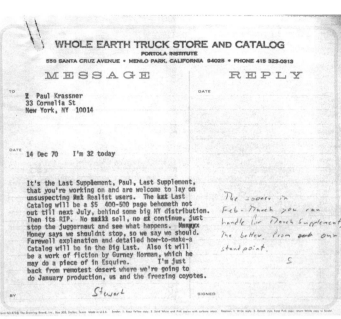

Figure 0.41 (top left)
Stewart Brand, letter to Paul Krassner, December 14, 1970. M1237, box 6a, fol. 1. Courtesy of Stewart Brand and the Department of Special Collections and University Archives, Stanford University Libraries.

Figure 0.42 (top right)
Steve Baer erecting a geodesic dome, preparing for the Alloy conference, spring 1969. M1237, box 26, fol. 12. Courtesy of Stewart Brand and the Department of Special Collections and University Archives, Stanford University Libraries.

Figure 0.43 (bottom)
Page layout for Stewart Brand's "A Report from Alloy," in *The Last Whole Earth Catalog* (Menlo Park, CA: Portola Institute, 1971), 111. The first print of this report was published in the *Difficult but Possible Supplement to the Whole Earth Catalog*, March 1969.

connected? How is it connected."[141] As Baer commented: "No one wants to be an audience.... It can't *get* out of control because it already is out of control."[142]

The *WEC* also sponsored research groups, channeled through the Point Foundation. As one example, the New Alchemy Institute on Cape Cod had an advanced program of research funded in part by the Point Foundation. Baer and his Zomeworks solar practice were also able to conduct research thanks to the help of Point Foundation grants.

As a result of the success of the catalog and events like the Alloy conference, members of the alternative movement close to the *WEC* would go on to assume official responsibilities in California. During his term as governor from 1975 to 1983, Jerry Brown appointed Baldwin as an adviser to the new Office of Appropriate Technology for a competition held in 1976.[143] As chief architect for the State of California, Sim Van der Ryn, intimately involved in the Berkeley student protests of 1965 while he was teaching at the College of Environmental Design, developed the Energy Efficient Office Building program. This program was the first initiative of energy conservation in architecture undertaken by a government body. The personal and professional connections between Baldwin, Brand, and Brown—Brand served as an adviser to the governor—explain in large part how countercultural ideas penetrated the political sphere. Nonetheless, Brand remained skeptical about the impact that state intervention could really have. He had already expressed his suspicion in 1970 of the ability of legislation to transform the world, when he was solicited to give his opinion in writing to John Brademas, head of the Educational Committee in the House of Representatives. Brand wrote:[144]

I am delighted by the spirit behind your Environmental Quality Education Act and depressed by every measure in it. I'm a former ecology student, and I can report that ecology as a science is pretty boring.... Ecology as a movement, as a religion, is tremendously exciting, and everyone can get a piece of the fervor.... However, this voluntary mass education could be poisoned by federal "help."... In my experience, the whole apparatus of application, approval, and funding commonly introduced dishonesty into an operation that can never be eradicated.[145]

According to Brand, only well-organized voluntary gatherings, such as the Alloy conference, could contribute to societal transformation. However, it was the encouragement of U.S. senator Gaylord Nelson for students to develop environmental awareness projects in their communities that resulted in the first Earth Day celebration on April 22, 1970 (text 12).[146]

A Performance: The *Difficult but Possible Supplement*, 1971

The importance of creating a particular ambience of risk taking and experimentation was underlined by the production of the 1971 supplement. In January 1971, the team temporarily moved its headquarters from Menlo Park to the Saline Valley, a broad and deep valley in the northern part of California's Mojave Desert. The valley's famous hot springs bubble to the surface in several basins.

The 23.31-minute film titled *Saline Valley Production*, shot in Super 8 by Brand, documents the production.[147] The camera captures the desert landscape some nine hours from Menlo Park.

The film's opening frames show a geodesic dome built by the Ant Farm group, transported on a vehicle of which we see only the wheels. The film continues with scenes of preparing the land to welcome the dome and digging in the anchor points. Then we see the deployment of the inflatable structure

Figures 0.44–0.46
Stills from *Saline Valley Production* (1971), filmed by Stewart Brand. http://www.youtube.com/watch?v=OqRdGfDPCDE (accessed August 4, 2014). Courtesy of Stewart Brand and the Department of Special Collections and University Archives, Stanford University Libraries.

where the production atelier would be set up. The camera lingers on the Airstream mobile home where Brand and his wife lived.

As the film continues, we see the men busy erecting the dome while the women cook around the geothermal springs. Despite the emergence of the women's liberation movement, traditional roles seem to have been maintained in many of these settlements.[148] Then, several panoramas document the installation of the furniture and equipment for producing the *WEC Supplement*, including the light table and the IBM Selectric. We also see the construction of a second geodesic dome to serve as a photographic darkroom.

In that winter month, snow covered the surrounding mountains, and the wind began to seriously shake the dome. An entire film sequence documents the intense physical struggle to hold the structure in place despite the wind. The installation of a second inflatable membrane was perilous, and production continued in Brand's mobile home. His pickup truck, filled with sand and debris, served to anchor the dome to the soil. The film's final images show the return to the Menlo Park office, where the photography and page layout work on the light table could continue.[149]

The equipment necessary for production—paper, glue, IBM Selectric, and a Polaroid—included the actors as much as the six members of the *WEC* team—Hal Hershey, Barbara and Bud DeZonia, Fred Richardson, Stewart and Lois Brand—and the members of Ant Farm present: Andy Shapiro, Joe Hall, and Curtis Schreier. We could ask a number of questions about this expedition. Why go so far to produce the catalog? Was it to test geodesic domes or inflatable structures? They no doubt played a role in the experimental character of the undertaking, implying a collective effort to erect them. Brand, however, imagined the desert adventure to be an event or a performance: "Why did we do it? My excuse was to prove that production could be anywhere—an early Maker impulse I suppose. The real reason was the pure joy of doing something difficult but possible in the desert. The whole story of the making of the Supplement was printed in the *Supplement* making it possible to share with the readers the anguishes and joys of teamwork."[150] One can equally see this approach again as a military technique of team building, also taken up by managers.

Reader Participation

Who were the readers? How did they perceive the catalog? Brand insisted on his responsibility toward the readership and aimed to enter into a dialogue with them. "I keep coming back to the reader/user because that's who the editor represents. I've had to feel that my obligations to Portola Institute, to staff, friends, relatives, and to myself are all secondary. So are our obligations to authors, suppliers, publishers, other editors. Usually [there] is no conflict, but when there is[,] the editor has to see that the reader wins."[151]

Brand thus sought reciprocity; the best means to know the readers was by inviting them to react. He occasionally challenged the subscribers directly. Introducing D'Arcy Thompson's *On Growth and Form* (text 6), Brand wrote: "Would one of you do a thorough review of D'Arcy Thompson's venerable book for the CATALOG?" Events organized at the Whole Earth Truck Store were also a way to meet readers and contributors. The supplements, which appeared between the different catalogs had, as a mission, to give the readers a voice. The sharing of knowledge, experience, and information can be understood as a challenge to institutions of knowledge and even to the important certified laboratories. Most importantly, Brand conscientiously responded to the readers' letters and often published their recommendations.

Figure 0.47

Publication of "'Whole Earth' Philosophy," *Paddington Journal* 4 (February 1971). "The Whole Earth philosophy believes that the individual has enormous power…power to conduct his own education, find his own inspiration, shape his own environment and share his adventure with whoever is interested." M1045, flatbox 37. Courtesy of Stewart Brand and the Department of Special Collections and University Archives, Stanford University Libraries.

Paddington Journal
With Sydney Notes

CONTENTS
Antiques .... Pg.2   Fashion Pg.5
Art Gallery .... Pg.3   House & Garden Pg.7
Bookshop .... Pg.4   Gifts & Specialty Shops Pg.4
Crafts .... Pg.4   Real Estate Pg.7
Food & Drinks .... Pg.6   Travel Theatre Pg.8
For Visitors Pg.8

EDITOR & PUBLISHER: Jannie Southan
209 Underwood St., Paddington. Tel. 32-6905
PRINTER: R. V. Byers, 59 Carnarvon St., Auburn
SUBS: 12 issues by post $1.50

VOL. 4 NO.4.   FREE   ISSUE   FEBRUARY 1971.

'WHOLE EARTH' PHILOSOPHY

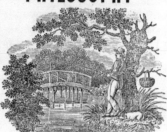

We have a talent in Australia for picking the eyes out of the worst kind of things done overseas, and importing them here as gospel. We've got all the hungry developers, the business expansionists and a plague of people. Survival hasn't reached the danger level yet, but with the help of people like Clutha Developments (see p.5) it will. In other places where danger level has been reached, reaction is setting in and in America, part of this reaction is the "alternative life revolution" described in the magazine "Cavalier" as "a return-to-basic nature kind of revolution that a minority of young people (not all long hairs either) is trying to bring about. It is a non-political revolution. It is a totally different way of looking at life, a dropout from standardised work patterns and family structures in what amounts to an increasingly totalitarian society, and finally an effort to create a whole new culture or at least sub-culture, a whole new society within the established one." It has spawned a publication called "Whole Earth Catalog" which sells for $(US)1.00 and which lists the "tools" for natural living. The Catalog describes its function as "an evaluation and access device. With it the user should know better what is worth getting and where and how to do the getting."

The Whole Earth philosophy believes that the individual has enormous power (something the Journal has been preaching too) … "power to conduct his own education, find his own inspiration, shape his own environment and share his adventure with whoever is interested". If this sounds outrageous, what could be more outrageous than the state to which organisation (in business, government, religion, leisure and everything else) has brought us? The earth won't support these organised systems, so something has to change. It isn't going to be the planet, that's for sure, unless it changes to a state where it won't support.life any longer. So for survival, people have to change.

This is one of the provocative, daring, fresh ideas from America that nobody is promoting outside in a big way. Probably because there's no money in it for some big operator. But there's life in it, and fun, and laughter, and involvement and all the things fundamental to man that can give him a sense of well-being. The Whole Earthers believe you should be involved with your living. Use lamps you have to make operate by tending them, eat food you grow, wear clothes you make … learn to make your own soap, cloth, guitars even. The Catalog not only publishes products, but users' comments about them as well, good and bad. So you know the facts and make up your own mind what you buy.

Bicycles are promoted for transport (what about that, bicycle makers — they're non-pollutant); someone has written in about the cultivation of bean sprouts" … (those wormy things in chow mien are sprouts) … an enjoyable and cheap way to add valuable vitamins and minerals to your diet. You can rig up a sprouter yourself with a glass dish, rack and towel and sprout beans sold at Health Food Stores." Even a person living in a single room could do that.

The wonderful thing about this kind of thinking is that people, no matter how old or young, need not feel helpless in the face of the overwhelming and lethal destruction they see all around in the name of progress and profits. There are things we all can do. For instance, our poverty-stricken old-age pensioners could grow bean sprouts to help their diets, and enjoy doing it. There ought to be more adventure in individual self-reliance that would not only pull exploitation of people and places into line, but offer a better kind of life to individuals, in which creativity would be encouraged to find new outlets. Busy, creative people aren't bored; they're not prowling round robbing banks, damaging peoperty, molesting other people and generally behaving like aimless souls with no hope.

The "Whole Earth Catalog", though related to American living and dealing with American problems, would have wide application here and for this reason we pass on the following information about it. It is a tabloid size book of 53 pages of articles, ideas and information (there was even a letter in it from Handweavers & Spinners' Guild of Australia at Bankstown, listing dyes and wools, so the sources are widespread). It is published 6 times a year with 2 big issues (their Spring and Autumn) and 4 smaller ones. The subscription is $9 for one year by surface mail and $14.60 (US currency) by air mail. But so you may check in case costs have changed (we have the March 1969 issue) the address is: WHOLE EARTH CATALOG, Portola Institute, 558 Santa Cruz, Menlo Park, California 94025, U.S.A. The Portola Institute "was established in 1966 on a non-profit basis to encourage, organise and conduct innovative educational projects." It relies for support on private foundations and it says because it is a private organisation with no need for profits or guaranteed "success", it can experiment with new and unusual educational projects. It seemed worth while knowing about.

One reader, Forest Shormer, from the Abundant Life Seed Foundation based in Port Townsend, Washington, which specialized in organic and bio-dynamic seeds, responded regarding the part of the catalog that he knew best. It is useful, however, to note that Shormer benefited from a review of one of his products, which could interfere with his objectivity: "Your coverage of 'Seeds' is impeccable—free of chaff. And a veil has lifted to show me why Catalog layout is so subtly compelling. We are honored to appear on page 100—may I suggest that pp. 100–101 printed on poster stock would be a valuable tool by itself—we'd order a few score copies for openers."[152]

Not all the readers were thrilled, however, with the regular catalog updates:

This "Joe Customer" will not buy your justification for a revised edition of the *Next Whole Earth Catalog*. It seems to me from this vantage point that you and your colleagues are falling or have fallen into the "better is big and big is better" business syndrome. Your weak rationalization for the need to update 10% of the old edition centers on the use of FIGURES. You speak of 30,000 this and 100,000 that, $230,000 for this and $75,000 for that but bypass the most important issue. Stewart, what about the energy consumption and the acres of trees that will sacrifice themselves for your business NEEDS? How do you rationalize that? Another 6,160 trees or 14 acres will fall for your 10% revision, to say nothing of the wasted trees in your pulped 1st edition will fall for your 10% revision.[153]

A collaborator on the *Whole Earth Catalog* added in the upper margin of this letter: "Interesting argument against 2nd Edition."[154]

Among the entries recounting the readers' testimonials, one finds, for example, on the page presenting the *Inflatocookbook* (text 30), an article by Charles Tilford (who calls himself an "inflatoenvironmentologist") titled "The Latest Inflatables." Tilford keeps the readers informed of his activities: "Here's what we've been doing with inflatables recently—at the Media Conference at Goddard, many of us (Cosmic Labs, Southcoast) lived in poly bubbles in the woods (300-foot extension cord) and stayed dry through three days of rain. Very functional and nice."[155]

The archives of the *CoEvolution Quarterly*—the magazine launched by Brand in 1974 as an intellectual sequel to the *WEC*, with an average circulation of 40,000 copies of its 150-page issues—contain much material that throws light on the *WEC*'s readership. The *CoEvolution Quarterly*'s articles were often written by readers, sometimes by well-known contributors, sometimes by both. The magazine was "read by scientists, scholars, government officials, engineers, and assorted varieties of Aquarian Argonauts, giving the magazine influence out of proportion to its circulation."[156] Bruce Black, a reader from Glendale, California, expressed his admiration for the quarterly: "One thing that is probably more important to me than anything is to always try to keep an open mind to 'everything.' I have read a number of other mags that are clinging like moss to a rock to the '60's/70's 'Organic Back to Nature' type of philosophy. This is completely backwards to what I thought we were trying to say back then—KEEP AN OPEN MIND....I see so much of the *modus operandi* of 'if it fits into our philosophy.'"[157] The added inscription "Fine Fan Mail" qualifies the spirit of the letter, probably set aside for Brand to read.

Not all the mail was positive. Here is an example of a reader who complains about the rebellious tone adopted by *CoEvolution Quarterly* in the late '70s:

Dear CoEvolution Staff: I enjoyed subscribing to your magazine a couple years ago and found many informative articles, as I am a geologist and environmentalist. I found, however, in the last issues of my subscription, attitudes of godlessness, hearty approval to lewd references and rebellion reflected in the articles. I have grown up since I subscribed to your magazine and have found a truth which may seem quite far off from what is falsely called knowledge exposed in articles by hot young scientists.[158]

Reception

Several reviews of the *Whole Earth Catalog* appeared in the professional press right from the beginning. In 1969, *Scientific American* highlighted the "well behaved child" aspect of the publication, far from the violent politics of the era: "The document is good reading: it assembles the input and output channels of a strangely attractive contemporary subculture, that of the dissenter and reformer who seeks to construct a philosophical, personal and economic refuge from this curious industrial society."[159]

The article continued by stressing the catalog's pragmatic and pedagogical ambition, including, for example, information about how to repair a car, where to find the best camping equipment catalog, or how to procure the best camera: "There is, for example, an attachment for converting your chain saw to a one-man lumber mill, a knowing account of the best kerosene lamps ('Coleman lamps are terrible—they hiss and clank and blind you, just like civilization') and information about automobile repair data, consumer guides, extrication of Government documents, discount houses and the better mail-order suppliers of camping gear, radio parts, optical components—even art reproductions."[160] The reviewer emphasized the standards for book selection but was quick to add: "There is apparently a whole list of unwritten, personal and arbitrary criteria including 'vibrations.'"[161]

The reviewer thus also identified the irrational and indulgent component and picks up on the sometimes vulgar or drug-associated language:

The personal side is conservationist, rural and iconoclastic; the books of Yoga and Tantra are here, with the journal of the Modern Utopians. Just as in the wider society, there is a strong component of the irrational and the indulgent, more from the supplement, which contains the letters of readers. ... There is a good stock of a few Anglo-Saxon words, and some touching visual exhibits of drug-induced mental deterioration. Yet even this ugly side of the subculture is more benign, one can argue, than the cancers of war and cruelty, which proliferate in the publications of our straight world and which are completely absent here.[162]

It is worth pointing out that in this passage the use of slang and information on marijuana did not disturb the *Scientific American* journalist.

The *New York Times Book Review* of February 15, 1970, examined readers' habits on both coasts by asking two journalists, Bennett Kremer and Peter Collier, to investigate bookstores. The catalog shows up in both reports. At Cody's, a Berkeley bookstore, one interviewee insisted that "the best-selling book here and probably in all northern California is 'the Whole Earth Catalog,' a Buckminster Fuller–inspired catalog of tools and devices leading man to self-sufficiency. ... 'We sell hundreds of them,' says Cody. 'We sell out every new issue as soon as we get it. People around here want to get free from mass-consumed objects and the sort of culture that grows up around them.'"[163]

That same year, *Publishers' Weekly* gave an analysis of the *Whole Earth Catalog* in a review that was both humorous and slightly cynical. The reviewer also commented on the specificity of the page layouts:

As a physical entity, the *Whole Earth Catalog* is something only a matrix could love. An awkward 11 × 14-1/4 inches, it looks like a nursery school cut-and-paste project, printed with pieces of potato and mud for ink on yellowish pulpwood paper. The design is a demented mishmash that blows the reader's mind and eye from right to left to right, to top to bottom to middle. It's hard to decide whether the typefaces are uglier than they are small or smaller than they are ugly. But, unlike *Science*, the *Whole Earth Catalog* is lovable. Giving off good vibes, it passes from hand to hand; it is used, cherished, laughed at, grooved on, quoted as Gospel, damned. It is not cut up and hung upon a nail. It's not free, either.[164]

Publishers' Weekly also said of page 48: "A sample page of the catalog shows its more and less lovable aspects: a lot of worthwhile information in an amateurish, confusing format."[165]

Publishers' Weekly noted that, of the two hundred works listed in the *WEC* in autumn 1969, only three were works of fiction:[166] Ayn Rand's *Atlas Shrugged*, a new science fiction novel by Frank Herbert, and Richard Brautigan's *Trout Fishing in America*. Publishers' Weekly asked: "'Why not *Stranger in a Strange Land*, a best-seller and a very admired cult book?' 'Fuzzy thinking,' a catalog editor told *PW* candidly. 'Stewart doesn't like fuzzy thinking.'"[167]

To conclude its assessment, *Publishers' Weekly* reported on the *WEC*'s intention to sink its own ship.[168] Brand, believing that he could go no further with the enterprise, had made it known that the catalog would cease publication in spring 1971. If he did distance himself for a time by delegating the editorship to others, the fact remains that he produced new editions in 1973, 1975, and 1980.

Recent Reception

The recent interest in the *Whole Earth Catalog* is partly a by-product of the return to ecological issues of the 1960s and 1970s and partly the work of cultural and media historians who are reassessing the period. The *WEC*'s place in the ecological movement has also been much discussed. In 2007 the American historian Andrew Kirk examined, in his book *Counterculture Green: The Whole Earth Catalog and American Environmentalism*, the emerging concept of "soft technology" in terms of counterculture. Kirk pointed out that the countercultural movement is not necessarily antitechnological. On the contrary, as the *WEC* frequently insisted, a good command of science and technology can be beneficial for producing tools—computers, for example—capable of increasing individuals' ability to act.[169] The catalog contributed to bringing together environmentalist movements by encouraging debates about technological progress versus the preservation of nature and the environment. Brand fueled discussion of appropriate technology by publishing the works of, among others, the economist E. F. Schumacher (text 40).[170] He also contributed to forging the apparently paradoxical idea that the return to a simpler, more ecologically conscious life did not necessarily mean repudiating technology.[171] Kirk also discussed how the awakening of environmental awareness would increase the outdoor recreation business. He analyzed the catalog's "Nomadics" section as the paradigm of this trend.

A growing number of scholars see the catalog as a fundamental milestone in the transition from a countercultural society to a "new capitalist" one. The already cited book by Fred Turner, *From Counterculture to Cyberculture: Stewart Brand, the Whole Earth Network, and the Rise of Digital Utopianism* (2006), shows the links between counterculture and technology that were instrumental in putting the computer within reach of a broader public. Turner was especially interested in understanding the evolution of digital culture on the West Coast, and he laid out a sociotechnical genesis of the Internet. He also described the cultural milieu of the times, especially from the point of view of media, tracing the development of new means of communication. Turner highlighted the fact that Brand, as an entrepreneur who knew how to grasp and develop communication tools, created one of the first online communication networks, which would lead to the World Wide Web.[172]

Along the same lines as Turner's book, an exhibition at the Haus der Kulturen des Welt in Berlin during the summer of 2013, accompanied by a catalog, explored the *WEC* as a "key player."[173] Like Turner, the exhibition's curators Diedrich Diederichsen and Anselm Franke sought to respond to the question "How is it that ecology and cybernetics led to a networked capitalist society." They referred to the book by Luc Boltanski and Eve Chiapello, *New Spirit of Capitalism*, published in French in 1999, which scrutinizes capitalism's transformation from Fordism to a networked organization.[174]

In the 1990s, with the emergence of the Internet, the "cyberculture utopia" was another way to express this ambivalence about capital. It took for its model the community ideal of the hippies who rebelled against the establishment. Turner, like Kirk, warned, however, against considering the *WEC* only from a countercultural perspective:

Far from rejecting mainstream culture, these two publications [*WEC* and *Radical Software*] actually embraced its technocratic ideals, its faith in expertise, and even its information technologies. In their view, mainstream America offered a cornucopia of high-technology tools; given the proper instructions, readers could use these tools to transform their collective consciousness and thereby build a new, more collaborative society. Though it is tempting today to recall the youth movements of the 1960s as a single antinomian uprising, these publications remind us that at a fundamental level, parts of the counterculture were not countercultural at all.[175]

Felicity Scott, for her part, has drawn attention to the proclaimed apolitical character of Brand and his *WEC* undertaking.[176] The *LWEC* did nevertheless recommend some works calling for direct action. For example, the "Community" section cites several engaged political texts, such as the OM Collective's *The Organizer's Manual* (text 53), and the book by Saul D. Alinsky, *Rules for Radicals: A Practical Primer for Realistic Radicals* (text 52). It seems that the *WEC* generation did manage to modify behavior on many levels even if their capitalism ran in the same vein as "natural capitalism" (capitalism with an emphasis on environmentalist thinking). Turner explained the group's reticence to adopt a political stance:

For Kesey and the Pranksters, and, ultimately, for an entire wing of the American counterculture, the confrontational politics favored by the New Left looked like a trap. To do politics was to become a politician; to change the world, you needed to begin with yourself, at home. More specifically still, you needed to begin by changing your state of mind. In his 1969 best seller, *The Making of a Counter Culture*, Theodore Roszak put it this way: "Building the good society is not primarily a social, but a psychic task."[177]

Simon Sadler has sought to understand the epistemological shift that took place with the *WEC*. In an article titled "An Architecture of the Whole,"[178] he considered the *WEC* as a sort of architecture, in spatial terms, as a meeting place where readers and contributors were led to think about their relationship to objects and to the world.[179]

The sociologist Sam Binkley, in his book *Getting Loose: Lifestyles Consumption in the 1970s*, notably in the chapter "Book as Tool: Lifestyle Print Culture and the West Coast Publishing Boom," has situated the *WEC* in the context of California's print culture.[180] U.S. publishers were traditionally situated on the East Coast, and the success of the *WEC* entailed, on the one hand, a redeployment of the publishing industry and, on the other, a "Californization" of East Coast periodicals. The underground style became mainstream.

David Senior, a librarian, curated a small exhibition titled Access to Tools: Publications from the *Whole Earth Catalog*, 1968–1974, at the Museum of

Modern Art (MoMA) in 2011. The exhibit encouraged a more detailed exploration of the books presented in the catalog.[181]

The *WEC* must be considered in the context of the 1960s, marked by a series of social and intellectual movements: demonstrations against the war in Vietnam, growing awareness of injustices against Native Americans, the civil rights movement, the feminist movement, and the founding of communities more or less dedicated to the hippie lifestyle or the ecology movement. In this context, the *Whole Earth Catalog* stands apart because of its holistic ambition. This implies perceiving and emphasizing the importance of the whole and the interdependence of its parts, within a set of integrative symbols. It furnished a basis of practical, material, intellectual, and cultural knowledge and "tools" that permitted a better and happier life. The emphasis was on taking control of one's own life rather than passively calling on aid from the store.

The Selection of Texts

More than a thousand items are referred to in the *LWEC*. The appendix lists about nine hundred titles, almost all of which can be found in the catalogs of the Library of Congress. It is, however, not comprehensive. The list does not include commercial catalogs of seeds or tools. It does include a selection of the journals advocated by *LWEC*, but these are lists of publications in a series, for which the Library of Congress does not have all the titles. Scanning the list reveals the extremely wide range of publication types included. It is worth noting that more than a third of the books were published between 1968 and 1971, which was the life span of the catalog up to that point. In part this reflects Brand's use of the R. R. Bowker publications *Publishers' Weekly* and *Forthcoming Books in Print*. The list also demonstrates the *WEC*'s global reach. Far from having a Californian bias, the catalog cited many works published in Washington, New York, London, and elsewhere.

The *Whole Earth Catalog* played an important role in publicizing books that went on to acquire a broad readership. Sam Binkley has researched the readership of some of these works. For example, John Muir's *How to Keep Your Volkswagen Alive* (text 57) sold 125,000 copies; Jeanie Darlington's *Grow Your Own: An Encounter with Organic Gardening* (text 16) sold 56,000 copies, and Alicia Bay Laurel's *Living on the Earth* (text 45) sold 10,000 copies in four months.[182]

The eighty items we have selected are intended to give a flavor of the *WEC* and show something of the range of material presented to the catalog's readers. We have provided introductions to each section and headnotes to help contextualize each piece. In our recommendations for further reading, we list one or two available works, either new editions of the texts or relevant secondary sources.

—Caroline Maniaque-Benton

Notes

1. The Stewart Brand papers in the Special Collections of Stanford University Libraries (SBP M1237) covers the period 1954–2000. It consists of 16 linear meters of documents, including 120 boxes of manuscript documents, 4 folders, 15 audiocassettes, 140 minicassettes, and some films. The archive is classified by year and theme. The *Whole Earth Catalog* archive (M1045, Special Collections, Stanford University Libraries, henceforth WEAP M1045) covers the period 1969–1986 and includes editorial files of Stewart Brand and Jay Kinney for *CoEvolution Quarterly*, as well as correspondence with readers and authors, a collection of photographs, memorandums, and publicity material.

2. *WEC* denotes the *Whole Earth Catalog* in general, and *L WEC* refers to the 1971 edition, the *Last Whole Earth Catalog. L(U)WEC* refers to the 1974 updated edition. Both the *LWEC* and *L(U)WEC* are 447 pages long. Minor changes in the same edition were also introduced from one printing to the next.

3. Typescript draft for the National Book Award ceremony at the Lincoln Center in New York on April 13, 1972, box 26, Stewart Brand Papers M1237, Special Collections, Stanford University Library, Stanford (henceforth SBP M1237).

4. Ibid. "In keeping with our practice, I owe you a brisk financial report. Actually it's very convenient: I was on the East Coast fund raising this week anyway. Your $1,000.00 donation will go through our usual non-profit channels to help out on events in Stockholm surrounding the United Nations Conference on the human environment this June. You have purchased trans-Atlantic passage for…"

5. Ibid.

6. For more on *CoEvolution Quarterly*, see David Armstrong, *A Trumpet to Arms: Alternative Media in America* (Cambridge, MA: South End Press, 1981), 287–290.

7. The different editions are explained at http://www.wholeearth.com/history-1968 -1988.php (accessed September 3, 2014). In 1975, "The *Whole Earth Epilog* was a continuation of the *Last Whole Earth Catalog*—in effect Volume II. It commences where the *Catalog* leaves off, with page 449, and comprises all new material, deeper research, and a more professional index covering both *Catalog* and *Epilog*" (*L[U]*

WEC Last (Updated) Whole Earth Catalog, 2). It contains 320 pages. The industrial designer Jay Baldwin was editor in chief of *The Next Whole Earth Catalog*, which appeared in 1980, revisiting the same principle as the *WEC*. In 1985, *CoEvolution Quarterly* changed its title to the *Whole Earth Review* (no. 44, January 1985–no. 89, spring 1996), then to *Whole Earth* (no. 90, summer 1997). *The Millennium Whole Earth Catalog: Access to Tools and Ideas for the Twenty-first Century* was published in 1994 with Howard Rheingold as editor and contained 384 pages.

8. The distribution of books and various objects was made possible by contractual agreements with publishers or manufacturers. See Sam Binkley, "The Sears of Menlo Park: The Discourse of Heroic Consumption in the *Whole Earth Catalog*," *Journal of Consumer Culture* 3, no. 3 (2003): 283–313.

9. Manuel Castells, *La société en réseaux*, vol. 1, *L'ère de l'information* (Paris: Fayard, 2001).

10. Extract from the National Book Prize evaluation by Digby Diehl and E. Harrison, journalists for *Contemporary Affairs*, cited by Stewart Brand, April 13, 1972, box 26, SBP M1237. The extract is highlighted in biro, indicating its importance to Brand.

11. Fred Turner, *From Counterculture to Cyberculture: Stewart Brand, the Whole Earth Network, and the Rise of Digital Utopianism* (Chicago: University of Chicago Press, 2006), 41.

12. Stewart Brand, notebooks, box 17, diary 1957, 7, SBP M1237.

13. The architect Louis Kahn designed its superb library and refectory in 1965.

14. The oil tycoon Edward Harkness made a substantial donation to the academy in 1930 to design a new way of teaching and learning. See "Harkness History," *Phillips Exeter Academy*, http://www.exeter.edu/ admissions/109_1220_11688.aspx (accessed July 10, 2014).

15. Andrée Goudot-Perrot, *Cybernétique et biologie* ("Que sais-je?") (Paris: Presses Universitaires de France, 1967), 8.

16. Brand kept in constant touch with Paul Ehrlich, who was a regular reader of the *WEC* and a contributor to *CoEvolution Quarterly* throughout the 1960s and 1970s. The correspondence between the two is held in the archive SBP M1237.

17. Felicity Scott adopts an ironic tone in her analysis of Brand's military training. See Scott, "The Environmental Game," in *Climats*, ed. Thierry Mandoul et al. (Gollion: Infolio, 2012), 125–126.

18. Brand drew attention to a small mistake in Turner's work: "I'll add here one micro-correction that I gather Turner plans to fix in the paperback edition. In the early 1960s I was not a draftee in the Army, but an officer on two years active duty in the Infantry. If I at times took a leadership role later on, I was just deploying what I'd been trained to do." http://www.amazon.fr/From-Counterculture -Cyberculture-Stewart-Utopianism/dp/ 0226817423 (accessed September 4, 2014).

19. Reserve Officers' Training Corps, "Principles of Military Organization," 2, box 26, fols. 1–12, SBP M1237.

20. Quoted in Dominique Cardon's introduction to Fred Turner, *Aux sources de l'utopie numérique: De la contreculture à la cyberculture, Stewart Brand, un homme d'influence* (Caen: C&F Éditions, 2012), 18. See also Loïc Petitgirard's review of Amy Dahan and Dominique Pestre, eds., *Les sciences pour la guerre: 1940–1960* (Paris: Éditions de l'École des Hautes Études en Sciences Sociales, 2004), in *La Revue pour l'Histoire du CNRS* 13 (2005), http://histoire -cnrs.revues.org/441 (uploaded January 22, 2007). Note in particular the pieces by the historian of sciences: Paul N. Edwards, *The Closed World: Computers and the Politics of Discourse in Cold War America* (Cambridge, MA: MIT Press, 1996). On the teamwork of the Manhattan Project, see Peter Galison, "The Legacy of Los Alamos," in *Critical Assembly: A Technical History of Los Alamos during the Oppenheimer Years, 1943–1946*, ed. Lillian Hoddeson, Paul W. Henriksen, and Roger A. Meade (Cambridge: Cambridge University Press, 2004), 477.

21. Stewart Brand, interview with Katherine Fulton, 1994, box 24, fol. 13, WEAP M1045.

22. Dennis J. Opatrny, "Governor's Part-Time Guru for Ideas: Stewart Brand," *San Francisco Sunday Examiner and Chronicle,*

October 2, 1977, 8. The press cutting is conserved in SBP M1237.

23. Reserve Officers' Training Corps, "Principles of Military Organization," 5.

24. Brand, interview with Fulton.

25. Stewart Brand, "Texture of Life," undated typescript, 2, box 26, fol. 1, note 8, SBP M1237. The text was written after 1975, since it refers to the article and book titled *Space Colonies.*

26. Ibid.

27. Brand, "First Jump," box 26, fol. 1, SBP M1237.

28. Robert Misrahi, *Les actes de la joie: Fonder, aimer, agir* (Paris: PUF, 1987). Overcoming fear is also a source of intense pleasure, and this is perhaps what Brand experienced. Brand had also had the experience of participating in an important scientific experiment on LSD at Menlo Park. See also Brand, notebooks, December 18, 1962, SBP M1237; and Turner, *Counterculture to Cyberculture*, 117.

29. Brand, "Alloy Report," in *LWEC* (Menlo Park: Portola Institute, 1971), 116.

30. Ibid.

31. Brand, interview with Fulton.

32. The value of evaluations of this kind lies in the fact that people are still aware of the motivations for their actions and will thus be less likely to self-censure. The information is collected swiftly, which also ensures a rapid response.

33. See Luc Boltanski and Eve Chiapello, "1. Management Discourse in the 1990s," in *The New Spirit of Capitalism* (London: Verso, 2007).

34. Headquarters and Service Co., *The Airborne Classbook*, graduation, February 10, 1961, box 26, fol. 7, SBP M1237, 1961, n.p. See also diploma: "For merit in photography: Stewart Brand has been named a winner in the Monthly Qualifying Competition of the *Popular Photography* magazine 1961," box 26, fol. 1, SBP M1237.

35. This is where USCO built the Tabernacle, a multimedia space for psychedelic meditation. See the interview with Gerd Stern regarding the exhibition *Traces du sacré*, Centre Pompidou, May 7–August 11, 2008, http://traces-du-sacre.centrepompidou .fr/exposition/paroles_artistes.php?id=50 (accessed September 14, 2014). See also

"Psychedelic Art," *Life* 61, no. 11 (September 9, 1966): 61. The light shows and cinematic experiments of the USCO group are described in this piece across four illustrated double-page spreads. This issue of *Life* also featured an article titled "Lunar View of a Socked-in Earth," 340. The planet appears on a very dark background, which may have influenced Brand in his cover for the *WEC.*

36. Interview 3 with the poet Gerd Stern (born in 1928), oral history conducted by Victoria Morris Byerly, April 16, 1996, in *From Beat Scene Poet to Psychedelic Multimedia Artist in San Francisco and Beyond, 1948–1978* (Regional Oral History Office, Bancroft Library, University of California, Berkeley, 2001), ix, 104, available in the Online Archive of California (OAC), https://archive.org/stream/beatscenepoet 00gerdrich#page/n0/mode/2up (accessed February 10, 2015). During the late 1960s, members of USCO started the Lama Foundation in New Mexico.

37. See Fred Turner, *The Democratic Surround: Multimedia and American Liberalism from World War II to the Psychedelic Sixties* (Chicago: University of Chicago Press, 2013), 287–289.

38. Brand, notebooks, January 12, 1961– January 29, 1962, quoted in Turner, *From Counterculture to Cyberculture*, 46.

39. Brand, "'We Haven't Noticed That We Are as Gods': Hans Ulrich Obrist in Conversation with Stewart Brand," *Electronics Beats*, July 2013, 66–72.

40. Kate Murphy, "Stewart Brand," *New York Times*, November 5, 2011, http://www .nytimes.com/2011/11/06/opinion/sunday/ what-stewart-brand-is-reading-and-watching .html (accessed September 4, 2014).

41. Stewart Brand, letter to "Lee," July 18, 1963, box 5, fol. 7, SBP M1237.

42. In 1967, Brand met Lois Jennings, a Native American Ottawa Indian, raised in Washington, DC. They met at a conference in Sheridan, Wyoming, during the period when Brand was photographing Native American Indians. "Through Ansel Adams, I got work with Stuart Udall, secretary of the interior, photographing Indians." Stewart Brand, interview with Andrew Brown, *Guardian*, August 4, 2001, http://www .theguardian.com/education/2001/aug/04/

artsandhumanities.highereducation (accessed November 4, 2014). See also Sherry L. Smith, *Hippies, Indians, and the Fight for Red Power* (Oxford: Oxford University Press, 2012), chap. 2.

43. Brand, notebooks, "Captions," box 20, fol. 2, SBP M1237.

44. Brand's role in the appreciation of Native Americans during the 1960s is outlined in Smith, *Fight for Red Power*. Brand wrote a substantial article on the Amerindian question, "Indians and the Counter-Culture," *Clark Creek* 18 (December 1972): 34–38.

45. Charles Perry, *The Haight-Ashbury: A History* (Rolling Stone Press, 1984), 19.

46. He created several "shows" between 1965 and 1968. See Turner, *From Counterculture to Cyberculture*, chap. 2.

47. Reserve Officers' Training Corps, "Principles of Military Organization."

48. Stewart Brand, notebooks, July 12, 1963, 5, box 17, fol. 2, SBP M1237.

49. Ibid.

50. Ibid.

51. Brand's commentary in "Understanding Whole Systems/*Geology Illustrated*," in *LWEC*, 10–11.

52. Turner, *From Counterculture to Cyberculture*, chapters 2 through 4.

53. For more on the different aspects of his career, see Turner, *From Counterculture to Cyberculture*, and Andrew G. Kirk, *Counterculture Green: The Whole Earth Catalog and American Environmentalism* (Lawrence: University Press of Kansas, 2007).

54. However, we must note Brand's relationship with countercultural designers. See Lorraine Wild and David Karwan, "Agency and Urgency: The Medium and Its Message," in *Hippie Modernism: The Struggle for Utopia*, ed. Andrew Blauvelt (Minneapolis: Walker Art Center, 2015), 44–57.

55. See Kevin Starr, "*Sunset Magazine* and the Phenomenon of the Far West," *Stanford University Library* (May 1998), http://web.stanford.edu/dept/SUL/library/sunset-magazine/html/influences_1.html (accessed August 10, 2014). The history of *Sunset: The Magazine of Western Living* (1898–1998) is recounted in this publication prepared for Stanford Library in *The American West*, a program developing cultural histories of information providers. In this article, Starr, a historian, points out that *Sunset* helped its readers—or at least the cultivated elite of the West Coast—to discover and define a "Californian" aesthetic and lifestyle. *Sunset* also helped form a particular ideal and taste. The magazine also promoted action, for example, against pollution caused by DDT.

56. In its editorial offices at Menlo Park, *Sunset* built a kind of "Laboratory of Western Living," including a kitchen in which recipes proposed by readers could be tried out and tested. There was also a garden for verifying readers' gardening tips.

57. Stewart Brand's manuscript notes of his meeting with Bill Lane (1919–2010), co-owner and chief editor of *Sunset*, box 26, SBP M1237.

58. Ibid.

59. See Sim Van der Ryn's letter to Stewart Brand, December 26, 1978, box 1, WEAP M1245.

60. Brand, meeting with Bill Lane.

61. Ibid.

62. Brand, manuscript notes of his meeting with William Flynn, *Newsweek*, box 26, SBP M1237.

63. Ibid.

64. Ibid.

65. *Harper's Magazine* has been published as a monthly without interruption since 1850. This general-interest magazine, often referred to simply as *Harper's*, covers politics, literature, culture, and the arts. The most venerable magazine published in the United States, it had a print run in 2014 of more than 200,000 copies.

66. Ibid., 9.

67. Beatriz Colomina and Craig Buckley, *Clip/Stamp/Fold: The Radical Architecture of Little Magazines, 196X to 197X* (Barcelona: Actar, 2010), 9–10.

68. *Harper's Magazine*, April 1974, 3.

69. Tony Jones (*Harper's Magazine*), letter to Stewart Brand, September 1973, box 1, WEAP M1045.

70. Brand, box 1, WEAP M1045.

71. The British economist Barbara Ward (1914–1981) became a well-known authority on the problems of developing countries with the publication of *Spaceship Earth* (1966) and another book, cowritten with René Dubos, *Only One Earth: The Care and Maintenance of a Small Planet* (1972). See also Barbara Ward, "Only One Earth," *Courrier Unesco*, January 1973, 8–10.

72. Stewart Brand, letter to Mr. Cornish (World Future Society), February 18, 1969, box 6, fol. 5, SBP M1237.

73. The Windsor typeface used in the L. L. Bean catalog was also used for the title page of the *Whole Earth Catalog*.

74. A point made by Jean-Louis Cohen in conversation, March 20, 2015.

75. The Portola Institute was a charitable foundation with an educational mission. Richard (Dick) Raymond founded the institute in Menlo Park in 1966 to encourage and help other institutions to work with music, simulation games, and computer technology in a learning environment. Raymond and Brand subsequently collaborated on the Point Foundation (1969–1975). The archives of the Point Foundation, also conserved at Stanford University (M1441), provide precise records of investment in laboratories and research groups working on appropriate technologies. Turner, *From Counterculture to Cyberculture*, 291.

76. Stewart Brand, "How to Do a *Whole Earth Catalog*," in *LWEC*, 438.

77. Ibid.

78. Joe Bronner and Annie Helmuth were named, with Brand, as editors of the *Difficult but Possible Supplement to the Whole Earth Catalog* for January, March, and July 1969. Joe Bronner, Jan Ford, Diana Shugart, and Annie Helmuth were named with Brand as editors for the *Difficult but Possible Supplement to the Whole Earth Catalog*, September 1969. Lloyd Kahn and Sarah Kahn were named for *Whole Earth Catalog*, spring 1970. Cappy McClure, Hal Hershey, Mary McCabe, and Fred Richardson were named for the supplement of January 1970. A list of the contributors is found in *LWEC*, 434.

79. Stewart Brand, "History: Some of What Happened around Here for the Last Three Years," *Whole Earth Catalog*, June 1971, http://www.wholeearth.com/issue/1150/article/322/history-some.of.what.happened.around.here.for.the.last.three.years (accessed August 10, 2014).

80. Brand, "How to Do a *Whole Earth Catalog*," 435–441.

81. Ibid., 435, quoted in Andrew Kirk, *Counterculture Green: The Whole Earth Catalog and American Environmentalism* (Lawrence: University Press of Kansas, 2007), 114.

82. Stewart Brand, "Function," *WEC*, fall 1968, 2.

83. Simon Sadler, "TEDification versus Edification," *Places*, January 2014, http://places.designobserver.com/feature/the-magical-thinking-and-many-contradictions-of-the-ted-talks/38293 (accessed August 27, 2014). In this article, Sadler analyzes the fascination of the Californian intelligentsia with a series of TED lectures. The rhetorical mode of address, themes, and invited speakers sum up what Silicon Valley residents like to hear. Sadler compares TED with the *WEC*. At first glance, the *WEC* is just a miscellaneous collection of publications; not all are of fundamental importance.

84. Brand, "How to Do a *Whole Earth Catalog*," 435.

85. Ibid.

86. Ibid.

87. "Short words, brief sentences and a concise text also.... Research has shown that the reader reads few articles after leafing through them. He prefers short articles. Magazines are no exception to this rule, and most success is achieved with short, well-illustrated articles." See Yves Agnès, *Manuel de journalisme* (Paris: La Découverte, 2008), 126.

88. Brand, "How to Do a *Whole Earth Catalog*," 435.

89. Ibid.

90. Wolfgang Langewiesche, *Stick and Rudder: An Explanation of the Art of Flying* (1944; New York: Whitlessey House/McGraw Hill, 1972). Special appendix, "The Dangers of the Air," by Leighton Collins. Illustrated by Jo Kotula.

91. Brand, "How to Do a *Whole Earth Catalog*," 435.

92. Stewart Brand, *WRAPAROUND, Harper's Magazine*, April 1974, 5.

93. Brand, "How to Do a *Whole Earth Catalog*," 435.

94 See Cathy D. Smith, "Handymen, Hippies and Healing: Social Transformation through the DIY Movement (1940s to 1970s) in North America," *Architectural Histories* 2, no. 1 (2014): 1–10, http://dx.doi.org/10.5334/ah.bd (accessed August 7, 2015).

95. The sociologist Antoine Hennion devoted much of his work to the study of amateur behavior, observing notably how amateurs respond to music and build a musical taste. See also Hennion, "Réflexivités: L'activité de l'amateur," *Réseaux*, no. 153 (2009): 55–78.

96. Brand, "How to Do a *Whole Earth Catalog*," 435–441.

97. Ibid.

98. Sam Binkley notes that "within a year and a half Bookpeople mushroomed into a $2 million business with twenty-five employees, carrying a list of mostly how-to titles, catalogs and manuals and drawing eager interest from the larger East Coast firms." Binkley, *Getting Loose: Lifestyle Consumption in the 1970s* (Durham, NC: Duke University Press, 2007), 117–118.

99. Walter Szykitka, ed. and comp., *Public Works: A Handbook for Self-Reliant Living* (Barcelona: Links Books, 1974).

100. Brand, "Texture of Life."

101. Psychological research on the workplace was important. See, among others, Kurt Lewin (1890–1947) on group dynamics and leadership, and Abraham Maslow (1908–1970) on the theory of motivation (*Motivation and Personality* [New York: Harper, 1954; 2nd ed., New York: Harper & Row 1970]). R. John Williams, "Technê-Zen and the Spiritual Quality of Global Capitalism," *Critical Inquiry* 38, no. 1 (autumn 2011): 17–70. In this article, Williams takes as his point of departure Pirsig's book *Zen and the Art of Motorcycle Maintenance* (1974) and explores the business world from the perspective of Zen Buddhism.

102. See Peter M. Senge, *The Fifth Discipline: The Art and Practice of the Learning Organization* (New York: Doubleday/Currency, 1990). The Center for Organizational Learning eventually changed its name to the Society for Organizational Learning.

103. For example, Art Kleiner, who collaborated with Senge on *The Fifth Discipline Fieldbook* (New York: Doubleday, 1994), also published a collection of articles with Brand: *News That Stayed News, 1974–1984: Ten Years of CoEvolution Quarterly* (San Francisco: North Point Press, 1986). Stephanie Mills, collaborator on the *CoEvolution Quarterly* from 1974, is well known for her interest in bioregionalism, ecological resuscitation, the economy of small communities, and the concept of voluntary simplicity. Among her publications are *Tough Little Beauties* (2007), *Epicurean Simplicity* (2002), and *Turning Away from Technology: A New Vision for the 21st Century* (Sierra Club Book, 1997).

104. Rita Fink, "Cosmic Celebration," and Stephanie Mills, "From the Encampment," *Pacific Sun*, September 1–4, 1978, 5, 23. Press cutting, box 24, WEAP M1045.

105. Art Kleiner and George Roth, *Field Manual for the Learning Historian* (Cambridge, MA: MIT-COL and Reflection Learning Association, 1996), 26.

106. Stewart Brand, letter to Cappy McClure, October 21, 1970, SBP M1237.

107. Ibid.

108. Brand, "How to Do a *Whole Earth Catalog*," 435.

109. Ibid.

110. On behavioral psychology, leadership, and the division into "directive," participatory," or "laissez-faire" groups, see Ralph White and Ronald Lippitt, *Autocracy and Democracy: An Experimental Inquiry* (New York: Harper, 1960).

111. Brand, "How to Do a *Whole Earth Catalog*," 435.

112. Kevin Kelly, "One Dead Media," *The Technium*, http://kk.org/thetechnium/2008/06/one-dead-media (accessed July 28, 2014).

113. Kathleen Musante DeWalt and Billie R. DeWalt, *Participant Observation: A Guide for Fieldworkers*, cited in Kelly, *The Technium*.

114. See Alex Wright, "The Web Time Forgot," *New York Times*, June 17, 2008, http://www.nytimes.com/2008/06/17/science/17mund.html?8dpc (accessed July 28, 2014). Cited in Kelly, *The Technium*. Article about the museum in Mons, Belgium, devoted to the Mundaneum, an early concept of a worldwide information and communication network, 1939. In his blog, Kelly mentions his book *Cool Tools: A Catalog of Possibilities* (2008), http://kk.org/books/cool-tools-a-catalog-of-possibilities.php (accessed July 28, 2014).

115. Kelly, "One Dead Media."

116. Brand, "How to Do a *Whole Earth Catalog*," 435.

117. During the 1980s, Brand set up the Whole Earth 'Lectronic Link (WELL), a first attempt to create a virtual community. See Fred Turner, "Where the Counterculture Met the New Economy: The WELL and the Origins of Virtual Community," *Technology and Culture* 46, no. 3 (July 2005): 485–512. See also Felicity D. Scott, "Networks and Apparatuses, circa 1971: Or, Hippies Meet Computers," in *Hippie Modernism*, 102–113.

118. Denis Cosgrove, "Contested Global Visions: One-World, Whole-Earth, and the Apollo Space Photographs," *Annals of the Association of American Geographers* 84, no. 2 (June 1994): 270–294.

119. Brand, "How to Do a *Whole Earth Catalog*," 435.

120. Wild and Karwan, "Agency and Urgency," 44–57.

121. The first model equipped with this new typewriter technology, invented by IBM, came onto the market in 1961. J. S. Morgan and J. R. Norwood, "The IBM Selectric Composer: Justification Mechanism," *IBM Journal of Research and Development* 12, no. 1 (January 1968): 68–75. The Selectric was designed by Eliot Noyes, chief designer at IBM for twenty-one years. See John Harwood, "The White Room: Eliot Noyes and the Logic of the Information Age Interior," *Grey Room*, no. 12 (summer 2003): 20.

122. The Univers font was designed for IBM typewriters by Adrian Frutiger around 1957 for the French type foundry Deberny and Peignot.

123. Brand, "How to Do a *Whole Earth Catalog*," 435. "We publish considerable detailed information—fine print. Sorting among that is aided by a consistent code of type-faces (reviews are always 'univers italic,' access is always 'teeny,' *Divine Right* is always 'bold teeny,' and so forth). The IBM Selectric Composer makes this an easy matter. Still we're not as consistent as we should be."

124. Ibid.

125. Ibid., 436.

126. Ibid.

127. Greg Castillo, "Counterculture Terroir: California's Hippie Enterprise Zone," in *Hippie Modernism*, 87–101.

128. Ibid., 94.

129. Ibid.

130. Ibid., 95.

131. Ibid.

132. Charles and Ray Eames, "Polaroid Ad," 1972, YouTube, https://www.youtube.com/watch?v=5jaiq_ZZ_eM (accessed September 3, 2014).

133. Brand, "How to Do a *Whole Earth Catalog*," 436.

134. Fred Richardson, *L WEC*, 436.

135. Stewart Brand took the opportunity of this break to build a cabin in Canada (SBP M1237).

136. Gurney Norman, letter to Stewart Brand, 1970, box 6, fol. 5, SBP M1237.

137. The novel serialized in the *L WEC* was published under the title *Divine Right's Trip: A Folk-Tale* (New York: Dial, 1972).

138. Hannah Arendt, *The Human Condition*, quoted by Brand in the *L WEC*, 17.

139. Stewart Brand, *The Last Whole Earth Catalog*, 112.

140. Stewart Brand, ed., "Report from Alloy," *Supplement Whole Earth Catalog*, March 1969, is reprinted in the *L WEC* and in the *L(U)WEC*, 111–117. The quote is on page 112.

141. Brand, "Report from Alloy," 116. For more on Alloy and technology, see Lionel Devlieger, "Alloy, la retraite dans le désert," *Faces: Journal d'Architecture* 71 (summer 2012): 56–59. On the environmental aspect of the Alloy conference, see Kirk, *Counterculture Green*.

142. Steve Baer, quote from Brand's notepad, March 1969, SBP M1237, box 30. In the same document, Steve Durkee added: "The mike is a ploy to create structure. It doesn't have to be plugged in."

143. A series of publications attests to the works undertaken by this group. See *Office of Appropriate Technology: Purpose, Organization, and Activities* (Sacramento: Appropriate Technology, 1976).

144. John Brademas, letter to Stewart Brand, April 14, 1970, box 30, SBP M1237.

145. Stewart Brand, reply to John Brademas, Chairman, Committee on Education and Labor, U.S. House of Representatives, April 1970, box 30, SBP M1237.

146. Wisconsin senator Gaylord Nelson proposed the first important demonstration on American soil to shake up the political arena and prompt a political reaction to the environmental issue. The first Earth Day led to the establishment of the Environment Protection Agency (EPA) and the adoption of the Clean Air Act in 1970, Clean Water Act in 1972, and Endangered Species Act in 1973.

147. *Saline Valley Production*, 1971 (film), produced in Saline Valley, CA, http://www.youtube.com/watch?v=OqRdGfDPCDE (accessed August 4, 2014).

148. For the selection of texts—for example, in the "Community" and "Craft" sections of the *WEC*—dealing with everyday life, see Alicia Bay Laurel, *Living on the Earth* (New York: Random House, 1970); on the kitchen, Ita Jones, *The Grubbag* (New York: Random House, 1971); and on health, Boston Women's Health Book Collective, *Our Bodies, Ourselves* (New York: Simon & Schuster, 1977). These works conform in many ways to gender stereotypes of the day.

149. Stewart Brand's commentary, in *Saline Valley Production*, 1971 (film).

150. Stewart Brand, quoted in Kevin Kelly's blog, http://www.youtube.com/watch?v=OqRdGfDPCDE (accessed August 4, 2014).

151. Brand, "How to Do a *Whole Earth Catalog*," 435.

152. Forest Shormer, letter to Stewart Brand, Port Townsend, WA, October 17, 1980, box 28, WEAP M1045.

153. David Brooks (Tin Mountain Conservation Center, NH), letter to the *Whole Earth Catalog* team, November 12, 1981, box 28, WEAP M1045.

154. Ibid.

155. Charles Tilford, letter to the *Whole Earth Catalog* editorial team (WEAP M1045). Tilford, trained as a mechanical engineer, started creating inflatable multimedia environments at Columbia University. Wanting to make more permanent fabric structures, he applied for an MS in Architectural Technology from Columbia University in 1972, working with Mario Salvadori. Tilford was a one-time Ant Farm collaborator for a structure at Antioch College, Ohio.

156. David Armstrong, *A Trumpet to Arms: Alternative Media in America* (Cambridge, MA: South End Press, 1981), 289.

157. Bruce Black, letter to the *Whole Earth Catalog* team, December 6, 1980, WEAP M1045.

158. Charles Tilford, letter to the *Whole Earth Catalog* editorial team, box 28, WEAP M1045.

159. *Scientific American*, June 1969, 142, box 24, WEAP M1045.

160. Ibid.

161. Ibid.

162. Ibid.

163. Bennett Kremer and Peter Collier, "Unrequired Reading: East and West," *New York Times Book Review*, February 15, 1970, 5.

164. "The Whole Earth Catalog," *Publishers' Weekly*, May 11, 1970, 21, box 24, WEAP M1045.

165. Ibid.

166. Fiction played an important role in the *LWEC*, with the serialization of Gurney Norman's novel on each page.

167. "The Whole Earth Catalog," *Publishers' Weekly*, 20.

168. Ibid., 21. "Perhaps the most interesting aspect of the *Whole Earth Catalog* is its intention to self-destruct. Having achieved its purpose, to aid 'the power of the individual to conduct his own education, find his own inspiration, shape his own environment, and share his adventure with whoever is interested,' the Catalog has announced that it will cease publication with the issue of spring, 1971."

169. Kirk, *Counterculture Green*.

170. *Whole Earth Catalog* and *CoEvolution Quarterly* helped to publicize the ideas of E. F. Schumacher (*Small Is Beautiful* [New York: Harper & Row, 1973]) and Amory Lovins ("Energy Strategy: The Road Not Taken," *Foreign Affairs* 55 [October 1976]). Stewart Brand and Jay Baldwin published *Soft-Tech* (New York: Penguin Books, 1978).

171. Andrew Kirk, "Appropriating Technology: *The Whole Earth Catalog* and Counterculture Environmental Politics," *Environmental History* 6, no. 3 (July 2001): 374–394.

172. Turner, *From Counterculture to Cyberculture*.

173. The exhibition Whole Earth: California and the Disappearance of the Outside was presented at the Haus der Kulturen der Welt (HKW), Berlin, during the summer of 2013. Diedrich Diederichsen and Amselm Franke, dir., *Whole Earth: California and the Disappearance of the Outside* (Berlin: Sternberg Press, 2013). The organizers drew on contemporary artists to reflect on the 1970s. The two curators, Diederichsen and Franke, depict this historic moment, considered today as being at the origin of the "Californian ideology," an alliance between hippie culture and cybernetics, a certain romantic vision of nature melded with a reverence for technology and the computational culture. The political question runs through the technological one. The political conservatism, or apoliticism, of the *WEC* heroes goes hand in hand with the rerouting of technology. However, this gives us cause to reflect on another alliance: that of cybernetics and power. This means state power, the power to make war, of course, but also to organize the territory and to think about and build the city.

174. Boltanski and Chiapello, *New Spirit of Capitalism*. The authors demonstrate capitalism's ability to incorporate criticism and build it into its structures.

175. Fred Turner, "Bohemian Technocracy and the Countercultural Press," in *Power to the People: The Graphic Design of the Radical Press and the Rise of the Counterculture, 1964–1974*, ed. Geoff Kaplan (Chicago: University of Chicago Press, 2013), 133.

176. Scott, "The Environmental Game," 99–127. Scott stresses that Brand not only promised access to technology and information but also succeeded in conceptualizing the demands of an increasingly larger group of individuals, which permitted the emergence of alternative lifestyles and a largely depoliticized environmental sensitivity. She observes that it was under the tutelage of Brand that the *WEC* seized the imagination of the young American generation and facilitated the shift of the New Left protesters' demands toward the desire for new lifestyles, and that, as a disciple of Buckminster Fuller, Brand always emphasized his unapologetic rejection of politics.

177. Turner, "Bohemian Technocracy," 136.

178. Simon Sadler, "An Architecture of the Whole," *Journal of Architectural Education* 61, no. 4 (May 2008): 108–129.

179. Despite his determinism and his interest in the science of systems (cybernetics), Sadler's article suggests that the holistic thinking present in the *WEC* opens the aesthetic and ideological pathway to a sort of indeterminism, appropriate to the idea of ecological sustainability. Sadler positively evaluates the contribution of the *WEC* as a crystallization of inventions and possibilities, as opposed to the regulatory requirements of sustainable development that were narrowly focused on energy performance, for example. He points out that today we neglect the intellectual project and the transformational, perhaps even revolutionary, possibilities that the *WEC* presupposed. See also from the same author, "Mandalas or Raised Fists? Hippie Holism, Panther Totality, and Another Modernism," in *Hippie Modernism*, 114–125.

180. Binkley, *Getting Loose*.

181. "Access to Tools: Publications from the *Whole Earth Catalog*, 1968–1974," April 18–December 10, 2011, http://www.moma.org/interactives/exhibitions/2011/AccesstoTools (Accessed July 18, 2013).

182. Binkley, *Getting Loose*, 118.

Understanding Whole Systems

"Understanding Whole Systems" was not the longest section of the catalog, but it was probably the most important, for two reasons. The *LWEC*'s opening section imaginatively pursued the principle of holistic thought, as opposed to specialization, across a range of metaphors, from the geological and cartographic view of the world, to the analysis of fundamental economic and biological systems, to the structure of human thought. Many of these topics would be picked up in subsequent sections. It is characteristic that the section began with Richard Buckminster Fuller's books, since he was seen as a visionary global thinker. Fuller was at once a charismatic spokesman for environmental concerns, an enthusiastic technocrat, and an example of American inventiveness. His contribution to the catalog was both intellectual and cultural.

The section was also important as an illustration of the methods Brand used in juxtaposing images and texts in double-page spreads. In *The Mechanical Bride: Folklore of Industrial Man* (1951), Marshall McLuhan analyzed newspaper layouts, showing how apparently unconnected events, presented on the same page, can create new levels of meaning. Brand deployed these tactics explicitly and with great skill. On pages 8 and 9 of the *WEC*, for example, we find images of geological formations viewed from above juxtaposed with images of the human body. Brand did not explicitly suggest the connection, but in his text the telling use of keywords revealed the intention. After describing Joseph Royce's book of photographs *Surface Anatomy* (1965) in terms that could be used for describing landscape, Brand then reversed the process when discussing Hanns Reich's *The World from Above* (1968). Here Brand's use of words like "glamor" and "anatomy" made the connection with the human body clear.

The cumulative effect was to provide unexpected and striking perceptions of the world and everything in it. In the end, the aim was to help us better understand ourselves. As Brand said of *Surface Anatomy*: "The book is included as a companion piece to the earth picture books. The whole lovely system of the human creature, seen without, surface by surface is here. One of its main revelations is how cliché ridden our usual views of ourselves are—we are still not good with mirrors" (*LWEC*, 10).

"Understanding Whole Systems" also placed an emphasis on biology. Biologists played a central role in the alternative culture of the 1960s and 1970s: Gregory Bateson (his father was a prominent British biologist, and he studied zoology, then anthropology), Paul Ehrlich, Stewart Brand, and Jay Baldwin all shared a training in biology. These men were sensitive to the risks to society of environmental changes.

As Howard T. Odum observed: "Bit by bit, the machinery of the macroscope is evolving in various senses and in the philosophical attitudes of students. ... Man already having a clear view of the parts in their fantastically complex detail, must somehow get away, rise above, step back, group parts, simplify concepts, interpose frosted glass, and thus somehow see the big patterns (Odum, *Environment, Power, and Society*, 8). This was the section's fundamental aim. Many of the authors also stressed the ecological dangers of industrialization.—CM

1 Richard Buckminster Fuller, *Utopia or Oblivion*, 1969

Richard Buckminster Fuller was a figure of inspiration for participants in the counterculture. As Stewart Brand said, "The insights of Buckminster Fuller initiated this catalog." This respect was mutual. In 1966 Fuller awarded the Drop City community his Dymaxion Award for "poetically economic" domed living structures. It was the breadth of Fuller's contribution that was striking, from technological invention, through environmental concerns, to a kind of transcendentalism and a belief in the essential goodness of human beings and nature when uncorrupted by politics and materialism. It was this, as much as his invention of the geodesic dome, that made him the father figure of the catalog. Critics of alternative architecture are quick to point out that Fuller had worked for many years in the service of the U.S. Marine Corps and designed geodesic domes for use as emergency shelters and equipment protection. A much-reproduced photograph shows one of his domes dangling from a helicopter.

Fuller was famous for his long and complex lectures, delivered all over the world, using a unique vocabulary and mixing highly technical descriptions with utopian schemes for changing the world. *Utopia or Oblivion* consists of a series of essays drawn from these lectures. One of the experiments described in the book is the World Peace Game, an ironic reversal of strategic war games, which was intended to search for means of redistributing natural resources across the globe. Played out on a Dymaxion world map and supported by a database of statistical information, the game was supposed to provide a radical alternative to market forces.

The World Peace Game had originally been conceived at Southern Illinois University as the "Inventory of World Resources, Human Trends, and Needs," and Fuller envisioned and expected it to be played by major individuals and teams. Based on an unrealistic idealism, he imagined a rational distribution of the world's resources without conflict or national boundaries. —CM

Source
R. Buckminster Fuller, *Utopia or Oblivion* (Baden, Switzerland: Lars Müller, 2008). © 1969, 2008, the Estate of R. Buckminster Fuller. (Referred to in *LWEC*, 3.)

Further Reading
Gabel, Medard. "Buckminster Fuller and the Game of the World." In *Buckminster Fuller: Anthology for a New Millennium*, ed. Thomas T. K. Zung, 122–127. New York: St. Martin's Griffin, 2001.

"R. Buckminster Fuller Retrospective." *Architectural Design* 42 (December 1972): 746–773.

"The World Game—How to Make the World Work," from *Utopia or Oblivion*

The fact is that now—for the first time in the history of man for the last ten years, all the political theories and all the concepts of political functions—in any other secondary roles as housekeeping organizations—are completely obsolete. All of them were developed on the you-or-me basis. This whole realization that mankind can and may be comprehensively successful is startling. ...

Here on Southern Illinois' campus we are going to set up a great computer program. We are going to introduce the many variables now known to be operative in economics. We will store all the basic data in the machine's memory bank; where and how much of each class of the physical resources; where are the people, what are the trendings and important needs of world man?

Next we are going to set up a computer feeding game, called "How Do We Make the World Work?" We still start playing relatively soon. We will bring people from all over the world to play it. There will be competitive teams from all around earth to test their theories on how to make the world work. If a team resorts to political pressures to accelerate their advantages and is not able to wait for the going gestation rates to validate their theory they are apt to be in trouble. When you get into politics you are very liable to get into war. War is the ultimate tool of politics. If war develops the side inducing it loses the game. ...

"The game" will be hooked up with the now swiftly increasing major universities' information network. This network's information bank will soon be augmented by the world-around satellite-scanned live inventory of vital data. Spy satellites are now inadvertently telephotoing the whereabouts and number of beef cattle around the surface of the entire earth. The exact condition of all the world's crops is now simultaneously and totally scanned and inventoried. The interrelationship of the comprehensively scanned weather and the growing food supply of the entire earth are becoming manifest.

In playing "the game" the computer will remember all the plays made by previous players and will be able to remind each successive player of the ill fate of any poor move he might contemplate making. But the ever-changing inventory might make possible today that which would not work yesterday. Therefore the successful stratagems of the live game will vary from day to day. The game will not become stereotyped. ...

The general-systems-theory controls of the game will be predicated upon employing within a closed system the world's continually updated total resource information in closely specified network complexes designed to facilitate attainment, at the earliest possible date, by every human being of complete enjoyment of the total planet earth, through the individual's optional traveling, tarrying, or dwelling here and there. This world-around freedom of living, work, study, and enjoyment must be accomplished without any one individual interfering with another and without any individual being physically or economically advantaged at the cost of another. ...

All the foregoing objectives must be accomplished not only for those who now live but for all coming generations of humanity. How to make humanity a continuing success at the earliest possible moment will be the objective. The game will also be dynamic. The players will be forced to improve the program—failure to improve also results in retrogression of conditions. Conditions cannot be pegged to accomplishment. They must also grow either worse or better. This puts time at a premium in playing the game.

Major world individuals and teams will be asked to play the game. The game cannot help but become major world news. As it will be played from a high balcony overlooking a football field-sized Dymaxion Airocean World Map with electrically illumined data transformations, the game will be visibly developed and may be live-televised the world over by a multi-Telstar relay system. ...

Ultimately its most successful winning techniques will become well known around the world and as the game's solutions gain world favor they will be spontaneously resorted to as political emergencies accelerate. ...

Comprehensive coordination of bookings, resource, and accommodation information will soon bring about a twenty-four hour, world-around viewpoint of society which will operate and think transcendentally to local "seasons" and weathers of rooted botanical life. Humanity will become emancipated from its mental fixation on the seven-day-week frame of reference.

2 Howard T. Odum, *Environment, Power, and Society*, 1971

"When the cosmic yum comes by, you get the ONE! all right, but that may not particularly help you work with connectedness," Stewart Brand noted in his 1971 review of Howard T. Odum's *Environment, Power, and Society*. "The terms and understandings in this book can." Odum, a trained zoologist, became interested in the concept of systems ecology while completing his PhD at Yale University; his particular interest, as may be observed in the following passage, was energetics, a viewpoint focused on the flow of energy through various types of systems, from coral reefs to human relations. Although he felt the detailed approach of many of his fellow systems theorists and scientists to be essential, Odum was concerned that, in adopting a microscopic perspective, they became unable to solve the larger problems of contemporary society, namely, those regarding "man's environment, his social systems, his economics, and his survival" (*Environment, Power, and Society*, 9). In response to this perceived difficulty, Odum took up a macroscopic view of larger systems, integrating the many parts into a simpler whole defined by energy language. This fundamental project of *Environment, Power, and Society* in some ways corresponded to Brand's own enterprise; both author and editor sought to promote a holistic worldview through systems theory. For readers, the book could be not only an educational text on ecology but also a reinforcement of the whole systems philosophy endorsed by the catalog.—MG

Source

Howard T. Odum, *Environment, Power, and Society* (New York: Wiley Interscience, 1971), 6–9. Reprinted with permission. (Referred to in *LWEC*, 8.)

Further Reading

Odum, Eugene, and Howard T. Odum. *Fundamentals of Ecology*. Philadelphia: W. B. Saunders, 1960.

Odum, Howard T. *Environment, Power, and Society for the Twenty-first Century: The Hierarchy of Energy*. New York: Columbia University Press, 2007.

Van Dyne, George M. *Ecosystems, Systems Ecology, and Systems Ecologists*. Oak Ridge, TN: Oak Ridge National Laboratory, Health Physics Division, 1966.

"Man Takes Over Nature with Fossil Fuels," from *Environment, Power, and Society*

In modern times man's energetic relation to his environment has changed, for he now has much more fuel under his control, and it enters through a different route. Man's new industrialized system now derives its energies from the flows of concentrated fossil fuels, coal, and oil. Much of this energy flow is put back into his environmental system so that his yield of food and critical materials is greater.

The changes have come so fast that man's customs, mores, ethics, and religious patterns may not have adapted to them. We may wonder whether the programs controlling the behavior of the individual human being, whether he is a leader or a person led, are realistic in providing for the best adaptation to the new systems of energy flow. Man's role in the environment is becoming so enormous that his energetic capacity to hurt himself by upsetting the environmental system is increasing. His system is becoming so large and complex and is changing so rapidly that it is more and more difficult to learn of the patterns in his own fuel supply. Money, energy flows, and causal actions can no longer be easily equated. The accumulation point of money in the network may be much removed from the origin of the work done, of the energy inflow, or of the forces causing stress.

Only a small part of the total controlled energy is now processed by the individual person or by work that is recognizably personal. More and more of the energy flows are in machines of the system. We may wonder whether the individual human being understands the real source of the bounty to him of the new energy support. How many persons know that the prosperity of some modern cultures stems from the great flux of oil fuel energies pouring through machinery and not from some necessary and virtuous properties of human dedication and political designs? It is not easy to fit into our thinking the concept that part of the energy for growing potatoes comes from fossil fuel. In the new system much of the higher agricultural yields are only feasible because fossil fuels are put back into the farms through the use of industrial equipment, industrially manufactured chemicals, and plant varieties kept in adaptation by armies of agricultural specialists supported on the fossil-fuel-based economy.

How many of the old landmarks in the relation of man and energy are gone? How often now is the economic purchasing power supplied a man given in relation to the work he processes? How often now is the energy processed by a group in proportion to the land area of forests and fields controlled by that group? How often now do the instructions on good and bad behavior given to the young have convincing reality for the best energetic flows of the system or the individual?

In a thousand ways, the accelerating changes accompanying an increasing budget of energy for man raise questions of his ultimate role and survival. The pessimist talks of man as the next in line of the extinct dinosaurs and other predominant types that once inherited the earth. Like the dinosaurs, man is developing a system of specialists and giant mechanisms. The extinction of the dinosaurs serves as a warning that our so-called progress may not be a safe plan for survival.

As the industrialized areas increase, how much longer will the biological cycles in the uninhabited environments be able to absorb and regenerate the wastes and thus prevent self-poisoning of waters and atmosphere? As man's system becomes large enough to control and prevail in the flows of the biosphere, will he understand it well enough to prevent disaster? If his energy sources begin to decline, can he return to a minor energetic position in the earth system without a collapse and extinction of culture as we now know it?

Critical issues in public and political affairs of human society ultimately have an energetic basis, and the increasing number of urgent energy demands measure the accelerating changes in the system containing man. The action programs needed on such public issues as birth control,

land ownership, man in space, war prevention costs, nuclear power, zoning of land space, human medical maintenance, and world economics must each be limited and fitted into the overall energetic budget for a successful system of nature. Energetic budgeting in simpler systems of nature that do not include man is readily discernible and has often been described. The same principles apply to the vast industrialized system. Yet how seldom is the energetic budget discussed or its rules applied to human society. Subconscious attempts to use economic data as a substitute for energy data may be misleading. . . .

Man's survival will probably depend on his being able to see what his vast human system has become in relation to preceding and possible earth systems. And he must acquire the necessary understanding rapidly enough to adapt his opinions, folkways, mores, and action programs to the great new systems and provide a continuing survival path for them.

Since decisions on such matters in the arena of public affairs are ultimately made according to the beliefs of the citizens, it is the citizens who must somehow include the energetics of systems in their education. In some way the behavior of the large and small systems must be understood and that knowledge must be communicated to the dispersed intelligence of the modern decision apparatus.

Can we learn the nature of macroscopic environmental systems including man, and can we clarify their functions so that they are visible to the citizen and his leaders? Is there an adequate old or new morality which is preadapted to guide individual actions for the survival of the group in the realities of the new energetic situations?

3 Hanns Reich, *The World from Above*, 1966

Stewart Brand's fascination with the power of photography to suggest hidden meanings explains his choice of Hanns Reich's astonishing book of photographs. The high-contrast photographs bring out strange and puzzling patterns on the earth's surface. As with the choice of the view of the earth from space, these images emphasize a kind of holistic knowledge of the earth, comparing the natural and the humanmade. As Reich explained: "In this book the reader is invited to view the world and man through the macroscope. We hope eventually to discern the great clanking wheels of the machinery in which man is such a small component.... The levels, folds, surface motions of the earth's crust belong to its inner dynamics. A stony primal landscape, curvaceous forms, tectonic figurations of rocks, tropics and jungles, steppes and deserts—the history book of the world, lies open, huge and readable, below us in the depths" (*The World from Above*, 11).

Brand was clearly struck by the technical information provided in the captions: altitude, make of camera, aperture, and exposure: "Close-up glamor shots of the Earth. Mystery shots (What is that? What's our altitude above it, 10 feet or 10,000?).... Good traffic flow pattern shots: surface anatomy of civilization. Not a bad compendium; it'll do until they reprint E. A. Gutkind's *Our World from the Air*." The blend of knowledge and aesthetics is characteristic of Brand's editorial approach. The book does not have a particularly ecological message, but it does prompt the reader to reflect on humanity's impact on nature. Later, photographers such as Yann Arthus-Bertrand (*La terre vue du ciel*, 1999) produced damning evidence of the pollution of the earth.—CM

Source

Hanns Reich, *The World from Above* (New York: Hill and Wang, 1966), plate 13. (Referred to in *LWEC*, 11.)

Further Reading

Boeke, Kees. *Cosmic View: The Universe in 40 Jumps.* New York: J. Day, 1957.

Gutkind, Erwin Anton. *Our World from the Air: An International Survey of Man and His Environment.* Garden City, NY: Doubleday, 1952.

Figure 1.1
Hanns Reich, "River currents in Iceland. Beneath the Vatnajökull Glacier, volcanic activity brings forth an enormous quantity of melted water that flows down to the sea every five years in a great number of streams. Rolleiflex. Focal 80b Altitude 1,000/1,200 ft." Hanns Reich, *The World from Above* (New York: Hill and Wang, 1966), plate 13.

4 Lao-Tzu and Archie Bahm, *Tao Teh King*, 1958

Of the *Tao Teh King*, Stewart Brand wrote: "It describes how the universe is and makes an excellent case for harmony as the only survival technique that works"; his inclusion of the American philosopher Archie Bahm's translation of the ancient Chinese text lent a spiritual dimension to the concept of whole systems set forth in the catalog and revealed, more generally, the allure of Eastern religion in the evolving countercultural worldview. Unique to Bahm's translation, and perhaps what led Brand to describe this version as "straightforward," is his definition of *Tao* as "Nature" and *Teh* as "Intelligence," terms that Bahm thought most appropriately expressed the original Chinese meaning within American culture. Although not mentioned by Brand, another crucial element to Bahm's translation is an extensive commentary by the author, found at the end of the book, which explained the methodology by which he arrived at his determination. A philosophy professor at the University of New Mexico, Bahm was able to clearly situate his chosen translation of the *Tao Teh King* within the context of traditional Chinese culture, the Western philosophical canon, and the history of Western interpretations of the venerable document. While the *Tao Teh King* addressed practical topics beyond the somewhat mystical-sounding principle of "how the universe is"—community structure, proper government, and ethical behavior, for instance—Brand's summation, that "harmony [is] the only survival technique that works," succinctly affirmed its relevance to a whole systems perspective.—MG

Source

Lao-Tzu and Archie Bahm, *Tao Teh King: Interpreted as Nature and Intelligence* (New York: F. Ungar, 1958), 11, 68–69. (Referred to in *LWEC*, 12.)

Further Reading

Jackson, Carl. "The Counterculture Looks East: Beat Writers and Asian Religion." *American Studies* 29, no. 1 (spring 1988): 51–70.

Lao-Tzu. *Tao Te Ching*. Translated by Stephen Addiss and Stanley Lombardo. Boulder, CO: Shambhala Publications, 2007.

Williams, R. John. "*Technē*-Zen and the Spiritual Quality of Global Capitalism." *Critical Inquiry* 38, no. 1 (autumn 2011): 17–70.

Chapters I, LXXX, and LXXXI from *Tao Teh King*

I

Nature can never be completely described, for such a description of Nature would have to duplicate Nature.

No name can fully express what it represents.

It is Nature itself, and not any part (or name or description) abstracted from Nature, which is the ultimate source of all that happens, all that comes and goes, begins and ends, is and is not.

But to describe Nature as "the ultimate source of all" is still only a description, and such a description is not Nature itself. Yet since, in order to speak of it, we must use words, we shall have to describe it as "the ultimate source of all."

If Nature is inexpressible, he who desires to know Nature as it is in itself will not try to express it in words.

To try to express the inexpressible leads one to make distinctions which are unreal.

Although the existence of Nature and a description of that existence are two different things, yet they are also the same.

For both are ways of existing. That is, a description of existence must have its own existence, which is different from the existence of that which it describes; and so again we have to recognize an existence which cannot be described.

LXXX

The ideal state is a small intimate community
Where all the necessities of life are present in
    abundance.
There everyone is satisfied to live and die without
    looking around for greener pastures.
Even if they have cars or boats, they do not use
    them for traveling abroad.
Even if they have police and fortifications, these
    are never put to use.
Business transactions are simple enough to be
    calculated on one's fingers rather than requir-
    ing complicated bookkeeping.
The people are satisfied with their food,
Contented with their clothing,
Comfortable in their dwellings,
And happy with their customs.
Even though neighboring communities are
    within sight,
And the crowing of the neighbor's cocks and
    barking of the neighbor's dogs are within
    hearing,
They grow old and die without ever troubling
    themselves to go outside of their own com-
    munities.

LXXXI

He who is genuine is not artificial;
He who is artificial is not genuine.
He who is intelligent is not quarrelsome;
He who is quarrelsome is not intelligent.
He who is wise is not pretentious;
He who is pretentious is not wise.
Therefore the intelligent man does not struggle to
    achieve for himself.
The more useful he is to others, the more he will
    be taken care of by others.
The more he yields to the wishes of others, the
    more his needs will be cared for *by* those
    repeatedly benefitted by his yielding.
Nature's way is to produce good without evil.
The intelligent man's way is to accept and follow
    Nature rather than to oppose Nature.

5 Aldo Leopold, *Sand County Almanac and Sketches Here and There*, **1949**

Aldo Leopold (1887–1948) was a forestry conservationist who became a champion for preserving wilderness areas. He developed the principle of a "land ethic" that saw soil, water, plants, and animals as members of our community rather than mere objects in the service of humankind. His approach marked a substantial shift of emphasis from man's needs to the interests of nature as a whole, which he referred to as a "biotic community." He claimed: "A thing is right when it tends to preserve the integrity, stability, and beauty of the biotic community. It is wrong when it tends otherwise." In 1935 Leopold decided to purchase a strip of impoverished land, which had suffered from intensive farming. His aim was both to study and improve the land but also to use it as a teaching tool. His essays charting the course of this experiment were published posthumously as *Sand County Almanac* in 1949. The book is still highly regarded for its insights into nature conservation.

Brand empathized with Leopold's mission to try to educate the city-dwelling, wage-earning "modern man" to understand the nature on which his comforts depended. As Leopold said: "Perhaps the most obvious obstacle impeding the evolution of a land ethic is the fact that our educational and economic system is headed away from, rather than toward, an intense consciousness of land. Your true modern is separated from the land by many middlemen, and by innumerable physical gadgets" (*LWEC*, 13). —CM

Source

Aldo Leopold, *Sand County Almanac and Sketches Here and There* (New York: Oxford University Press, 1949). (Referred to in *LWEC*, 13.)

Further Reading

The full text is accessible at http://oregonstate.edu/instruct/phl201/modules/texts/text3/leopold.html.

Freyfogle, Eric T. "Entry to Land Ethic." In *Encyclopedia of Environmental Ethics and Philosophy*, ed. J. Baird Callicott and Robert Frodeman, 22–26. Detroit: Macmillan Reference USA, 2009.

"The Land Ethic," from *Sand County Almanac and Sketches Here and There*

The Ethical Sequence

This extension of ethics, so far studied only by philosophers, is actually a process in ecological evolution. Its sequences may be described in ecological as well as in philosophical terms. An ethic, ecologically, is a limitation on freedom of action in the struggle for existence. An ethic, philosophically, is a differentiation of social from anti-social conduct. These are two definitions of one thing. The thing has its origin in the tendency of interdependent individuals or groups to evolve modes of co-operation. The ecologist calls these symbioses. Politics and economics are advanced symbioses in which the original free-for-all competition has been replaced, in part, by co-operative mechanisms with an ethical content. The extension of ethics to this third element in human environment is, if I read the evidence correctly, an evolutionary possibility and an ecological necessity. It is the third step in a sequence. The first two have already been taken. Individual thinkers since the days of Ezekiel and Isaiah have asserted that the despoliation of land is not only inexpedient but wrong. Society, however, has not yet affirmed their belief. I regard the present conservation movement as the embryo of such an affirmation. An ethic may be regarded as a mode of guidance for meeting ecological situations so new or intricate, or involving such deferred reactions, that the path of social expediency is not discernible to the average individual. …

The Community Concept

All ethics so far evolved rest upon a single premise: that the individual is a member of a community of interdependent parts. His instincts prompt him to compete for his place in the community, but his ethics prompt him also to co-operate (perhaps in order that there may be a place to compete for). The land ethic simply enlarges the boundaries of the community to include soils, waters, plants, and animals, or collectively: the land.

This sounds simple: do we not already sing our love for and obligation to the land of the free and the home of the brave? Yes, but just what and whom do we love? Certainly not the soil, which we are sending helter-skelter downriver. Certainly not the waters, which we assume have no function except to turn turbines, float barges, and carry off sewage. Certainly not the plants, of which we exterminate whole communities without batting an eye. Certainly not the animals, of which we have already extirpated many of the largest and most beautiful species. A land ethic of course cannot prevent the alteration, management, and use of these "resources," but it does affirm their right to continued existence, and, at least in spots, their continued existence in a natural state.

In short, a land ethic changes the role of Homo sapiens from conqueror of the land-community to plain member and citizen of it. It implies respect for his fellow-members, and also respect for the community as such. In human history, we have learned (I hope) that the conqueror role is eventually self-defeating. Why? Because it is implicit in such a role that the conqueror knows, ex cathedra, just what makes the community clock tick; and just what and who is valuable, and what and who is worthless, in community life. It always turns out that he knows neither, and this is why his conquests eventually defeat themselves.

6 D'Arcy W. Thompson, *On Growth and Form*, 1917

D'Arcy W. Thompson's book *On Growth and Form*, based on his doctoral dissertation from 1917, crossed boundaries, attracting the interest of architects, artists, engineers, and biologists. Brand selected a montage of the illustrations with which Thompson, a Scottish mathematical biologist and professor of natural history at Dundee, demonstrated geometrical structures in nature. The images seem to have been chosen more to display the rich variety of forms in the book than to constitute a specific argument. The book inspired an influential exhibition produced by Richard Hamilton at the Institute of Contemporary Arts (ICA), London, in 1951, as part of the Festival of Britain.[1] Thompson's work confirmed what many philosophers had claimed, that nature contains many natural forms based on pure geometric structures. The regular polyhedrons found in microorganisms could be compared with geodesic domes.

The book reveals a preoccupation with pattern similar to that of Christopher Alexander, whose *Notes on the Synthesis of Form* (1964) was presented on the facing page in the *LWEC*. Alexander developed a method of analyzing the design process. As Brand wrote: "Christopher Alexander is a design person that other design people refer to a lot. This book deals with the nature of current design problems that are expanding clear beyond any individual's ability to know and correlate all the factors."[2] — CM

1. See Anne Massey, *The Independent Group: Modernism and Mass Culture, 1945–59* (Manchester: Manchester University Press, 1995).

2. Stewart Brand, on *Synthesis of Form, LWEC*, 15.

Source
D'Arcy W. Thompson, *On Growth and Form* (Cambridge: Cambridge University Press, 1917; 1943; abridged ed., 1961). (Images featured in *LWEC*, 14.)

Further Reading
Bonnemaison, Sarah, and Philip Beesley, eds. *On Growth and Form: Organic Architecture and Beyond*. Halifax: Tuns Press and Riverside Architectural Press, 2008.

Upitis, Alise. "Alexander's Choice: How Architecture Avoided Computer-Aided Design c. 1962," in *A Second Modernism: MIT, Architecture, and the "Techno-Social" Moment*, ed. Arindam Dutta, 474–505. Cambridge, MA: MIT Press, 2013.

Figure 1.2
D'Arcy Thompson, "On Growth and Form." *LWEC*, 14.

On Growth and Form

A paradigm classic. Everyone dealing with growth or form in any manner can use the book. We've seen worn copies on the shelves of artists, inventors, engineers, computer systems designers, biologists.

—SB

On Growth and Form
D'Arcy Wentworth Thompson
Two volume edition
1917, 1952

$27.50 postpaid

Abridged paper edition
1917, 1961; 346 pp.

$2.75 postpaid

from:
Cambridge University Press
510 North Avenue
New Rochelle, N. Y. 10801

or WHOLE EARTH CATALOG

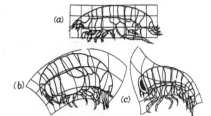

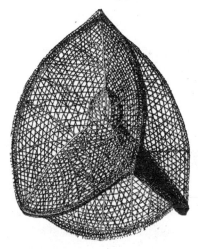

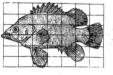

Fig. 150. *Polyprion.*

Fig. 151. *Pseudopriacanthus altus.*

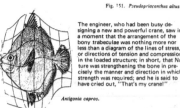

The engineer, who had been busy designing a new and powerful crane, saw in a moment that the arrangement of the bony trabeculae was nothing more nor less than a diagram of the lines of stress, or directions of tension and compression in the loaded structure; in short, that Nature was strengthening the bone in precisely the manner and direction in which strength was required; and he is said to have cried out, "That's my crane!"

Antigonia capros.

Fig. 143. (*a*) *Harpinia plumosa* Kr.; (*b*) *Stegocephalus inflatus* Kr.; (*c*) *Hyperia galba.*

When Plateau made the wire framework of a regular tetrahedron and dipped it in soap-solution, he obtained in an instant a beautifully symmetrical system of six films, meeting three by three in four edges and these four edges running from the corners of the figure to its centre of symmetry. Here they meet, two by two, at the Maraldi angle; and the films meet three by three, to form the re-entrant solid angle which we have called a "Maraldi pyramid" in our account of the architecture of the honeycomb. The very same configuration is easily recognized in the minute siliceous skeleton of *Callimitra.* There are two discrepancies, neither of which need raise any difficulty. The figure is not rectilinear but a *spherical tetrahedron,* such as might be formed by the boundary edges of a tetrahedral cluster of four co-equal bubbles; and just as Plateau extended his experiment by blowing a small bubble in the centre of his tetrahedral system, so we have a central bubble also here.

This bubble may be of any size; but its situation (if it be present at all) is always the same, and its shape is always such as to give the Maraldi angles at its own four corners. The tension of its own walls, and those of the films by which it is supported or slung, all balance one another. Hence the bubble appears in plane projection as a curvilinear equilateral triangle; and we have only got to convert this plane diagram into the corresponding solid to obtain the spherical tetrahedron we have been seeking to explain.

The geometry of the little inner tetrahedron is not less simple and elegant. Its six edges and four faces are all equal. The films attaching it to the outer skeleton are all planes. Its faces are spherical,

(*a*)　　　　　(*b*)

Fig. 63. Diagrammatic construction of *Callimitra.* (*a*) A bubble suspended within a tetrahedral cage; (*b*) another bubble within a skeleton of the former bubble.

and each has its centre in the opposite corner. The edges are circular arcs, with cosine $\frac{1}{3}$; each is in a plane perpendicular to the chord of the arc opposite, and each has its centre in the middle of that chord. Along each edge the two intersecting spheres meet each other at an angle of $120°$.[1]

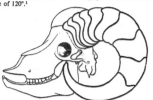

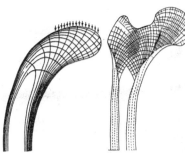

Fig. 101. Crane-head and femur. After Culmann and J. Wolff.

Purposive Systems

You're a purposive system. So am I. We're very good at it, and not as good as we'd like to be. Humanity, as a whole, is lousy at it, and worried. This collection of recent cybernetic thoughts can cheer you up and give you better concepts to worry with.

—SB

Purposive Systems

Ed.: Heinz von Foerster, J. D. White,
L. J. Peterson, J. K. Russell
1968; 179 pp. from
Spartan Books
432 Park Avenue South
New York, N. Y. 10016
or
WHOLE EARTH CATALOG

$10.00

We have not yet built into our educational system any recognition of the points where precision is essential, and yet we are living in a society where one mistake can dislocate the lives of thousands of people, wreck distribution systems, and distort life-history data, and subsequent career lines.

•

There is no basic reason why one cannot design a control memory with a different technology, a technology which would allow the computer itself to alter the information stored in the control memory. Thus we would have a computer that could alter its own character as required. To my knowledge very little conceptual work has been done in thinking through the implications of this extremely powerful possibility. The possibilities are so staggering and deep. The poor harried souls responsible for trying to understand the classical computer as we now know it wish this idea would go away.

The act of choosing a representation for a problem involves the specification of a space where the search for solution can take place. Such a specification involves the choice of a language—and its use—for expressing problem conditions, properties of solutions, and knowledge of regularities in the search space.

If we detach the concern of survival from computers, as is generally the case, they can learn abilities more useful to man than the struggle for existence.

•

A friend of mine once gave what I regard as a nice shorthand formula. When in a dilemma, introduce novelty.

•

Exaggerated politeness is a powerful source of misunderstanding.

•

Regularity seeking activities that seem to be generally useful include the detection of symmetries, the identification of 'critical points' (key points through which the search must go to get a solution), and the recognition of redundant information in problem descriptions.

Aspects of Form

This is a well-used collection of insights by venerable initiates of form study.

—SB

Aspects of Form

Lancelot Law Whyte, ed.
1951; 249 pp.

$1.95 postpaid from
Indiana University Press
10th and Morton Streets
Bloomington, Ind. 47401
or
WHOLE EARTH CATALOG

Fig. 2 Copperhead Snake——illustrating the effectiveness of disruptive contrast in relation to background configuration

A man may learn by experience to associate two series of events between which any connection seemed at first wildly improbable. For such associations to be possible, provision must be made for every signal entering the nervous system to be relayed to every part, not merely to the specialised receiving zone. Thus from the knot of an event is generated a web of speculation; when two series of events are perceived together they form the warp and woof of a shimmering fabric into which is woven the pattern of the probability that the two events are significantly related.

An 'image' in this biological sense, then, is not an imitation of an object's external form but an imitation of certain privileged or relevant aspects. It is here that a wide field of investigation would seem to open.

We know that there are certain privileged motifs in our world to which we respond almost too easily. The human face may be outstanding among them. Whether by instinct or by very early training, we certainly are ever disposed to single out the expressive features of a face from the chaos of sensations that surrounds it and to respond to its slightest variations with fear or joy.

14 Form
Whole Systems

7 Herbert Simon, *The Sciences of the Artificial,* 1969

Herbert A. Simon, Nobel laureate and a pioneer in the study of artificial intelligence (AI) and organization theory, delivered the Karl Taylor Compton lecture series at the Massachusetts Institute of Technology in 1968 on the subject of artificial phenomena; he then compiled his lectures, as well as one previous essay, "The Architecture of Complexity," into a single text to suggest the possibility of a science of the artificial, or a science of design, and how that science might be defined. A true polymath who over the course of his career won the Nobel Memorial Prize in Economics, the Association for Computing Machinery's Turing Award, and the American Psychological Association's Award for Outstanding Contributions to Psychology, Simon believed that a science of the artificial—"a body of intellectually tough, analytic, partly formalizable, partly empirical, teachable doctrine about the design process"—was essential to the curriculum of any field concerned with contingency and problem solving, including engineering, medicine, business, and art (*Sciences of the Artificial,* 58).

Simon's conception of subsystem stability—discussed in the passages that follow—seemed of particular interest to Brand, perhaps because it was somewhat analogous to his organizational strategy for the catalog; but the implications of the entire text for the field of computer science must also have captured Brand's attention at a time when he was just beginning to engage with engineers at the Stanford Artificial Intelligence Laboratory and develop his own perspective on personal computing.—MG

Source
Herbert Simon, *The Sciences of the Artificial* (Cambridge, MA: MIT Press, 1969), 90–92, 97–99. © Herbert Simon. Reprinted with permission of the MIT Press. (Referred to in *LWEC,* 15.)

Further Reading
Alexander, Christopher. *Notes on the Synthesis of Form.* Cambridge, MA: Harvard University Press, 1964.

Baber, Walter F. "The Arts of the Natural: Herbert Simon and Artificial Intelligence." *Public Administration Quarterly* 12, no. 3 (fall 1988): 329–347.

Simon, Herbert A. "How Complex Are Complex Systems?" In *PSA: Proceedings of the Biennial Meeting of the Philosophy of Science Association 1976,* vol. 2, *Symposium and Invited Papers,* 507–522. Chicago: University of Chicago Press, 1976.

Selections from *The Sciences of the Artificial*

The Evolution of Complex Systems

Let me introduce the topic of evolution with a parable. There once were two watchmakers, named Hora and Tempus, who manufactured very fine watches. Both of them were highly regarded, and the phones in their workshops rang frequently—new customers were constantly calling them. However, Hora prospered, while Tempus became poorer and poorer and finally lost his shop. What was the reason?

The watches the men made consisted of about 1,000 parts each. Tempus had so constructed his that if he had one partly assembled and had to put it down—to answer the phone, say—it immediately fell to pieces and had to be reassembled from the elements. The better the customers liked his watches, the more they phoned him and the more difficult it became for him to find enough uninterrupted time to finish a watch.

The watches that Hora made were no less complex than those of Tempus. But he had designed them so that he could put together subassemblies of about ten elements each. Ten of these subassemblies, again, could be put together into a larger subassembly; and a system of ten of the latter subassemblies constituted the whole watch. Hence, when Hora had to put down a partly assembled watch to answer the phone, he lost only a small part of his work, and he assembled his watches in only a fraction of the man-hours it took Tempus....

...A straightforward calculation shows that it will take Tempus on the average about four thousand times as long to assemble a watch as Hora.

We arrive at the estimate as follows:

1. Hora must make 111 times as many complete assemblies per watch as Tempus; but
2. Tempus will lose on the average 20 times as much work for each interrupted assembly as Hora (100 parts, on the average, as against 5); and
3. Tempus will complete an assembly only 44 times per million attempts ($0.99^{1}000 = 44 \times 10^{-6}$), while Hora will complete nine out of ten ($0.99^{10} = 9 \times 10^{-1}$). Hence Tempus will have to make 20,000 as many attempts per completed assembly as Hora....

The Sources of Selectivity

When we examine the sources from which the problem-solving system, or the evolving system, as the case may be, derives its selectivity, we discover that selectivity can always be equated with some kind of feedback of information from the environment.

Let us consider the case of problem solving first. There are two basic kinds of selectivity. One we have already noted: various paths are tried out, the consequences of following them are noted, and this information is used to guide further search. In the same way in organic evolution various complexes come into being, at least evanescently, and those that are stable provide new building blocks for further construction. It is this information about stable configurations, and not free energy or negentropy from the sun, that guides the process of evolution and provides the selectivity that is essential to account for its rapidity.

The second source of selectivity in problem solving is previous experience. We see this particularly clearly when the problem to be solved is similar to one that has been solved before. Then, by simply trying again the paths that led to the earlier solution, or their analogues, trial-and-error search is greatly reduced or altogether eliminated.

What corresponds to this latter kind of information in organic evolution? The closest analogue is reproduction. Once we reach the level of self-reproducing systems, a complex system, when it has once been achieved, can be multiplied indefinitely. Reproduction in fact allows the inheritance of acquired characteristics, but at the level of genetic material, of course; that is, only characteristics acquired by the genes can be inherited. We shall return to the topic of reproduction in the final section of this essay....

Conclusion: The Evolutionary Explanation of Hierarchy

We have shown thus far that complex systems will evolve from simple systems much more rapidly if there are stable intermediate forms than if there are not. The resulting complex forms in the former case will be hierarchic. We have only to turn the argument around to explain the observed predominance of hierarchies among the complex systems nature presents to us. Among possible complex forms, hierarchies are the ones that have the time to evolve.

The hypothesis that complexity will be hierarchic makes no distinction among very flat hierarchies, like crystals and tissues and polymers, and the intermediate forms. Indeed in the complex systems we encounter in nature examples of both forms are prominent. A more complete theory than the one we have developed here would presumably have something to say about the determinants of width of span in these systems.

8 Wendell Berry, "Think Little," 1970

Wendell Berry is an American novelist, poet, environmental activist, and farmer, as well as an important supporter of the *WEC*; he helped to edit the "Land Use" section. After publishing his first novel in 1960, Berry taught English and creative writing in New York and Kentucky. In 1965 he moved to a 125-acre farm in Kentucky and charted his experiences in a series of articles for the Rodale Press. He became an authority on organic gardening and farming. He took an aggressive stance against the Vietnam War in 1968 and took part the following year in protests against a proposed nuclear power plant in Indiana.

In the piece selected by Brand, Berry presents the protest movements in civil rights, the war, and the environment as having a common cause against the greed and exploitation of the market economy. Berry makes the important point that environmental issues affect every individual in his or her daily choices. It is not enough to blame government or industry for the abuse of the environment.

By coining the phrase "think little," Berry aligned himself with the wide base of the alternative movement. Constructing a healthy, ethical, and satisfying life depends on the individual and his or her community. The power of Berry's writing reached beyond the back-to-the-land counterculture to inspire many urban readers.—CM

Source

Originally presented as an Earth Day speech, April 1970, and then in *Blue-Tail Fly* 7 (1970), a Kentucky underground newspaper, and *LWEC* (1971), 24–25. The essay was republished in Wendell Berry, *A Continuous Harmony: Essays Cultural and Agricultural* (New York: Harcourt Brace Jovanovich, 1972). © 2012 by Wendell Berry, from *A Continuous Harmony*. Reprinted by permission of Counterpoint.

Further Reading

Berry, Wendell. *The Long-Legged House.* New York: Harcourt, Brace & World, 1969.

Berry, Wendell. "Some Difficulties to Think About before Buying a Farm." *Organic Farming and Gardening* 24 (July 1977): 59–61.

Smith, Kimberly K. *Wendell Berry and the Agrarian Tradition.* Lawrence: University Press of Kansas, 2003.

"Think Little"

To me, one of the most important aspects of the environmental movement is that it brings us not just to another public crisis, but to a crisis of the protest movement itself. For the environmental crisis should make it dramatically clear, as perhaps it has not always been before, that there is no public crisis that is not also private. To most advocates of civil rights, racism has seemed mostly the fault of someone else. For most advocates of peace the war has been a remote reality, and the burden of the blame has seemed to rest mostly on the government. I am certain that these crises have been more private, and that we have each suffered more from them and been more responsible for them, than has been readily apparent, but the connections have been difficult to see. Racism and militarism have been institutionalized among us for too long for our personal involvement in those evils to be easily apparent to us. Think, for example, of all the Northerners who assumed—until black people attempted to move into their neighborhoods—that racism was a Southern phenomenon.

But the environmental crisis rises closer to home. Every time we draw a breath, every time we drink a glass of water, every time we eat a bite of food we are suffering from it. And more important, every time we indulge in, or depend on, the wastefulness of our economy—and our economy's first principle is waste—we are causing the crisis. Nearly every one of us, nearly every day of his life, is contributing directly to the ruin of this planet. A protest meeting on the issue of environmental abuse is not a convocation of accusers, it is a convocation of the guilty. That realization ought to clear the smog of self-righteousness that has almost conventionally hovered over these occasions, and let us see the work that is to be done....

What we are up against in this country, in any attempt to invoke private responsibility, is that we have nearly destroyed private life. Our people have given up their independence in return for the cheap seductions and the shoddy merchandise of so-called "affluence." We have delegated all our vital functions and responsibilities to salesmen and agents and bureaus and experts of all sorts. We cannot feed or clothe ourselves, or entertain ourselves, or communicate with each other, or be charitable or neighborly or loving, or even respect ourselves, without recourse to a merchant or a corporation or a public-service organization or an agency of the government or a style-setter or an expert. Most of us cannot think of dissenting from the opinions or the actions of one organization without first forming a new organization. Individualism is going around these days in uniform, handing out the party line on individualism. Dissenters want to publish their personal opinions over a thousand signatures....

If we are to hope to correct our abuses of each other and of other races and of our land, and if our effort to correct these abuses is to be more than a political fad that will in the long run be only another form of abuse, then we are going to have to go far beyond public protest and political action. We are going to have to rebuild the substance and the integrity of private life in this country. We are going to have to gather up the fragments of knowledge and responsibility that we have parceled out to the bureaus and the corporations and the specialists, and we are going to have to put those fragments back together again in our own minds and in our families and households and neighborhoods. We need better government, no doubt about it. But we also need better minds, better friendships, better marriages, better communities. We need persons and households that do not have to wait upon organizations, but can make necessary changes in themselves, on their own.

For most of the history of this country our motto, implied or spoken, has been Think Big. I have come to believe that a better motto, and an essential one now, is Think Little. That implies the necessary change of thinking and feeling, and suggests the necessary work. Thinking Big has led us to the two biggest and cheapest political dodges of our time: plan-making and law-making. The lotus-eaters of this era are in Washington, D.C., Thinking Big. Somebody comes up with a problem, and somebody in the government comes up with a plan or a law. The result, mostly, has been the persistence of the problem, and the enlargement and enrichment of the government.

But the discipline of thought is not generalization; it is detail, and it is personal behavior. While the government is "studying" and funding and organizing its Big Thought, nothing is being done. But the citizen who is willing to Think Little, and, accepting the discipline of that, to go ahead on his own, is already solving the problem. A man who is trying to live as a neighbor to his neighbors will have a lively and practical understanding of the work of peace and brotherhood, and let there be no mistake about it—he is doing that work. A couple who make a good marriage, and raise healthy, morally competent children, are serving the world's future more directly and surely than any political leader, though they never utter a public word. A good farmer who is dealing with the problem of soil erosion on an acre of ground has a sounder grasp of that problem and cares more about it and is probably doing more to solve it than any bureaucrat who is talking about it in general. A man who is willing to undertake the discipline and the difficulty of mending his own ways is worth more to the conservation movement than a hundred who are insisting merely that the government and the industries mend their ways....

Odd as I am sure it will appear to some, I can think of no better form of personal involvement in the cure of the environment than that of gardening. A person who is growing a garden, if he is growing it organically, is improving a piece of the world. He is producing something to eat, which makes him somewhat independent of the grocery business, but he is also enlarging, for himself, the meaning of food and the pleasure of eating. The food he grows will be fresher, more nutritious, less contaminated by poisons and preservatives and dyes than what he can buy at a store. He is reducing the trash problem; a garden is not a disposable container, and it will digest and re-use its own wastes. If he enjoys working in his garden, then he is less dependent on an automobile or a merchant for his pleasure. He is involving himself directly in the work of feeding people.

If you think I'm wandering off the subject, let me remind you that most of the vegetables necessary for a family of four can be grown on a plot of forty by sixty feet. I think we might see in this an economic potential of considerable importance, since we now appear to be facing the possibility of widespread famine. How much food could be grown in the dooryards of cities and suburbs? How much could be grown along the extravagant right-of-ways of the interstate system? Or how much could be grown, by the intensive practices and economics of the small farm, on so-called marginal lands?

9 Lewis Mumford, *The Pentagon of Power*, 1970

"The only thing that slows me on Mumford is his occasional pious ought-to-be approach to this or that, which is ineffective. Fuller does it too,"[1] Brand wrote after reading *The Pentagon of Power*. The book is visually powerful, with arguments based on contrasting black-and-white images, with large captions. The image of the astronaut in his closed-system space capsule is, for Mumford, "the archetypal proto-model of Post-Historic Man, whose existence from birth to death would be conditioned by the megamachine, and made to conform, as in a space capsule, to the minimal functional requirements by an equally minimal environment—all under remote control."

The cultural historian and critic Lewis Mumford (1895–1990) became increasingly concerned about the effects of industrialization on society. His successful *Technics and Civilization* (1934) traced the development of technology into the era of electrification, which appeared to offer more satisfying roles in localized, small-scale production. Mumford became aware, however, that the problems lay deeper, in the structures of the marketplace and the basic concepts of progress associated with it. In *The Pentagon of Power*, volume 2 of *The Myth of the Machine*, Mumford expands his analysis, now seeing the very metaphor of the machine as the root cause of the problem. Thinking like a machine could only lead to catastrophe and war. What was needed was a return to the human values and wisdom contained in earlier civilizations: "With this extirpation of earlier cultures went a vast loss of botanical and medical lore, representing many thousands of years of watchful observation and empirical experiment whose extraordinary discoveries—such as the American Indian's use of snakeroot (reserpine) as a tranquillizer in mental illness—modern medicine has now, all too belatedly, begun to appreciate" (*The Pentagon of Power*, quoted in *LWEC*, 29).—CM

1. Stewart Brand, letter to Steve Baer, December 4, 1970, M1045.

Source
Lewis Mumford, *The Pentagon of Power* (New York: Harcourt Brace and Jovanovich, 1970), plates 14–15. (Referred to in *LWEC*, 29.)

Further Reading
Hughes, Thomas P., and Agatha C. Hughes, eds. *Lewis Mumford: Public Intellectual*. New York: Oxford University Press, 1990.

Novak, Frank G., ed. *Lewis Mumford and Patrick Geddes: The Correspondence.* New York: Routledge, 1995.

"Encapsulated Man," from *The Pentagon of Power*

Behold the astronaut, fully equipped for duty: a scaly creature, more like an over-sized ant than a primate—certainly not a naked god. To survive on the moon he must be encased in an even more heavily insulated garment, and become a kind of faceless ambulatory mummy. While he is hurtling through space the astronaut's physical existence is purely a function of mass and motion, narrowed down to the pinpoint of acute sentient intelligence demanded by the necessity for coordinating his reactions with the mechanical and electronic apparatus upon which his survival depends. Here is the archetypal proto-model of Post-Historic

Man, whose existence from birth to death would be conditioned by the megamachine, and made to conform, as in a space capsule, to the minimal functional requirements by an equally minimal environment—all under remote control.

Dr. Bruno Bettelheim reports the behavior of a nine-year-old autistic patient, a boy called Joey, who conceived that he was run by machines. "So controlling was this belief that Joey carried with him an elaborate life-support system made up of radio tubes, light bulbs, and a 'breathing machine.' At meals he ran imaginary wires from a wall socket to himself, so his food could be digested. His bed was rigged up with batteries, a loud-speaker, and other improvised equipment to keep him alive while he slept."

But is this just the autistic fantasy of a pathetic little boy? Is it not rather the state that the mass of mankind is fast approaching in actual life, without realizing how pathological it is to be cut off from their own resources for living, and to feel no tie with the outer world unless they are connected with the Power Complex and constantly receive information, direction, stimulation, and sedation from a central external source, via radio, discs, and television, with the minimal opportunity for reciprocal face-to-face contact? The stringent limitations of the space capsule have already been extended to other areas. Technocratic designers proudly exhibit furniture planned solely to fit rooms as painfully constricted as a rocket chamber. Even more ingenious minds, equally subservient to the Power Complex, have already conceived a hospital bed in which every function from the taking of temperature to intravenous feeding will be automatically performed within the limits of the bed. Solitary confinement thus becomes the last word in "tender and loving care."

Except for meeting emergencies, as with an iron lung or a space rocket, such mechanical attachment and encapsulation present a definitely pathological syndrome. Increasingly, the astronaut's space suit will be, figuratively speaking, the only garment that machine-processed and machine-conditioned man will wear in comfort; for only in that suit will he, like little Joey, feel alive. This is a return to the womb, without the embryo's prospect of a natal delivery. As if to emphasize this point, *the actual position of the astronaut here shown, under working conditions, is on his back*, the normal position of the fetus.

14–15: Encapsulated Man

Behold the astronaut, fully equipped for duty: a scaly creature, more like an oversized ant than a primate—certainly not a naked god. To survive on the moon he must be encased in an even more heavily insulated garment, and become a kind of faceless ambulatory mummy. While he is hurtling through space the astronaut's physical existence is purely a function of mass and motion, narrowed down to the pinpoint of acute sentient intelligence demanded by the necessity for coordinating his reactions with the mechanical and electronic apparatus upon which his survival depends. Here is the archetypal proto-model of Post-Historic Man, whose existence from birth to death would be conditioned by the megamachine, and made to conform, as in a space capsule, to the minimal functional requirements by an equally minimal environment—all under remote control.

Dr. Bruno Bettelheim reports the behavior of a nine-year-old autistic patient, a boy called Joey, who conceived that he was run by machines. "So controlling was this belief that Joey carried with him an elaborate life-support system made up of radio tubes, light bulbs, and a 'breathing machine.' At meals he ran imaginary wires from a wall socket to himself, so his food could be digested. His bed was rigged up with batteries, a loud-speaker, and other improvised equipment to keep him alive while he slept."

But is this just the autistic fantasy of a pathetic little boy? Is it not rather the state that the mass of mankind is fast approaching in actual life, without realizing how pathological it is to be cut off from their own resources for living, and to feel no tie with the outer world unless they are connected with the Power Complex and constantly receive information, direction, stimulation, and sedation from a central external source, via radio, discs, and television, with the minimal opportunity for reciprocal face-to-face contact? The stringent limitations of the space capsule have already been extended to other areas. Technocratic designers proudly exhibit furniture planned solely to fit rooms as painfully constricted as a rocket chamber. Even more ingenious minds, equally subservient to the Power Complex, have already conceived a hospital bed in which every function from the taking of temperature to intravenous feeding will be automatically performed within the limits of the bed. Solitary confinement thus becomes the last word in 'tender and loving care.'

Except for meeting emergencies, as with an iron lung or a space rocket, such mechanical attachment and encapsulation presents a definitely pathological syndrome. Increasingly, the astronaut's space suit will be, figuratively speaking, the only garment that machine-processed and machine-conditioned man will wear in comfort; for only in that suit will he, like little Joey, feel alive. This is a return to the womb, without the embryo's prospect of a natal delivery. As if to emphasize this point, *the actual position of the astronaut here shown, under working conditions, is on his back*, the normal position of the foetus.

10 Barry Commoner, *Science and Survival*, **1966**

Since the early 1950s, the American biologist, educator, and environmental activist Barry Commoner (1917–2012) warned the public about the dangers of modern technology and its effect on the environment. The issue of atomic fallout occupied Commoner for many years. He believed that scientists have a social responsibility to inform the public about the scientific facts and the experiments they are conducting, particularly when their experiments could affect the health of future generations. His reputation as a "serious" scientist was an important recommendation for Brand: "Scientific and technological critics are significantly rare and much needed. Barry Commoner is a good one — prestigious scientist, active administrator, capable writer, committed critic."

In the selected text, Commoner stresses the often invisible effect of technology on human beings. He was among the first to warn of the effects of the buildup of carbon dioxide in the atmosphere and the dangers of the greenhouse effect.

In the heightened atmosphere of the Cold War, Commoner's book prompted morbid thoughts. Brand wrote: "Reading his anthology of technological blunders, I am led to wonder if the success of death-fear-driven science merely changes the size of the package that death comes in. There's less of piecemeal local, 'natural' dying, and more of massive 'caused' dying. I think I prefer the old way. Is anybody working on establishing friendly relations with Death? Maybe he's a good guy who's had bad press." — CM

Source

Barry Commoner, *Science and Survival* (New York: Viking Press, 1966), 3–4, 16–19. (Referred to in *LWEC*, 33.)

Further Reading

"Barry Commoner." In *Environmental Activists*, ed. John Mongillo and Bibi Booth, 61–64. Westport, CT: Greenwood Press, 2001.

Egan, Michael. *Barry Commoner and the Science of Survival: The Remaking of American Environmentalism*. Cambridge, MA: MIT Press, 2007.

Figure 1.3 (opposite)
"Encapsulated Man." Lewis Mumford, *The Pentagon of Power* (New York: Harcourt Brace and Jovanovich, 1970), plate 14. Reprinted with permission.

"Is Science Getting Out of Hand?" from *Science and Survival*

Sorcerer's Apprentice

We are surrounded by the technological successes of science: space vehicles, nuclear power, new synthetic chemicals, medical advances that increase the length and usefulness of human life. But we also see some sharp contrasts. While one group of scientists studies ways to provide air for the first human visitors to the moon, another tries to learn why we are fouling the air that the rest of us must breathe on earth. We hear of masterful schemes for using nuclear explosions to extract pure water from the moon; but in some American cities the water that flows from the tap is undrinkable and the householder must buy drinking water in bottles. Science is triumphant with far-ranging success, but its triumph is somehow clouded by growing difficulties in providing for the simple necessities of human life on the earth.

Our Polluted Environment

For about a million years human beings have survived and proliferated on the earth by fitting unobtrusively into a life-sustaining environment, joining a vast community in which animals, plants, micro-organisms, soil, water and air are tied together in an elaborate network of mutual relationships. In the pre-industrial world the environment appeared to hold an unlimited store of clean air and water. It seemed reasonable, as the need arose, to vent smoke into the sky and sewage into rivers in the expectation that the huge reserves of uncontaminated air and water would effectively dilute and degrade the pollutants—perhaps in the same optimistic spirit that leads us to embed slotted boxes in bathroom walls to receive razor blades. But there is simply not enough air and water on the earth to absorb current man-made wastes without effect. We have begun to merit the truculent complaint against the works of the paleface voiced by Chief Satinpenny in Nathanael West's *A Cool Million:* "Now even the Grand Canyon will no longer hold razor blades."

Fire, an ancient friend, has become a man-made threat to the environment through the sheer quantity of the waste it produces. Each ton of wood, coal, petroleum, or natural gas burned contributes several tons of carbon dioxide to the earth's atmosphere. Between 1860 and 1960 the combustion of fuels added nearly 14 per cent to the carbon dioxide content of the air, which had until then remained constant for many centuries. Carbon dioxide plays an important role in regulating the temperature of the earth because of the "greenhouse effect." Both glass and carbon dioxide tend to pass visible light but absorb infra-red rays. This explains why the sun so easily warms a greenhouse on a winter day. Light from the sun enters through the greenhouse glass. Within, it is absorbed by soil and plants and converted to infra-red heat energy which remains trapped inside the greenhouse because it cannot pass out again through the glass. Carbon dioxide makes a huge greenhouse of the earth, allowing sunlight to reach the earth's surface, but limiting re-radiation of the resulting heat into space.

11 Paul Ehrlich, *The Population Bomb*, 1968

On the sensationalistic cover of the first edition of Dr. Paul R. Ehrlich's *The Population Bomb* (1968), just beneath the title, appeared the following warning: "While you are reading these words four people will have died from starvation. Most of them children." This harrowing statistic was just one of many that Ehrlich, an entomologist and population biologist, offered in his work, which cautioned readers that the exponential growth of human populations worldwide was ecologically unsustainable and, according to his calculations, threatened to exhaust Earth's finite natural resources in mere decades. Yet *The Population Bomb* was not simply a dystopian prophecy shrouded in scientific data; after explicating the problem in the first chapter, Ehrlich posed a solution—population control—and spent much of the following text discussing what was currently being accomplished toward this goal, what policies should be put into place to support it, and even what individuals could do to stem the crisis.

As Stewart Brand mentioned in his review, Ehrlich "freaked out of his lab and into the media with the bad news," even appearing multiple times on *The Tonight Show Starring Johnny Carson* after the book was released, to inform a wider audience of both the perils of unchecked growth and the benefits of population control. The following passages represent these two dimensions of Ehrlich's message.—MG

Source

Paul Ehrlich, *The Population Bomb* (New York: Ballantine, 1968), 14–18, 118. Courtesy of Paul Ehrlich. (Referred to in *LWEC*, 34.)

Further Reading

Hardin, Garrett, ed. *Population, Evolution, and Birth Control: A Collage of Controversial Ideas*. San Francisco: W. H. Freeman, 1969.

Meadows, Donella. *The Limits of Growth: A Report for the Club of Rome's Project on the Predicament of Mankind*. New York: Universe Books, 1972.

Simon, Julian Lincoln. *The Ultimate Resource*. Princeton, NJ: Princeton University Press, 1981.

Selections from *The Population Bomb*

Too Many People

It has been estimated that the human population of 6000 B.C. was about five million people, taking perhaps one million years to get there from two and a half million. The population did not reach 500 million until almost 8,000 years later—about 1650 A.D. This means it doubled roughly once every thousand years or so. It reached a billion people around 1850, doubling in some 200 years. It took only 80 years or so for the next doubling, as the population reached two billion around 1930. We have not completed the next doubling to four billion yet, but we now have well over three billion people. The doubling time at present seems to be about 37 years. Quite a reduction in doubling times: 1,000,000 years, 1,000 years, 200 years, 80 years, 37 years. Perhaps the meaning of a doubling time of around 37 years is best brought home by a theoretical exercise. Let's examine what might happen on the absurd assumption that the population continued to double every 37 years into the indefinite future.

If growth continued at that rate for about 900 years, there would be some 60,000,000,000,000,000 people on the face of the earth. Sixty million billion people. This is about 100 persons for each square yard of the Earth's surface, land and sea. A British physicist, J. H. Fremlin, guessed that such a multitude might be housed in a continuous 2,000-story building covering our entire planet. The upper 1,000 stories would contain only the apparatus for running this gigantic warren. Ducts, pipes, wires, elevator shafts, etc., would occupy about half of the space in the bottom 1,000 stories. This would leave three or four yards of floor space for each person. I will leave to your imagination the physical details of existence in this ant heap, except to point out that all would not be black. Probably each person would be limited in his travel. Perhaps he could take elevators through all 1,000 residential stories but could travel only within a circle of a few hundred yards' radius on any floor. This would permit, however, each person to choose his friends from among some ten million people! And, as Fremlin points out, entertainment on the worldwide TV should be excellent, for at any time "one could expect some ten million Shakespeares and rather more Beatles to be alive." ...

Enough of fantasy. Hopefully, you are convinced that the population will have to stop growing sooner or later and that the extremely remote possibility of expanding into outer space offers no escape from the laws of population growth. If you still want to hope for the stars, just remember that, at the current growth rate, in a few thousand years everything in the visible universe would be converted into people, and the ball of people would be expanding with the speed of light! Unfortunately, even 900 years is much too far in the future for those of us concerned with the population explosion. As you shall see, the next *nine* years will probably tell the story.

Of course, population growth is not occurring uniformly over the face of the Earth. Indeed, countries are divided rather neatly into two groups: those with rapid growth rates, and those with relatively slow growth rates. The first group, making up about two-thirds of the world population, coincides closely with what are known as the "undeveloped countries" (UDCs). The UDCs are not industrialized, tend to have inefficient agriculture, very small gross national products, high illiteracy rates and related problems. That's what UDCs are technically, but a short definition of undeveloped is "starving." Most Latin American, African, and Asian countries fall into this category. The second group consists, in essence, of the "developed countries" (DCs). DCs are modern, industrial nations, such as the United States, Canada, most European countries, Israel, Russia, Japan, and Australia. Most people in these countries are adequately nourished.

Doubling times in the UDCs range around 20 to 35 years. Examples of these times (from the 1968 figures just released by the Population Reference Bureau) are Kenya, 24 years; Nigeria, 28; Turkey, 24; Indonesia, 31; Philippines, 20; Brazil, 22; Costa Rica, 20; and El Salvador, 19. Think of what it means for the population of a country to double in 25 years. In order just to keep living standards at the present inadequate level, the food available for the people must be doubled. Every structure and road must be duplicated. The amount of power must be doubled. The capacity of the transport system must be doubled. The number of trained doctors, nurses, teachers, and administrators must be doubled. This would be a fantastically difficult job in the United States— a rich country with a fine agricultural system, immense industries, and rich natural resources. Think of what it means to a country with none of these.

What Needs to Be Done

A general answer to the question, "What needs to be done?" is simple. We must rapidly bring the world population under control, reducing the growth rate to zero or making it go negative. Conscious regulation of human numbers must be achieved. Simultaneously we must, at least temporarily, greatly increase our food production. This agricultural program should be carefully monitored to minimize deleterious effects on the environment and should include an effective program of ecosystem restoration. As these projects are carried out, an international policy research program must be initiated to set optimum population-environment goals for the world and to devise methods for reaching these goals. So the answer to the question is simple. Getting the job done, unfortunately, is going to be complex beyond belief—if indeed it can be done.

12 Garrett De Bell, *The Environmental Handbook*, 1970

By the late-1960s, many Americans had witnessed firsthand the environmental degradation of their nation, yet little had been accomplished politically to stem the crisis. US senator Gaylord Nelson, a Democrat from Wisconsin, sought to upset this political stagnation. In 1970 he founded Earth Day, a day of nation-wide environmental teach-ins and peaceful demonstrations that he hoped would educate citizens and consequently activate them to provoke policy change at the national, state, and local levels and inspire them to make eco-logically responsible choices in their daily lives.

De Bell's book, compiled at the behest of David Brower, one of the founders of Friends of the Earth (FOE), a San Francisco–based environmentalist organiza-tion, sought to provide "ideas and tactics" for the first Earth Day teach-in. The edited volume included essays by significant voices in the environmental move-ment, including René Dubos (the piece selected here), Paul Ehrlich, Lynn White Jr., and Kenneth Boulding, as well as practical information on individual and group activist techniques and an extensive bibliography for further reading.

In his review of the text, Gurney Norman called *The Environmental Hand-book* "'the bible' of New Conservation," its essayists "prophets," and environ-mental activists "young disciples working zealously to save [the Earth]." The following excerpts, which begin the book's first part, "The Meaning of Ecol-ogy," provide just a small fragment of the collage of "evangelical" ideas that informed both the first national environmental teach-in and countless readers of the *Whole Earth Catalog.*—MG

Source
Garrett De Bell, ed., *The Environmental Handbook* (New York: Ballantine Books, 1970), 1–3, 8–9. (Referred to in *LWEC*, 40.)

Further Reading
Bookchin, Murray. "Ecology and Revolu-tionary Thought." *Anarchy* 69, no. 6 (1966): 18.

Hays, Samuel P. *A History of Environ-mental Politics since 1945*. Pittsburgh, PA: University of Pittsburgh Press, 2000.

"The Phenomenon of Earth Day (or How I Learned to Stop Worrying about Pollution and Love Con Ed)." *Progressive Architec-ture*, June 1970.

Rome, Adam. "The Genius of Earth Day." *Environmental History* 15, no. 2 (April 2010): 194–205.

Selections from *The Environmental Handbook*

Smokey the Bear Sutra

Once in the Jurassic, about 150 million years ago, the Great Sun Buddha in this corner of the Infinite Void gave a great Discourse to all the as-sembled elements and energies: to the standing beings, the walking beings, the flying beings, and the sitting beings—even grasses, to the number of thirteen billion, each one born from a seed, as-sembled there: a Discourse concerning Enlight-enment on the planet Earth.

"In some future time, there will be a conti-nent called America. It will have great centers of power called [names] such as Pyramid Lake, Walden Pond, Mt. Rainier, Big Sur, Everglades, and so forth; and powerful nerves and channels such as Columbia River, Mississippi River, and Grand Canyon. The human race in that era will get into troubles all over its head, and practically wreck everything in spite of its own intelligent Buddha-nature."

"The twisting strata of the great mountains and the pulsings of great volcanoes are my love burn-ing deep in the earth. My obstinate compassion is schist and basalt and granite, to be mountains, to bring down the rain. In that future American Era I shall enter a new form: to cure the world of loveless knowledge that seeks with blind hunger; and mindless rage eating food that will not fill it."

And he showed himself in his true form of

SMOKEY THE BEAR.

A handsome smokey-colored brown bear standing on his hind legs, showing that he is aroused and watchful.

Bearing in his right paw the Shovel that digs to the truth beneath appearances; cuts the roots of useless attachments, and flings damp sand on the fires of greed and war;

His left paw in the Mudra of Comradely Display—indicating that all creatures have the full right to live to their limits and that deer, rabbits, chipmunks, snakes, dandelions, and lizards all grow in the realm of the Dharma;

Wearing the blue work overalls symbolic of slaves and laborers, the countless men oppressed by civilization that claims to save but only destroys;

Wearing the broad-brimmed hat of the West, symbolic of the forces that guard the Wilderness, which is the Natural State of the Dharma and the True Path of man on earth; all true paths lead through mountains—

With a halo of smoke and flame behind, the forest fires of the kali-yuga, fires caused by the stupidity of those who think things can be gained and lost whereas in truth all is contained vast and free in the Blue Sky and Green Earth of One Mind;

Round-bellied to show his kind nature and that the great earth has food enough for everyone who loves and trusts her;

Trampling underfoot wasteful freeways and needless suburbs; smashing the worms of capitalism and totalitarianism;

Indicating the Task: his followers, becoming free of cars, houses, canned food, universities, and shoes, master the Three Mysteries of their own Body, Speech, and Mind; and fearlessly chop down the rotten trees and prune out the sick limbs of this country America and then burn the leftover trash.

Wrathful but Calm, Austere but Comic, Smokey the Bear will Illuminate those who would help him; but for those who would hinder or slander him,

HE WILL PUT THEM OUT.
Thus his great Mantra:
Namah samanta vajranam chanda maharoshana
Sphataya hum traka ham mam

"I DEDICATE MYSELF TO THE UNIVERSAL DIAMOND BE THIS RAGING FURY DESTROYED"

And he will protect those who love woods and rivers, Gods and animals, hobos and madmen, prisoners and sick people, musicians, playful women, and hopeful children;

And if anyone is threatened by advertising, air pollution, or the police, they should chant

SMOKEY THE BEAR'S SPELL:

DROWN THEIR BUTTS
CRUSH THEIR BUTTS
DROWN THEIR BUTTS
CRUSH THEIR BUTTS
And SMOKEY THE BEAR will surely appear to put the enemy out with his vajra-shovel.

Now those who recite this Sutra and then try to put it in practice will accumulate merit as countless as the sands of Arizona and Nevada, Will help save the planet Earth from total oil slick, Will enter the age of harmony of man and nature, Will win the tender love and caresses of men, women, and beasts Will always have ripe blackberries to eat and a sunny spot under a pine tree to sit at,

AND IN THE END WILL WIN
HIGHEST PERFECT ENLIGHTENMENT.
thus have we heard.

The Environmental Teach-In

René Dubos

The world is too much with us. We know this intuitively. But our social and economic institutions seem unable to come to grips with this awareness. The most they do is to appoint blue-ribbon committees and organize symposia which endlessly restate what everybody knows in a turgid prose that nobody reads and that leads to no action.

I do not believe that the environmental teach-in will provide new insight or factual knowledge, but I do hope that it will help alert public opinion to the immediacy of the ecologic crisis. The teach-in should point to action programs that can be developed *now* in each particular community. It should try also to define the areas of concern where knowledge is inadequate for effective action, but could be obtained by pointed research. I know that many scientists and technologists would welcome a form of public pressure that would provide them with the opportunity to work on problems of social importance.

The colossal inertia and rigidity—if not indifference—of social and academic institutions make it unlikely that they will develop effective programs of action or research focused on environmental problems. Two kinds of events, however, may catalyze and accelerate the process. One is some ecological catastrophe that will alarm the public and thus bring pressure on the social, economic, and academic establishments. Another more attractive possibility is the emergence of a grassroots movement, powered by romantic emotion as much as by factual knowledge, that will give form and strength to the latent public concern with environmental quality. Because students are vigorous, informed, and still uncommitted to vested interests, they constitute one of the few groups in our society that can act as spearheads of this grassroots movement. I wish I were young enough to be a really effective participant in the Environmental Teach-In and to proclaim in action rather than in words my faith that GNP and technological efficiency are far less important than the quality of the organic world and the suitability of the environment for a truly human life.

Land Use

The second section of the *Whole Earth Catalog*, "Land Use," highlighted books and products pertaining to gardening, animal husbandry, passive solar and wind technology, sanitation systems, agricultural tools, and even, on the concluding page, the work of Paolo Soleri. "Land Use" was perhaps most directly applicable to the many readers who, in the late 1960s, were moving "back to the land," both individually and communally, but lacked the expansive knowledge needed to begin anew in rural America; countercultural homesteaders could depend on the texts selected by the catalog's editors to guide them, in an ecologically responsible way, through almost every challenge they might face during the establishment and successful maintenance of an agrarian lifestyle. Yet the section did not entirely cater to fledgling farmers; it also included resources that could empower residents of urban or suburban spaces to implement similar projects at a smaller scale.

Beginning with the foundational texts *Living the Good Life* (1954) and *Walden* (1854), through which Scott and Helen Nearing and Henry David Thoreau, respectively, inspired countless youths to adopt a simpler or more environmentally mindful way of life, the passages and images that follow represent a cross section of "Land Use." The Nearings' and Thoreau's criticisms of political corruption, pollution, and the general exploitation of the earth and human race resonated within the counterculture, and their solution, to return to nature, renouncing the mores of modern society, prompted this new generation to do the same. *Living the Good Life* and *Walden*, however, were more philosophical

than the instructional manuals of which this second section was predominantly composed; *The Oxford Book of Food Plants* (1969) and *Grow Your Own* (1970), for instance, presented more detailed, practical information regarding individual plants, as well as general approaches to organic gardening for both novices and more seasoned gardeners. Growing at least a portion of one's own food without the use of harmful chemical fertilizers or pesticides was an attainable goal for almost every reader, be it producing herbs in an apartment window, working a small plot in a community garden, or planting acres of crops, and the catalog's editors recognized the necessity for resources at every level.

Another component of ethical land use applicable to readers of all situations was the production of energy in situ by way of sustainable resources, including the sun and wind. The seven-volume set *New Sources of Energy* (1961), an excerpt of which is included herein, provided what Brand described as a "trove" of information on geothermal, solar, and wind energy systems for urban, suburban, and rural projects internationally, although these solutions were, perhaps, more suitable for expert readers. Similarly, readers were directed to manufacturers of wind-generating equipment or texts such as John Reynolds's *Windmills and Watermills*, which, while historical in intent, conveyed a great deal of information pictorially. Between these energy resource selections and the final pages of the "Land Use" section, which concentrated on landscape planning and the designs of Paolo Soleri, lay a multitude of tools, technical guides,

and informational brochures that users of the catalog might need to undertake tasks as disparate
as prospecting for gold, foraging for edible plants,
and acquiring tax-delinquent property at a low
cost; yet all these resources approached the natural environment, or "land," more generally, in a way
that was unique from other sections. Less abstract
than "Whole Systems," but still ecologically
inclined, "Land Use" considered human beings'
physical relationship with the natural environment
and explored the ways in which one might engage
and use Nature without dominating it or contributing to its demise. —MG

13 Helen and Scott Nearing, *Living the Good Life*, 1954

Scott Nearing (1883–1983), an American leftist radical, and his partner Helen Knothe (1904–1995) left New York City in 1932 to build a new life for themselves on sixty-five acres of farmland in Vermont. Although neither Nearing, an economist trained at the Wharton School of Business, nor Knothe had any previous agricultural experience, they developed a strategy of self-subsistence that proved effective for their entire tenure on the property. After marrying in 1947 and subsequently moving to another farm in Harborside, Maine, the Nearings coauthored *Living the Good Life*, a chronicle of their years in New England.

For countercultural pioneers who embraced the back-to-the-land movement in the 1960s and '70s, the Nearings' book proved essential; it offered both the practical advice and solid philosophical framework that many inexperienced prospective farmers would require to achieve success in their endeavors.—MG

Source

Helen and Scott Nearing, *Living the Good Life: How to Live Sanely and Simply in a Troubled World* (New York: Schocken Books, 1954), 47–50. © 1954 and renewed 1982 by Helen Nearing. Used by permission of Schocken Books, an imprint of the Knopf Doubleday Publishing Group, a division of Penguin Random House LLC. All rights reserved. (Referred to in *LWEC*, 47.)

Further Reading

Bromfield, Louis. *Malabar Farm*. New York: Harper, 1948.

Kaufman, Maynard. "The New Homesteading Movement: From Utopia to Eutopia." *Soundings: An Interdisciplinary Journal* 55, no. 1 (spring 1972): 63–82.

Nearing, Helen, and Scott Nearing. *The Good Life: Helen and Scott Nearing's Sixty Years of Self-Sufficient Living*. New York: Schocken Books, 1989.

"We Build a Stone House" from *Living the Good Life*

We began this book by describing our trek to Vermont and our buying first the Ellonen place and later the Hoard place. It was on the latter farm that we planned to build a home—a forest farmhouse.

The Hoard place had on it a half-dozen run-down but still usable buildings—a house and woodshed, a cow and hay barn with out-buildings, a horse stable, a pig pen and chicken-house. These buildings were designed to provide for the needs of a general Vermont farm, based upon animal husbandry. Since we had decided to keep no animals, most of the buildings were unsuited to our purposes, quite aside from their ramshackle appearance and condition. At the earliest possible moment, therefore, we gave the cow

and hay barn, pig pen and chicken-house to John Korpi, who lacked these buildings and wanted them as much as we wanted to get rid of them. He and his son took them down and hauled away all of the usable material. The horse barn we kept as a temporary tool shed. The old house was used as a carpenter shop and storage space for lumber and cement during part of our building operations, before it too was torn down and given away.

We decided to replace these old Hoard structures with a group of functional buildings, all to be constructed of stone. We chose stone for several reasons. Stone buildings seem a natural outcropping of the earth. They blend into the landscape and are a part of it. We like the varied

"We built a house of stone."

Forest Farm, Vermont.

Figure 2.1
"We Built a House of Stone." Helen and Scott
Nearing, *Living the Good Life: How to Live
Sanely and Simply in a Troubled World*
(New York: Schocken Books, 1954), n.p.
Photo by Helen Nearing, 1944. Courtesy of
Good Life Center.

color and character of the stones, which are lying around unused on most New England farms. Stone houses are poised, dignified and solid—sturdy in appearance and in fact, standing as they do for generations. They are cheaper to maintain, needing no paint, little or no upkeep or repair. They will not burn. They are cooler in summer and warmer in winter. If, combined with all these advantages, we could build them economically, we were convinced that stone was the right material for our needs.

We are not trained architects and know next to nothing of the details of that profession. We have read on the subject, and have put up over a dozen buildings. In view of this experience, we take our courage in our hands, and state four

general rules which we think should govern the architecture of domestic establishments.

Rule I: *Form and function should unite in the structure…*
Rule II: *Buildings should be adapted to their environment…*
Rule III: *Local materials are better adapted than any other to* create the illusion that the building was a part of the environment from its beginnings and has been growing up with the environment ever since…
Rule IV: *The style of a domestic establishment should express the inmates and be an extension of themselves.*

14 Henry David Thoreau, *Walden*, 1854

On July 4, 1845, Henry David Thoreau (1817–1862) began a two-year residence in a small cabin he had built at the edge of Walden Pond, just outside Concord, Massachusetts, his hometown. Thoreau's intent was to live simply and independently, immersed in nature yet not so far removed from Concord as to be entirely separate from society. Eight years after quitting his sylvan retreat, he recounted his experiment in *Walden; or, Life in the Woods*, an aphoristic text in which he imbued the prosaic tasks of self-subsistence or home economics with spiritual value and espoused the benefits of a simpler lifestyle.

While Walden did not enjoy great popularity upon its original publication in 1854, by the time the counterculture embraced the book, it was considered to be a classic of American literature, and Thoreau's reflections on the natural environment, American societal values, civil disobedience, proper government, and spirituality had inspired preservationists, political activists, authors, and artists for many decades. In the 1960s, the text became especially significant for members of the counterculture engaged in the back-to-the-land movement, whose yearning for an unaffected lifestyle, stripped of modern minutiae, closely paralleled their forefather's. In the passage that follows, Thoreau expresses the purpose of his personal movement back to the land.—MG

Source

Henry David Thoreau, *Walden; or, Life in the Woods* (1854; reprint, Garden City, NY: Anchor Press, 1973), 80–81, 84–86. (Referred to in *LWEC*, 47.)

Further Reading

Buell, Lawrence. *The Environmental Imagination: Thoreau, Nature Writing, and the Formation of American Culture.* Cambridge, MA: Belknap Press, 1995.

Skinner, B. F. *Walden Two.* New York: Macmillan, 1948.

Thoreau, Henry David. *The Journal: 1837–1861.* Ed. Damion Searls. New York: New York Review of Books, 2009.

"Where I Lived, and What I Lived For," from *Walden*

I went to the woods because I wished to live deliberately, to front only the essential facts of life, and see if I could not learn what it had to teach, and not, when I came to die, discover that I had not lived. I did not wish to live what was not life, living is so dear, nor did I wish to practise resignation, unless it was quite necessary. I wanted to live deep and suck out all the marrow of life, to live so sturdily and Spartan-like as to put to rout all that was not life, to cut a broad swath and shave close. To drive life into a corner, and reduce it to its lowest terms, and, if it proved to be mean, why then to get the whole and genuine meanness of it, and publish its meanness to the world; or if it were sublime, to know it by experience, and be able to give a true account of it in my next excursion. For most men, it appears to me, are in a strange uncertainty about it, whether it is of the devil or of God, and have *somewhat hastily* concluded that it is the chief end of man here to "glorify God and enjoy him forever."

Still we live meanly, like ants; though the fable tells us that we were long ago changed into men; like pygmies we fight with cranes; it is error upon error, and clout upon clout, and our best virtue has for its occasion a superfluous and evitable wretchedness. Our life is frittered away by detail. An honest man has hardly need to count more than his ten fingers, or in extreme cases he may add his ten toes, and lump the rest. Simplicity, simplicity, simplicity! I say, let your affairs be as two or three, and not a hundred or a thousand; instead of a million count half a dozen, and keep your accounts on your thumb nail. In the midst of this chopping sea of civilized life, such are the clouds and storms and quicksands and thousand-and-one items to be allowed for, that a man has to live, if he would not founder and go to the bottom and not make his port at all, by dead reckoning, and he must be a great calculator indeed who succeeds. Simplify, simplify. Instead of three meals a day if it be necessary eat but one; instead of a hundred dishes: five; and reduce other things in proportion....

Shams and delusions are esteemed for soundest truths, while reality is fabulous. If men would steadily observe realities only, and not allow themselves to be deluded, life, to compare it with such things as we know, would be like a fairy tale and the Arabian Nights' Entertainments. If we respected only what is inevitable and has a right to be, music and poetry would resound along the streets. When we are unhurried and wise, we perceive that only great and worthy things have any permanent and absolute existence, that petty fears and petty pleasures are but the shadow of the reality. This is always exhilarating and sublime. By closing the eyes and slumbering, and consenting to be deceived by shows, men establish and confirm their daily life of routine and habit every where, which still is built on purely illusory foundations. Children, who play life, discern its true law and relations more clearly than men, who fail to live it worthily, but who think that they are wiser by experience, that is, by failure. I have read in a Hindoo book, that "there was a king's son, who, being expelled in infancy from his native city, was brought up by a forester, and, growing up to maturity in that state, imagined himself to belong to the barbarous race with which he lived. One of his father's ministers having discovered him, revealed to him what he was, and the misconception of his character was removed, and he knew himself to be a prince. So soul," continues the Hindoo philosopher, "from the circumstances in which it is placed, mistakes its own character, until the truth is revealed to it by some holy teacher, and then it knows itself to be *Brahme*." I perceive that we inhabitants of New England live this mean life that we do because our vision does not penetrate the surface of things. We think that that *is* which *appears* to be. If a man should walk through this town and see only the reality, where, think you, would the "Mill-dam" go to? If he should give us an account of the realities he beheld there, we should not recognize the place in his description.

Look at a meeting-house, or a court-house, or a jail, or a shop, or a dwelling-house, and say what that thing really is before a true gaze, and they would all go to pieces in your account of them. Men esteem truth remote, in the outskirts of the system, behind the farthest star, before Adam and after the last man. In eternity there is indeed something true and sublime. But all these times and places and occasions are now and here. God himself culminates in the present moment, and will never be more divine in the lapse of all the ages. And we are enabled to apprehend at all what is sublime and noble only by the perpetual instilling and drenching of the reality that surrounds us. The universe constantly and obediently answers to our conceptions; whether we travel fast or slow, the track is laid for us. Let us spend our lives in conceiving then. The poet or the artist never yet had so fair and noble a design but some of his posterity at least could accomplish it.

Let us spend one day as deliberately as Nature, and not be thrown off the track by every nutshell and mosquito's wing that falls on the rails. Let us rise early and fast, or break fast, gently and without perturbation; let company come and let company go, let the bells ring and the children cry,—determined to make a day of it. Why should we knock under and go with the stream? Let us not be upset and overwhelmed in that terrible rapid and whirlpool called a dinner, situated in the meridian shallows. Weather this danger and you are safe, for the rest of the way is down hill. With unrelaxed nerves, with morning vigor, sail by it, looking another way, tied to the mast like Ulysses. If the engine whistles, let it whistle till it is hoarse for its pains. If the bell rings, why should we run? We will consider what kind of music they are like. Let us settle ourselves, and work and wedge our feet downward through the mud and slush of opinion, and prejudice, and tradition, and delusion, and appearance, that alluvion which covers the globe, through Paris and London, through New York and Boston and Concord, through church and state, through poetry and philosophy and religion, till we come to a hard bottom and rocks in place, which we can call *reality*, and say, This is, and no mistake; and then begin, having a *point d'appui*, below freshet and frost and fire, a place where you might found a wall or a state, or set a lamp-post safely, or perhaps a gauge, not a Nilometer, but a Realometer, that future ages might know how deep a freshet of shams and appearances had gathered from time to time. If you stand right fronting and face to face to a fact, you will see the sun glimmer on both its surfaces, as if it were a cimeter, and feel its sweet edge dividing you through the heart and marrow, and so you will happily conclude your mortal career. Be it life or death, we crave only reality. If we are really dying, let us hear the rattle in our throats and feel cold in the extremities; if we are alive, let us go about our business.

Time is but the stream I go a-fishing in. I drink at it; but while I drink I see the sandy bottom and detect how shallow it is. Its thin current slides away, but eternity remains. I would drink deeper; fish in the sky, whose bottom is pebbly with stars. I cannot count one. I know not the first letter of the alphabet. I have always been regretting that I was not as wise as the day I was born. The intellect is a cleaver; it discerns and rifts its way into the secret of things. I do not wish to be any more busy with my hands than is necessary. My head is hands and feet. I feel all my best faculties concentrated in it. My instinct tells me that my head is an organ for burrowing, as some creatures use their snout and fore-paws, and with it I would mine and burrow my way through these hills. I think that the richest vein is somewhere hereabouts; so by the divining rod and thin rising vapors I judge; and here I will begin to mine.

15 *The Oxford Book of Food Plants*, **1969**

This apparently simple book provides historical and nutritional information about everyday food plants: "In prehistoric times, food crops, after their first domestication, spread slowly outwards from their centres of origin to such other areas as were easily accessible by land travel. Thus before written history began, even such outlying areas of western Europe as Britain had received cereal crops like wheat and barley which originated in western Asia. Field beans and peas, originating in Asia and the Mediterranean, have been discovered in excavations of Swiss lake dwellings of the Bronze Age" (*The Oxford Book of Food Plants*, **195**).

With detailed botanical illustrations on each page, the accompanying text explains the origins of each plant in an accurate and precise way, always considering the complex circulation process of ordinary plants: "By contrast with the voyages of Columbus, the opening up of sea communications with eastern Asia by the great voyage of Vasco da Gama in 1497 was not followed by many plant introductions, though rhubarb, already exploited in Asia as a food plant, was in the course of time brought to plant in the gardens of Europe" (*The Oxford Book of Food Plants*, **197**).

The plants are arranged in groups, backed up by an index, according to the kind of food they produce: fruits, herbs, spices, cereal, sugar crops, oil crops, nuts, vegetables, and root crops. The beautiful illustrations, rare in an affordable book such as this, designed by the botanical artist Barbara Nicholson, trained at the Royal College of Art, were intended to encourage people to take an interest in the use and nutritional value of food plants. Reasonably priced books with good quality color illustrations were still rare in the 1960s, and one of the attractions of the book was its appearance. For town people discovering the country, the book provided not only practical information but also an interesting historical account from prehistoric times about the migration of food plants across the world. —CM

Source

G. B. Masefield, M. Wallis, and S. G. Harrison, with illustrations by B. E. Nicholson, *The Oxford Book of Food Plants* (Oxford: Oxford University Press, 1969), plate 167. (Referred to in *LWEC*, 49.)

Further Reading

United Nations. *Report of the United Nations Conference on the Human Environment, Stockholm, 5–16 June, 1972.* New York: United Nations, 1973.

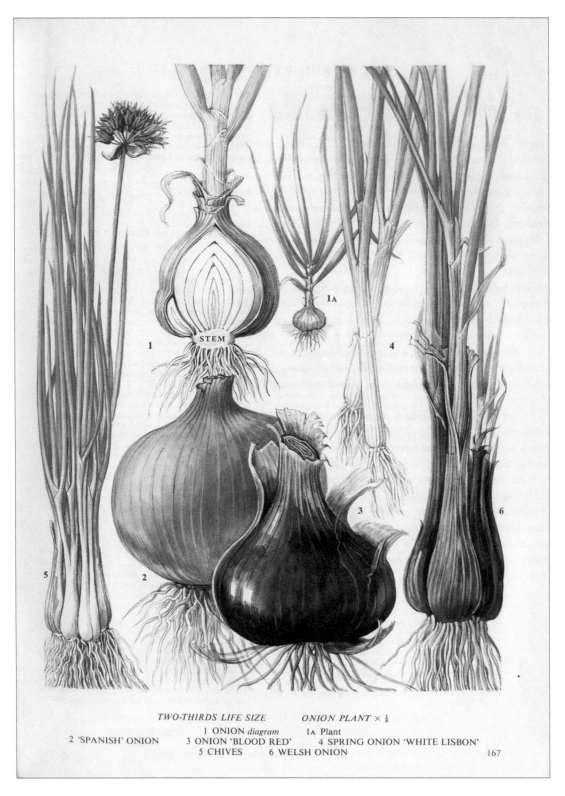

TWO-THIRDS LIFE SIZE   ONION PLANT × ⅛

1 ONION *diagram*   1A Plant
2 'SPANISH' ONION   3 ONION 'BLOOD RED'   4 SPRING ONION 'WHITE LISBON'
5 CHIVES   6 WELSH ONION

167

Figure 2.2
"Onion Plant." G. B. Masefield, M. Wallis,
S. G. Harrison, and B. E. Nicholson, *The
Oxford Book of Food Plants* (Oxford: Oxford
University Press, 1969), plate 167.

16 Jeanie Darlington, *Grow Your Own*, 1970

Jeanie Darlington began gardening in earnest in 1968, on a 10 x 10 foot plot in the backyard of her Albany, California, home. Inspired to grow tomatoes, Darlington faced a quandary: would she employ the chemical fertilizers and pesticides she promoted at the nursery where she worked, or follow the advice of her few customers who relied on only organic materials? After discovering *Organic Gardening and Farming Magazine*, she committed herself to the latter way, even quitting her job at the poison-pushing nursery shortly thereafter. Although Darlington faced setbacks along her path to create a thriving garden—for instance, she initially applied sulfate of ammonia to the soil to burn off grass and weeds, killing earthworms and creating a nutrient-depleted, hard clay patch in the process—through the use of homemade compost and other natural strategies, she eventually found great success. In her book *Grow Your Own*, she shared these victories and missteps, as well as her extensive knowledge of gardening; her friendly, easygoing style of writing about gardening techniques endeared the book to beginners and won the support of the editors of the *Whole Earth Catalog* and *Mother Earth News*.

In the *Whole Earth Catalog*, Gurney Norman lauded Darlington as a "competent freak," one of those figures who, by gaining professional knowledge "without sacrificing the joyous spirit of the amateur...may be the most important people in the entire culture at this nervous point in its development." Darlington's lessons span from proper compost production to seasonal growth patterns to indoor gardening and are interspersed with line drawings like "Manty," a praying mantis that lived in her garden. The excerpt found in the following pages, from the book's introduction, demonstrates the intimacy of her prose, her keen understanding of organic gardening strategies, and an ecological worldview shared by the editors and readers of the *Whole Earth Catalog*.—MG

Source

Jeanie Darlington, *Grow Your Own: An Introduction to Organic Gardening* (Berkeley, CA: Bookworks, 1970), 47–50. Courtesy of Jeanie Darlington. (Referred to in *LWEC*, 50.)

Further Reading

Philbrick, John, and Helen Philbrick. *Gardening for Health and Nutrition: The Organic Methods*. San Francisco: Harper & Row, 1963.

Rodale, J. I. *The Encyclopedia of Organic Gardening*. Emmaus, PA: Rodale Books, 1959.

Stout, Ruth. *How to Have a Green Thumb without an Aching Back*. New York: Exposition Press, 1955.

Introduction to *Grow Your Own*

Meanwhile I had been thinking about how I could try to restore the ecological balance in the backyard. The birds had come back and were already busy eating lots of insects on the fruit tree branches. One night while out snail hunting, I encountered a lizard. It scared the wits out of me, until I remembered that they ate slugs. And I knew I could order ladybugs and praying mantis egg cases.

In this case, when I speak of "restoring the ecological balance," I mean that I wanted to cut down on the plant eating bugs without resorting to bug sprays. To do that, you invite certain predatory, carnivorous insects into your garden, such as ladybugs and mantids. This is a little hard in a small backyard, because a neighbor's spray program could defeat the effort. But it was worth a try and it only cost me $4.00. The ladybugs and mantids arrived in early April. After eating their fill of aphids, the ladybugs mated, laid eggs and died because it was the end of their life cycle. By the end of May, the baby mantids had hatched and the new ladybugs had come out of their larvae stage.

The ladybugs ate aphids, mealybugs, scale and many other tiny insects. There were plenty of them around. We never did figure out exactly what the mantids ate, but they looked fat and well fed and were very tame. I only saw one or two Mexican bean beetles, and it is said that mantids like them. All I know is that my bean leaves weren't eaten to a lacey remain of veins like they were the year before. And every once in a while, little pint-sized birds would hop among the rose bushes and gobble all the aphids off each new shoot.

The peas grew 6 feet high, despite the fact they were dwarf gray sugar peas (2 and 1/2 feet maximum said the package). We were eating them from the middle of May through the end of June. Beets and carrots soon followed. Then broccoli and chard; Italian and Kentucky Wonder beans, white corn, artichokes, zucchini, greyzini, Italian cocozelle; oak leaf and ruby lettuce; escarole and endive; shallots, onions and leeks; and finally Spring Giant, Pearson and yellow pear tomatoes. The tomatoes ripened very late, but that was because I had rotated the crop from last year's spot and it wasn't as hot and sunny in the new place. Now I know that rotation isn't really necessary as long as you replenish the organic matter in the soil. The soil bacteria working on the humus will destroy any disease organisms connected with tomatoes if they are there.

We loved watching the ladybugs, who especially liked to live in the upper leaves of the sunflowers. Often they would fly down and land on our shoulders and walk along for a bit and then go back to their roost. A friendly hello.

The mantids seemed to like the parsley and dusty miller best for their homes, although one even migrated to the long row of potted plants on our front porch in late August, where he contemplated a tiny piece of chicken we offered him for four hours before deciding not to eat it. We named him Manty and he stayed on the porch until early December, living in a pot of basil. He shed his skin three times and ate baby leafhoppers and a worm Sandy once brought him. On his last skin moult, he acquired a pair of long brown wings, and soon after he wandered away. To mate, then die? The Bay Area's weather is not mild enough for mantids to survive the winter.

I hope this book will help people get started growing their own vegetables and flowers organically. Having a garden is such a wonderful experience.

Some people still wonder why go to all the trouble to do it organically. I think it's much simpler to garden organically, at least on a small backyard scale. People are beginning to be aware of ecology. Organic gardening is something each of us can do to help. I'm quite sure it's cheaper to garden organically than with synthetic chemicals. You don't have to buy five different types of poison sprays and several different fertilizer mixes. Compost can be made for free or for a very little bit of money. For less than $20.00, I bought 100 lbs. each of blood meal (N), phosphate (P), and granite dust (K). That will last me several years.

It's without a doubt more fun to garden organically. It's nice to have living things like Manty and the ladybugs around. They become friends. And it's a good influence on your children. It's a pleasure to dig into rich soil, full of fat, happy earthworms. I love watching the birds splash around in our improvised bird bath and knowing that any bread I toss out to them will be gone in several hours.

Organically grown food really does taste better. Unfortunately I have seen some pretty sad looking organic produce at some health food stores. I don't know if this is due to the problems of large scale farming, or bad shipping and storage methods or what. My vegetables almost always look beautiful enough to be photographed for seed catalogs and I'm no veteran farmer.

Most importantly, it's better for your soul to garden organically. If you use chemical fertilizers, you are disregarding the fact that soil is a living breathing thing. Soil becomes only a medium which supports plants upright. Chemical fertilizers destroy many life forms such as beneficial soil bacteria and earthworms. Poison sprays not only pollute the atmosphere, but also kill many harmless insects and many helpful predators, thus destroying the balance of nature. Gardening organically is working in harmony with nature.

17 United Nations, *New Sources of Energy*, 1962

Held in Rome in August 1961, the United Nations Conference on New Sources of Energy gathered more than four hundred scientists, engineers, health officials, and other related professionals from seventy-four countries to create a productive space for

the exchange of ideas and experiences in the applications of solar, wind, and geothermal energy; consideration of how techniques can be brought into wider use, particularly for the benefit of less developed areas; [and] provision of up-to-date information on progress achieved and on the potentialities of as well as the limitations in utilizing these three sources of energy, especially in those areas lacking conventional energy sources or facing high energy costs. ("General Review of Proceedings," in *General Sessions*, vol. 1 of *New Sources of Energy and Energy Development*, 5)

The proceedings of the conference were published in seven volumes, which contained papers and reports, written by experts in their field, relating to geothermal energy (vols. 2–3), solar energy (vols. 4–5), and wind power (vol. 7). Although the proceedings were not wholly intended for a popular audience, Brand recognized their educational potential, especially for readers who had exhausted the more straightforward texts on solar and wind power. — MG

Source

J. Geoffroy, "Use of Solar Energy for Water Heating," in *Solar Heating II*, vol. 5 of *New Sources of Energy and Energy Development: Report on the United Nations Conference on New Sources of Energy; Solar Energy, Wind Power, Geothermal Energy, Rome, 21 to 31 August, 1961* (New York: United Nations, 1962), 45–47. Reprinted with the permission of the United Nations. (Referred to in *LWEC*, 68.)

Further Reading

United Nations. *Report of the United Nations Conference on the Human Environment, Stockholm, 5–16 June, 1972.* New York: United Nations, 1973.

"Use of Solar Energy for Water Heating," from *New Sources of Energy*

Theoretically, the ideal zone for the use of solar energy is the zone with the maximum annual number of days of sunshine and where the daily duration of sunshine is constant and the conventional forms of energy scarce and therefore expensive. Commercial experience, however, proves that *all the regions* located between 45° N. lat. and 45° S. lat. and having over 2,000 hours of sunshine a year may be regarded as suitable for the rational use of solar energy to heat water for domestic purposes, without fearing the competition of the conventional forms of energy.

It is well known that solar energy may be collected *with or without concentration*, and that either of these methods may be used for water heating.

Concentration may be accomplished either by a parabolic mirror or by a cylindro-parabolic mirror.

The parabolic mirror requires a delicate and expensive mechanism to follow the sun during the day, and it takes a good deal of space. On the other hand, it does collect the maximum energy during the entire period of sunshine, at temperatures of the order of several hundred degrees centigrade.

The cylindro-parabolic mirror requires periodic adjustment of its inclination and is delicate, expensive and bulky. It collects no more energy than a plane collector, but it gives working temperatures over 100°C.

The flat collector is neither delicate nor very expensive. When properly oriented and tilted, according to the site, it collects as much energy as the cylindro-parabolic collector, but the operating temperature is below 100°C.

Conditions to Be Met by a Solar Water-Heater

It must provide continuous service. Consequently, it must be a storage water-heater of capacity sufficient to meet the 24-hour needs of the consumer. It must also be provided with a stand-by heating system to take care of any interruptions of sunlight at the site.

It must not raise architectural problems:

Because of its bulk: its storage tank must, if required, fit into a single room, while the collector is installed on the front of the building;

Because of its weight, which must not exceed the permissible loads on roof, balconies and floors;

Because of its unaesthetic appearance: the water heater installed must be compact in appearance and compatible with the general style of the building.

Its installation must not require a specialist, and, like any other household equipment, the solar water-heater must be simple enough to operate after simply reading detailed instructions.

It must not be more liable to scaling than a water-heater using a conventional form of energy. The water supply being what it is, the solar water-heater must deliver water no hotter than 70°C, since scaling is more rapid above this temperature. The use of a heating circuit and a heat-exchanger eliminates the danger of scaling in the sensitive parts of the equipment.

It must not be liable to freeze. The use of a heating circuit independent of the distribution circuit makes it possible to protect the sensitive parts of the equipment by using an antifreeze mixture (water + alcohol or water + antifreeze in proper proportions). The storage tank and the distribution circuit must be very thoroughly heat-insulated.

It must be sturdy and require neither maintenance nor inspection. A solar water-heater is usually installed on the roof or balcony for optimum exposure. Access is thus inconvenient. The heater should have no moving parts, which might get out of order or need lubrication, and the materials must be resistant to weather conditions (wind, rain, sand, sea air).

Finally, it must be competitive in cost with water heaters using conventional forms of energy. But the cost of conventional water heaters should include not only the total capital investment but also the cost of energy for at least three years.

Figure 2.3
"Figure 4. Schematic diagram of a Radiasol hot-water service." J. Geoffroy, "Use of Solar Energy for Water Heating," in *Solar Heating II*, vol. 5 of *New Sources of Energy and Energy Development: Report on the United Nations Conference on New Sources of Energy; Solar Energy, Wind Power, Geothermal Energy, Rome, 21 to 31 August, 1961* (New York: United Nations, 1962), 45. Reprinted with the permission of the United Nations.

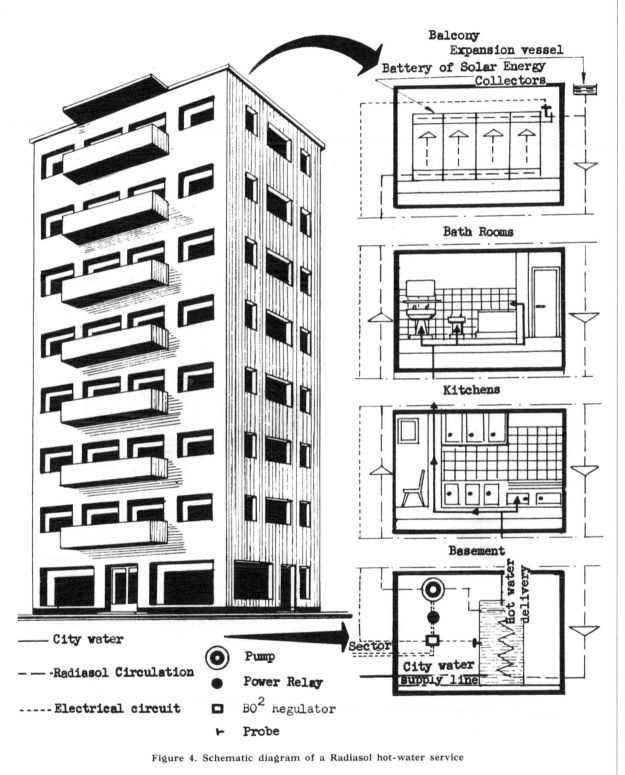

Balcony
Expansion vessel
Battery of Solar Energy
Collectors

Bath Rooms

Kitchens

Basement

Hot water delivery

Sector

City water
supply line

—— City water

— — Radiasol Circulation

· · · · Electrical circuit

⊚ Pump

● Power Relay

□ BO^2 Regulator

⊢ Probe

Figure 4. Schematic diagram of a Radiasol hot-water service

18 Ian McHarg, *Design with Nature*, 1969

The attraction of Ian McHarg's (1920–2001) approach was the way in which it brought a wide range of scientific knowledge and design considerations into relationship with one another. McHarg's aim was to discover, in any particular landscape configuration, the best potential value for various purposes, whether agricultural, recreational, or urban. His method was to draw a series of analytical maps identifying geological formation over time, soil and subsoil layers, vegetation, hydrographic flows, microclimate, human activities, and transportation networks. Unlike some idealist commentators, McHarg was a realist. As the reviewer Vic Conforti noted: "He also realizes that growth is going to take place. To meet this growth with only derisive criticism will do little. Only by developing the attitude of 'steward' towards Earth and acquiring the tools and knowledge that are needed to properly care for her, will we survive." Founder of the Department of Landscape Architecture at the University of Pennsylvania, McHarg, a Harvard-educated landscape architect, developed, from 1957 onward, a course titled "Man and Environment."

One of the case studies he analyzed was the Potomac River Basin. As he explained: "It is our intention to understand the Potomac River Basin as an interacting process, to interpret this as a value system and to designate appropriate land uses." Lewis Mumford, who wrote the introduction to the book, ended with this comment: "In presenting us with a vision of organic exuberance and human delight, which ecology and ecological design promise to open up for us, McHarg revises the hope for a better world. Without the passion and courage, and confident skill of people like McHarg, that hope might fade and disappear forever."—CM

Source

Ian McHarg, *Design with Nature* (Garden City, NY: American Museum of Natural History/Natural History Press, 1969), 127–128. (Referred to in *LWEC*, 82.)

Further Reading

McHarg, Ian. "An Ecological Method for Landscape Architecture." In *The Subversive Science: Essays toward an Ecology of Man*, ed. Paul Shepard and Daniel McKinley, 328–332. Boston, MA: Houghton Mifflin, 1969. The essay first appeared in *Landscape Architecture*, January 1967.

Olgyay, Victor. *Design with Climate: Bioclimatic Approach to Architectural Regionalism*. New York: Princeton Architectural Press, 1963.

Spirn, Anne Whiston. "Ian McHarg, Landscape Architecture, and Environmentalism: Ideas and Methods in Context." In *Environmentalism in Landscape Architecture*, ed. Michel Conan, 97–114. Washington, DC: Dumbarton Oaks Research Library and Collection, 2000.

"The River Basin," from *Design with Nature*

It is our intention to understand the Potomac River Basin as an interacting process, to interpret this as a value system and to designate appropriate land uses. Now this is not a plan—a plan is a determination to achieve certain social goals, related to the power of society to accomplish these. No: this exercise seeks only to reveal nature as a working storehouse, with implications for land use and management. This information is an indispensable ingredient to a plan, but is not the plan itself.

The first considerations are historical—geology and climate which, in conjunction, have interacted upon the river basin, for they have created the basic form. When this is understood, the various physiographic regions become clearly evident.

The current morphology, with climate and lithology, can be invoked to explain the pattern of rivers and streams, the distribution of groundwater, relative quantities and physical properties. The pursuit of this information on the movements of sediments, some by fluvial processes, others from deposition, will reveal the pattern, distribution and properties of soils. When climate, topography, the water regimen and soils are known, the incidence of plants as individuals and as communities becomes clearer. As animals are all either directly or indirectly plant-related—whether in terrestrial or aquatic environments—knowledge of the plant communities, their age and condition, will tend to explain the distribution of animals. We have seen that coal, iron, limestone and rich, productive soils occur where they do for reasons that derive from physical and biological processes deep in geological history. They are where they are … because. So too with the Fall Line on the major rivers, where the watercourses cut deep trenches through the gravels of the Coastal Plain after leaving the Piedmont crystalline rocks. Here is the break-of-bulk point where cities were located. There are transportation routes, but these are likely to follow river courses and passes; there are fitting places for cities, level, well-drained sites adjacent to abundant water and surrounded by productive soils, and so the same method can be used to follow land use over time and to see the march and growth of men upon the land. Indeed, even battles can be better understood if the facts of physiography are known. From this same method, the presence of unique sites, limestone caves or garnet beaches—home of oyster and clam, trout and bass—can be found.

An examination of this sort will reveal the most productive soils, the presence of coal and limestone deposits, the relative abundance of water in rivers and aquifers, the great forests, oyster banks, areas of wilderness or relative accessibility, historic forts or areas of great natural beauty. As the regions vary from one another because of their geological history, there will be regional variation in all the resources that are considered. There will also be a relative consistency within each region. So, having acquired this information, it is not difficult to interpret it in terms of the dominant, intrinsic resource or resources. After all, coal exists in only one region, limestone is extensive in only one; the great agricultural soils are concentrated in a single region. From this view can be seen the dominant prospective land use for each region, and this is the first overview. Where is the major recreational opportunity in the basin? Where the agricultural heartland? Where are the best forest locations? Where the best sites for urbanization? This preliminary investigation can answer such questions, deriving the information from the place itself. So now when we know something of the inventory of the storehouse, we can turn and ask, "what do you want?" The questions can vary. "Where can I find 15,000 acres of land of less than five per cent average slope for a new city, within one hour's

travel distance from Washington?" "Where is there coal to be stripped with an overburden of less than 12 to 1?" "Where in West Virginia can I drill and find 600 gallons of water a minute?" "Where can I find a large wilderness area or a wild river, a trout stream or a ski slope?" These questions can be answered.

We have become accustomed to think of single-function land use and the concept of zoning has done much to confirm this—a one-acre residential zone, a commercial or industrial zone—but this is clearly a most limiting concept. If we examine a forest, we know that there are many species—and, thus, that many cooperative roles coexist. In the forest there are likely to be dominant tree species, subdominants and a hierarchy of species descending to the final soil microorganisms. The same concept can apply to the management of resources—that there be dominant or codominant land uses, coexisting with subordinate, but compatible ones.

It takes only a moment of reflection to realize that a single area of forest may be managed either for timber or pulp; it may be simultaneously managed for water, flood, drought, erosion control, wildlife and recreation; it may also absorb villages and hamlets, recreational communities and second homes.

Now we have a program; we seek to find the highest and best uses of all the land in the basin, but in every case we will try to identify the maximum conjunction of these. This, then, is the image of nature as an interacting and living storehouse—a value system.

Shelter

For the *Whole Earth Catalog*, which was essentially about changing consciousness and finding better ways to live, the most pressing challenge was that of the dwelling. The twenty-six-page "Shelter" section offered everything to satisfy the home builder and the do-it-yourself *bricoleur*, presenting manuals that explained how to build lightweight structures, tipis, and sheds, as well as how to repair and maintain existing structures. The editors' skill was therefore how to present a wide range of publications—from the lyrical to the highly technical—in such a way as to attract the readers' attention and stimulate them to take action. Included were technical manuals for carpenters, electricians, and plumbers. The first big step toward empowerment and responsibility lay in demystifying the cult of expertise. The *Whole Earth Catalog* constantly implored people to take responsibility for their own lives by taking control of the technology they depended on.

The section opened with a discussion of Frank Lloyd Wright and Antoni Gaudí, held up as examples of inventive and unconventional creators. By implication, the practice of contemporary architects was dismissed. Surprisingly, the geodesic domes and lightweight structures of alternative architecture were placed toward the end of the section, after the two books by Steve Baer, *Dome Cookbook* and *Zome Primer*, as well as Lloyd Kahn's *Domebook 2*. With Paul Oliver's *Shelter and Society*, the reader was encouraged to look carefully at vernacular architecture, and even conventional house types, but to adapt them to new modes of living, an organic approach to materials, and a response to climate and available materials. It was the Zen Mountain Center that suggested Heinrich Engel's book *The Japanese House* for inclusion. Brand saw in it a mine of information on the art of building in Japan derived from the spiritual and social structures of the Japanese way of life.

Attention to climate, microclimate, or the effects of extreme weather—such as tornadoes and hurricanes—was covered by several books such as Frank Lane's *The Elements Rage*, illustrated with striking photographs.

Nevertheless, the selection did include some avant-garde architects such as Archigram, Moshe Safdie, Richard Buckminster Fuller, Paolo Soleri, and Christopher Alexander. Most attention was given to Frei Otto, probably because his innovative approach to engineering took into account experimentation, energy conservation, and ephemeral structures, as well as for the graphic and photographic quality of his two books devoted to lightweight structures. For similar reasons, the editors devoted a third of the page to the collective Ant Farm's pneumatic structures and their explanatory guide, the *Inflatocookbook* (1971).

The overall effect was to provide home builders with everything they needed to know and think about when planning a home, from choosing a site, selecting materials, and seeking technical help to much wider questions about how to live.—CM

19 C. L. Duddington, *Evolution and Design in the Plant Kingdom*, 1969

Opening the "Shelter" section with a book on the analysis of plants made an impressive claim for the biological analogy as a source of architectural inspiration. It is a recurring theme in Western architecture, but rarely had it been taken so literally. Not that C. L. Duddington was at all interested in architecture; rather, his analysis suggested how buildings might "grow" naturally just as plants do.

Duddington, a British biologist, studied the structures of plants to demonstrate principles of engineering. He showed how the microstructures in wood adapt to tension and compression. He explained the passage of water, drawn up through the trunks of trees and plant stems.

Brand's take on the book was fanciful: "'Live dwellings—how soon? Houses of living vegetable tissue. The walls take up your CO_2 and return oxygen. They grow or diminish to accommodate your family changes. Add a piece of the kitchen wall to the stewpot. House as friend. Dweller and dwelling domesticate each other."

Part of the interest of books like this one was not only to understand how natural forms function but to learn how to design like nature. Both Frank Lloyd Wright and Antoni Gaudí have been described, in different ways, as organic architects. For Gaudí, the close study of natural forms was reflected directly in his constructions. The piers in his model of the nave of the Sagrada Familia divide like branches of a tree, and the buttresses of terraces at the Park Güell actually imitate palm trees.—CM

Source
C. L. Duddington, *Evolution and Design in the Plant Kingdom* (New York: Thomas Y. Crowell, 1969), 36, 51–53. (Referred to *LWEC*, 84.)

Further Reading
Duddington, C. L. *The Living World*. London: Pan Books, 1966.

Selections from *Evolution and Design in the Plant Kingdom*

Plants as Structural Engineers

One of the first problems that plants had to solve on emerging from the ocean was that of raising their stems up to the light without the supporting embrace of the sea around them. For a seaweed it is easy. Supported by the water that bathes it on every side, the most it needs for maintaining the buoyancy of its fronds is a few air bladders scattered here and there. The bladders float and hold up the fronds where the light can get at them and carry out its vital role in photosynthesis. Even the giant kelp *Macrocystis* can keep its six hundred foot branches floating gracefully in the ocean currents in this way.

Not so with a land plant. No plant has so far been able to evolve bladders filled with hydrogen that float in the air like barrage balloons, trailing the plant beneath them. It is unlikely that one ever will—though I would hesitate to say that anything is impossible to a plant. A land plant must have built-in rigidity, so that its stem will not only stand erect as a result of its own solid structure, but also resist the efforts of the wind to blow it down.

A very small herbaceous plant can stand erect with the aid of hydrostatic pressure, by blowing its cells out with water. We see the same principle in the inner tube of a bicycle tire. When deflated it flops; it can be bent into any shape, and is quite incapable of supporting itself. Blow it up with air and it is another story. The tube takes on a circular form and resists strongly any attempt to bend it out of shape. If it is inflated hard enough it is almost as rigid as the wheel itself....

Weight-Supporting Problems

Plants live under varying loads. Some of these are problems added by the plant breeder, such as too-heavy wheat ears and branches of the Victoria plum which break under the weight of their own fruit. Even without these risks of civilization, however, there are more inherent weight-supporting difficulties to be solved. Why does the weeping willow weep—or, rather, why do not all trees weep? A tree grows from a sapling, often without much change of form. The young branches of the sapling tend to jut out at certain angles, and these angles are more or less retained as the tree grows from a sapling into a giant. As more and more wood is added to a branch it naturally gets heavier and heavier. The engineer would regard a branch of a tree as a cantilever loaded with its own weight; as it gets heavier it ought to bend more and more, until finally, if it grows long enough, it would hang vertically downward under the load. Yet, with few exceptions, this does not happen.

The mechanical stresses that occur in the branch of a tree may be presumed to be the same as in any other cantilever. The material toward the top of the branch is in tension, while the material near the bottom is in compression. If it is to keep a straight course the wood must resist both these stresses equally. How it maintains its apparently predetermined course we do not know.

If we study the wood on the lower sides of branches of coniferous trees, or on the upper sides of branches in broad-leaved trees, we find that it has a distinct structural difference from the rest of the wood. This wood is called "reaction wood," and it differs from normal wood both anatomically and chemically. In conifers, where it is called "compression wood," it is darker than normal wood, and the walls of the cells are more heavily impregnated with lignin. The tracheids are shorter than usual, and the wood is denser. The secondary lignified cell wall which is laid down inside the primary wall consists of three layers in normal wood, but in compression wood there are only two layers, the innermost layer missing.

In branches of broad-leaved trees there is no compression wood, but the wood laid down on the upper side of the branch, known as "tension wood," shows modifications in structure. The vessels are narrower, and the fibers have a thick inner layer—the gelatinous layer—which refracts light very readily and is composed mainly of cellulose instead of lignin.

20 "Shelter"

The section's first double-page spread, pages 84 and 85, compared the natural structures of plants, above (C. L. Duddington), with the serpentine benches of Antoni Gaudí's Park Güell in Barcelona. The vernacular structures of Bernard Rudofsky, published in *Architecture without Architects* (1964), were discussed below. Although Gaudí's work was illustrated, Wright was represented only by a text and a vignette showing interlocked fingers.

Brand took the opportunity to praise Gaudí as a stick to beat conventional architects: "God Damn architects who bore and stifle and preen their chickenyard feathers. Give us more jungle birds like this one. If their buildings are inconvenient, good." Here, he was following the advice of the editor of *Newsweek*: "Be impudent and get them mad!"[1] Brand commented on Wright's texts: "Get the habit of analysis,—analysis will in time enable synthesis to become your habit of mind,"[2] emphasizing as always the need for firsthand experience. Brand welcomed Wright's insistence on considered thought as a principle motor for action. Reflection was particularly necessary for the judicious choice of materials and a deep understanding of their nature: "Beware of the architectural school except as the exponent of engineering. Go into the field where you can see the machines and methods at work that make the modern buildings, or stay in construction direct and simple until you can work naturally into building-design from the nature of construction. Immediately begin to form the habit of thinking 'why' concerning any effects that please or displease you."—CM

1. Stewart Brand, *Notes*, box 26, M1237.

2. Frank Lloyd Wright, *Writings and Buildings: Frank Lloyd Wright*, selected by Edgar Kaufmann and Ben Raeburn (New York: New American Library, 1960), quoted by Stewart Brand, *LWEC*, 85.

Source
LWEC, 84–85.

Further Reading
Conrads, Ulrich, and Hans Günther Sperlich. *The Architecture of Fantasy: Utopian Building and Planning in Modern Times*. New York: Praeger, 1962.

Rudofsky, Bernard, and Museum of Modern Art (New York). *Architecture without Architects: A Short Introduction to Non-pedigreed Architecture*. 1964. Albuquerque: University of New Mexico Press, 1987.

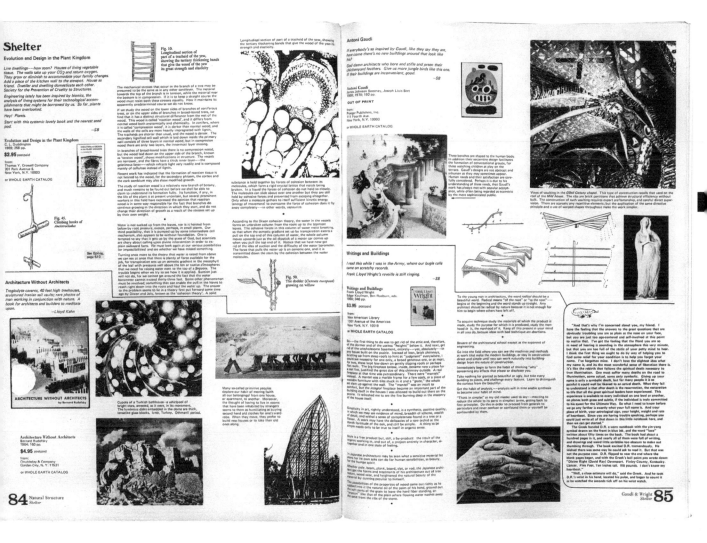

Figure 3.1
The "Shelter" section's first double-page spread. *LWEC*, 84–85.

21 Paul Oliver, *Shelter and Society*, 1969

"Seldom do you see anything on buildings by the people, of local materials and in simple harmony with the surrounding landscape. Here's an exception," wrote Lloyd Kahn of Paul Oliver's edited volume *Shelter and Society*, which examined a diversity of vernacular constructions—from the Norwegian *laftehus* to the Navajo hogan to the geodesic domes of Drop City—and the cultural and environmental conditions from which they emerged. That many of the contributors to the book were Oliver's colleagues at the Architectural Association in London demonstrated the burgeoning interest in vernacular studies within the discipline, one that was mirrored by the editors and readers of the *Whole Earth Catalog*, among others.

Insofar as the ecologically sensitive, humanistic approach to design offered by vernacular architecture complemented their own worldviews, countercultural builders embraced its possibilities. By including Bill Voyd's essay "Funk Architecture," an analysis of Drop City's "new geodesic vernacular" (*Shelter and Society*, 167), Oliver legitimized the idiosyncratic structures of similar communities, physical responses to their particular customs, material limitations, and climatic requirements. Oliver's text was certainly less technical than many manuals found in the "Shelter" section, but no less educational: it motivated a generation of New Communalists and other countercultural builders to consider construction in a new way. —MG

Source

Paul Oliver, ed., *Shelter and Society* (New York: A. Praeger, 1969), 27–28. Reprinted with the kind agreement of Paul Oliver. (Referred to in *LWEC*, 86.)

Further Reading

Moholy-Nagy, Sibyl. *Native Genius in Anonymous Architecture.* New York: Horizon Press, 1957.

Rudofsky, Bernard. *The Prodigious Builders: Notes toward a Natural History of Architecture with Special Regard to Those Species That Are Traditionally Neglected or Downright Ignored.* New York: Harcourt Brace Jovanovich, 1977.

Introduction to *Shelter and Society*

As the anthropologist may often reveal patterns of behavior in our own society which go unnoticed precisely because they are so familiar but become apparent when exposed to another and alien culture, so the study of vernacular architecture can reveal either directly, or by analogy, some of the responses which the occupants may have to the buildings erected by architects. They may illuminate aspects of our own reactions to buildings within our culture but, more important, as the spread of Western technology and architectural forms increases, those of other peoples. Such responses may be as much to the forms as to the enclosure; many architects are aware of the inspirational sensations awoken by church spires and cathedral roofs, of the gestural significance of pillars, columns and obelisks, of the symbolic associations of Roman ceremonial arch and Japanese *torii*, for they are familiar forms of considerable potency. Less easily recognized are the values attached to seemingly arbitrary forms or the psycho-social implications of the proximity of masses which exert their influences in subtle ways. Only a study of the building in relation to the culture that produced it may illumine these shadowy areas of architectural knowledge.

In these instances the intimate relationship between the society and its shelter is self-evident but it is no less significant in those aspects which are more frequently the concern of the architect in studying vernacular shelter—the situation of buildings, the resources from which they are constructed, the technology employed in their fabrication and their functionalism in meeting the conditions of climate or use. Such considerations are fundamentally interdependent and are all related to the total structure of the society. Whether the basic economy is one of hunting or of herding, of industry or trade, will affect precisely the nature of the buildings that are erected; whether a society lives in harmony with its neighbors, whether it has established commercial ties or whether defense of the community is a matter of major concern; proximity to water and to pasture, to fishing grounds and to overland routes—these are factors that determine the placement of buildings. Community, tribal, extended family and nuclear family structures or the complex reciprocity of "dual organization" are frequently mirrored in planning which superficially appears formless. What roles are assigned to the men and to the women, what social behavior is required of children and youths, what respect is accorded to the aged, what responsibilities fall on the shoulders of councils and elders, of headmen and chiefs, must ultimately affect the nature of the shelter that is provided for the members of the community and the lodges, council houses and ceremonial houses of its leaders.

If the spaces created must serve these social functions, the means whereby they are produced are conditioned by the materials available, whether rock or small stone, clay for daubing or baking, timber for frame or mass structure, bamboo and reed, grass and fiber. How they are quarried, felled, harvested, cured, fired or otherwise prepared must in turn depend on how labor is distributed and this in turn upon community organization. Methods of combating the weather—burning sun and monsoon rains, penetrating snow and tropical storm, seasonal flood waters and unpredictable land tremors alike—are the result of long modification and adaptation, experience shared and innovations approved. For the very congruity and harmony which is so frequently a source of admiration and comment in architectural writings on vernacular shelter is evidence of the integration of the building in the life of the community as a whole. The appreciation of the one without the understanding of the other must necessarily remain superficial. ...

Today, Western-trained architects are cheerfully prepared to embark with confidence, and often incompetence, on the resettlement of diverse ethnic groups and cultures in many countries; they are planning, replanning, rehousing, rehabilitating people throughout the world in buildings often alien to the way of life and in totally inappropriate structures for the climate and environment. At one end of the scale they build the successors to Imperial Delhi in Chandigarh and Brasilia; at the other, the pill-box rows of native housing in Kampala or the empty, inhospitable slab-blocks of the new quarters of Casablanca. Perhaps the most depressing aspect of this is the rapidly increasing tendency in many countries which are accelerating in their economic and technological development, to adopt all things western, including immense blocks of steel and glass, as necessary symbols of their achievement.

So far, however, there is little evidence that these new developments have contributed to the sum of human happiness. They may have added to the quality of "architecture"; they have added little to the quality of life. It would seem self-evident that a better understanding of the vernacular communities being resettled and replaced would help substantially in the better application of Western technology to building in exotic cultures. But have vernacular studies anything to reveal to us of our approach to building within our own society? Can shelter give any indication of the needs of people in a modern, industrialized, capitalistic, acquisitive, consumer-directed society? Can it even reveal the needs of a minority within it?

"It may be romantic to search for the salves of society's ills in slow-moving, rustic surroundings, or among innocent, unspoiled provincials, if such exist, but it is a waste of time," wrote Jane Jacobs in her memorable study of *The Death and Life of Great American Cities*; "does anyone suppose that, in real life, answers to any of the great questions that worry us today are going to come out of homogenous settlements?" But many cities—vernacular cities—display a homogeneity that is lacking in the metropolis, though in America, as in Great Britain, homogenous sub-cities of migrants from the country struggle to preserve their humanity and identity in the metropolitan areas. For many of them the city is hostile and frightening, presenting a granite face to the people who breath its fume-laden air. Human beings are the life of village and city alike; it is their societies which make both possible to live in. The architect cannot disengage himself from part of the responsibility for creating the urban wilderness.

It is painfully evident that the architects and planners who have unwittingly created a new mental oppression and have indirectly added the "new town blues" to the sociologist's vocabulary, who have ignored the widely expressed anxiety over living in high-rise flats and dismissed it as prejudice, have failed to offer developments which give the deepest sense of belonging within the community which vernacular ones commonly do. They are still more in touch with "architecture" than with society.

Figure 3.2 (opposite)
The "Libre" double-page spread: "Realization of the dwelling place as interrelating completely with the total environment, with the greater part of our lives being lived outdoors." "Libre," *Architectural Design* 12 (December 1971): 728–729. Courtesy of Dean Fleming.

22 "Libre," *Architectural Design*

The countercultural movement was closely related to the back-to-the-land movement and communes—there were several thousand in the 1960s and 1970s—and followed a wide variety of philosophies, social forms, and housing, making good use of traditional houses. A few communes, though, attained distinctive architectural forms, as was the case with Libre, "a growing-living thing in space & consciousness" ("Libre," 727) founded in 1966 near Gardner, Colorado. Libre was featured in *Architectural Design* in 1971, along with Drop City and the Lama Foundation. The anonymous author, in all probability Dean Fleming, gives a day-by-day account of life in the commune and the building process: "We began living in the house as soon as it vaguely looked like a shelter. With 2 out of 8 walls it at least served as a windbreak. Although living in a construction site can be both chaotic and hazardous, we found this was the only way the house could grow organically to fulfill our living needs. There are no blueprints and very little conception of what the end product will be. The house builds itself" ("Libre," 727). Fleming designed the main domes after studying those at Drop City.

The mixture of intimate details of everyday life and the work and construction that made it possible demonstrates the unity of life and work that was a feature of *WEC* values. The countercultural discourse put great stress an authenticity found through the secret nature of the site, populated by Native Americans, and living in communion with the silence of nature.—CM

Source

"Libre," *Architectural Design* 12 (December 1971): 728–729. Courtesy of Dean Fleming. (*Architectural Design* is referred to in *LWEC*, 88.)

Further Reading

Fairfield, Richard. *Communes USA: A Personal Tour.* Baltimore: Penguin, 1972.

Gordon, Alastair. *Spaced Out: Radical Environments of the Psychedelic Sixties.* New York: Rizzoli, 2008.

Miller, Timothy. *The 60s Communes: Hippies and Beyond.* Syracuse, NY: Syracuse University Press, 1999.

It's hard to say what architecture is for us — it seems we needed a way to keep warm — keep the snow & rain off, and the houses? THEN came about as individual responses to the four elements.

Having expended most of our "Architectural "energy in the physical building of the structures, it now becomes hard to explain the structures in terms of words

The pictures are our best means of relaying information.

GENERAL BUILDING FACTS AND CONSIDERATIONS:

Temperature ranges from 100° + to –50°

Water run-off collection (water is our scarcest commodity)

Sunlight is our greatest and free-est resource

Scavaging of materials from abandoned buildings & mines

Use of native materials in great abundance : logs, rocks, adobe

Use of "second-hand" or used materials and tools, where the concept of NEW and OLD has no meaning

Realization of the dwelling place as interrelating completely with the total environment, with the greater part of our lives being lived outdoors

The descriptions of individual houses are thoughts of the builders.

We live, eat, work, play, worship, build — no one is seperate from the other.

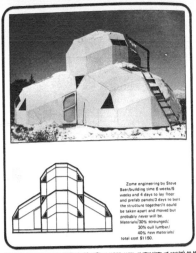

Zome engineering by Steve Baer/building time 6 weeks/5 weeks and 4 days to lay floor and prefab panels/3 days to bolt the structure together/it could be taken apart and moved but probably never will be. Materials/30% scrounged/ 30% cull lumber/ 40% new materials/ total cost $1150.

WE BUILT A 40FT DOME THINKING, BECAUSE HE'D LIVED IN 100 FT. LOFTS IN NEW YORK, WE COULDN'T GO ANY SMALLER. 40 FT. WAS BIG → A CERTAIN MAD EQUISTRATION DEVELOPS IN THE SEEMINGLY ENDLESS REPETITION OF SOAKING UP TRIANGLES. IN 3 MOS. WE MOVED IN BUT IT TOOK A YEAR TO FINISH SHEETROCKING ON THE INSIDE (WE FOUND WE'D AVERAGE 3 PANELS BEFORE INSANITY SET IN AND WE'D HAVE TO STOP AND ENJOY LIFE INSTEAD).

THE DOME SEEMED ESPECIALLY FIT FOR CHANGES, BEING AN EMPTY UNIMPOSING SHELL. IN FACT, IN 3½ YEARS, THE INTERIOR DESIGN HAS NOT SETTLED IN, EACH MONTH IT CHANGES AGAIN, MAY NEVER STOP SHIFTING AS NEW NEEDS ARISE, NEW IMAGES ARE LAUNCHED.

40 foot, 4 phase geodesic dome using 2 x 4 structure, plastic Nubs, plywood exterior, rock wool insulation and short rock interior, 2 x 4 floor, plastic skylights.

22 foot, 3 phase 3/8 semi-geodesic dome using essentially the same materials.

AND SHE LEAKS LIKE A SIEVE, ALWAYS HAS. BUT THE LIGHT AND LOFTINESS OF A DOME IS MAGNIFICENT. THE SPACE IS MORE SUITED FOR COMMUNAL ACTIVITY AND THAT'S THE WAY ITS GONE. WITH A POT SHOP, KILNS, WELDING, WOODWORKING, JEWELRY, SEWING & A KITCHEN FOR ALL.

THE LITTLE DOME , A 22 FT. BEDROOM, WAS CHEAP, QUICK & EASY —AND IS SEALED! WHILE THE 40 FOOTER COST $700 IN ITS PRIMAL STATE, THE BEDROOM WAS ABOUT $100 AND TOOK SCARCELY A WEEK TO BUILD.

IT'S A CONTINUOUS STRUGGLE TO MAINTAIN DOME SPACE BUT THE DANCES AND MEETINGS ALWAYS HAPPEN AT OUR HOUSE!

LIBRE IS A SOCIAL PHENOMENA, THE EXTENDED FAMILY, A CHANCE TO LOVE AND WORK IT OUT TO GETHER.
LIBRE IS A CREATIVE CENTER WITH TIME AND SPACE FOR IMAGINATIONS FREE FLOW THROUGH ALL FORMS.
LIBRE IS A RELATIONSHIP TO THE LAND THRU PLOWING, COMPOSTING, PLANTING AND EATING HER SPROUTING.
BUT LIBRE IS ALSO THE INCREDIBLE SILENCE OF THIS ANCIENT MOUNTAIN WHOSE RECEIVING OF THE INFINITE FORCES SO FAR TRANSCENDS EVEN OUR OWN CAPACITY TO UNDERSTAND THAT WE FEEL THE GENTLE HAND OF GOD ON OUR HEADS TEACHING US SOME /HOLY HUMILITY.
LIFE AND DEATH PASS BY THIS MOUNTAIN AS EASILY AS THE HUMMING OF THE WIND IN PINE AND FIR AND WE SMALL FOLK MUST ACCEPT IT ALL AND GIVE OUR DAILY THANKS.

AQUÍ COMPRAMOS DIOS

We live on 360 acres on the south facing slope of a mountain above 9000' / rainfall 16-20 inches a year / we seem preoccupied with water & water problems/ sun shines over 3000 hours a year / perfect for solar energy conversion / we're messing with it but haven't gotten it together yet.

23   Steve Baer, *Dome Cookbook*, 1968

In 1969, in a letter to the architectural historian Robin Middleton, technical editor of *Architectural Design*, Brand praised Baer's publication: "Have you seen Baer's *Dome Cookbook*—enclosed—; over here it's become an underground classic, yet no review beside the Catalog's has appeared; especially no one has attempted a critique of Baer's geometry and what it promises for fluidity in dome design."[1] In the same letter, Brand informed Middleton that the *WEC*'s July supplement would be prepared at Baer's place in New Mexico, which was the "best possible environment for us: there's new dome forms and connector forms and closed system solar trackers and much else here." Baer was in fact extremely active with the builders of New Mexico communes such as Drop City and the Lama Foundation. He aimed to develop systems and products that contributed to the self-sufficiency of a building and its occupants.

Dome Cookbook (1968) is a practical handbook, partly handwritten, for the construction of geodesic domes. In a punchy and lively style, Baer explains the geometry and practical problems involved, including some reflections of a more personal nature. As Brand put it: "A new art is evolving in the south west desert. Multi-colored cartop domes, put together with whatever's lying around. Free heat from the sun. Behind much of the innovating stands Steve Baer, a young inventor who generates enough energy to get others moving too."—CM

1. Stewart Brand, letter to Robin Middleton, July 17, 1969, SBP M1237, box 6.

Source

Steve Baer, *Dome Cookbook* (Corrales, NM: Cookbook Fund/Lama Foundation, 1968), 20–21. Courtesy of Steve Baer. (Referred to in *LWEC*, 91.)

Further Reading

Maniaque, Caroline. "In the Footsteps of the Counter-Culture." In *Sorry, Out of Gas: Architecture's Response to the 1973 Oil Crisis*, ed. Giovanna Borasi and Mirko Zardini. Montreal: Canadian Centre for Architecture; Montova, Italy: Corraini Edizioni, 2007.

"Steve and Holly Baer: Dome Home Enthusiasts; A Plowboy Interview with Steve and Holly Baer Who Showed the World That Very Inexpensive Dome Housing Could Be Fabricated from the Tops of Junked Automobiles." *Mother Earth News*, July–August 1973.

the angle iron was 1¾" x 1¾" x ⅛" and the ⅜" holes for the hinge bolts were drilled .083' (one inch) from the edge of the angle iron. The obvious place to run the sacred edges lines was through the centers of the hinge holes. With this relation between the sacred lines and the panels plus the decision to always place the leading face of the hinge .600' back from the intersection of the lines the following relations result between the hinges and the corners of the panels.

.600'
.517'
.511'
90°

.600'
.455'
60°

.600'
.483'
.483'
70° 32'

.600'
.541'
109° 28'

21

CAR TOPS

Car tops are a good building material. They are cheap, strong, have an excellent paint job and are available almost everywhere. The thickness of the tin varies from car to car some areonly about 20 guage (butter tops) others 18 and 19 guage. The tops can be cut into huge shingles and nailed onto a wood frame or they can be made int structural panels themselves which can be bolted, screwed, riveted or welded together to form a dome made of only car tops.

Most tops taken off a car with an axe will be between 45 and 52 inches wide and 50 to 70 inches long, a station wagon will give as much as 8 feet and a van or mini-bus even more. An experienced man with a good axe can easily chop 5 or 6 tops an hour and when the cars are packed close together so you don't have to touch ground you can go faster than this. Dave Park of General Salvage in Albuquerque has a stockpile of near 1000 car tops he has cut with various implements including a kind of giant can opener made from a spring leaf and driven with blows of a 3 lb machinist's hammer. Power saws with carborundum blades, electric unlashers, electric nibblers, acetylene torches, pneumatic chisels all of these will take tops off. The advantage of using an axe is that it's cheap and after some practice it can become a real pleasure to chop the top out of a car.

Chop along the sides first then the front and back. Throw open the doors- one foot on an open door and one foot on the car is a good stance for chopping the sides.

On cars with missing back glasses be careful of standing on the shelf in back of the rear seat—some of these have gaps spanned only by cardboard and its easy to step through— old fords are made this way.

Don't swing the axe hard once you have a slot going. Out swinging the head almost parallel to the top, if you hit flat, as you would a log you'll only smash the metal in. You have to first go after tops with a good deal of ambition, it takes a while before even the best get the hang of it.

Recently Mandel Bell and I rented an air compressor, marked off the shapes we needed right on the cars and then cut them out easily with a pneumatic chisel. Cutting the trimmed shapes right out of the junk yard eliminated at least two thirds of the work in the previous routine. It also meant we could haul twice as many tops in a load.

25 cents a top has been the going rate in New Mexico and colorado. The tin can only be sold as junk and in many places junking car bodies is so unprofitable that the yards give them to who ever will haul them away. Not all junkyards will let you at their cars some junk dealers just don't like the idea of it. The best people to work with are the men who haul the bodies to a railhead or a smelter, they don't care what you take.

every man needs a gun but that's not enough- what can you do with a rifle gun today. We need bigger artillas bullholds to every law abiding citizens There should be some very long phone numbers that you could dial after work gives you a bad time your drive into the next area where the book are kept secret from youngsters, criminals and people with poor judgement;

what is ? ll to what the blade? the edge? the handle?

you swing the axe - or what the head of the axe is moving more or less in the plane of the car top

Figure 3.3
Dome Cookbook double-page spread.
Steve Baer, *Dome Cookbook* (Corrales, NM:
Cookbook Fund/Lama Foundation, 1968),
20–21. Courtesy of Steve Baer.

24 Steve Baer, *Zome Primer*, 1970

"The geodesic dome, if it is large and composed of many different edges and joints, has many different edge lengths. It is complicated in structure and simple in shape. Zomes are simple in structure and complicated in shape," explained Steve Baer in his *Zome Primer* (1970), which built on the concepts he had laid out in *Dome Cookbook* two years earlier. Baer's "zome"—a portmanteau word combining "dome" and "zonohedron"—employed fivefold symmetry toward creating trusslike building components that offered greater design flexibility than did the geodesic construction techniques discussed in his previous work.

In *Zome Primer*, Baer, who had studied mathematics and physics at Amherst College and the Swiss Federal Institute of Technology in Zürich, presented technical drawings for zome frame fabrication, as well as photographs of completed residential zomes and playground climbers. His corporation, Zomeworks, patented the thirty-one-zone structural system in figure included here and developed from it the "Zometoy," an instructional modeling kit, alluded to by Brand in his review, which is still available today. Zomes represented the ingenuity of skilled design outlaws such as Baer, whose company went on to make major strides in passive solar technology. Including the *Zome Primer* in the *LWEC* also showed desire, on Brand's part, to disseminate such new inventions to a wider audience. —MG

Source
Steve Baer, *Zome Primer: Elements of Zonohedra Geometry* (Albuquerque, NM: Zomeworks Corporation, 1970), 16, 34. Courtesy of Steve Baer. (Referred to in *LWEC*, 91.)

Further Reading
Baer, Steve. *Sunspots: Collected Facts and Solar Fiction.* Albuquerque, NM: Zomeworks Corporation, 1975.

Prenis, John. *The Dome Builder's Handbook.* Philadelphia: Running Press, 1973.

Figure 3.4
"Symmetries of the Regular Thirty-one Zone Star." Steve Baer, *Zome Primer: Elements of Zonohedra Geometry* (Albuquerque, NM: Zomeworks Corporation, 1970), 16. Courtesy of Steve Baer.

Symmetries of a Regular Thirty-one Zone Star

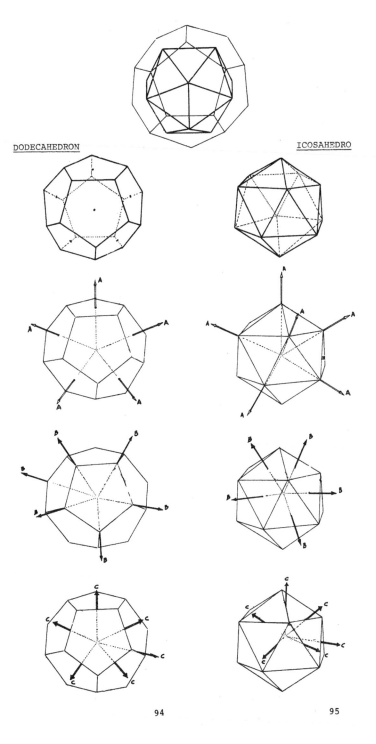

DODECAHEDRON

ICOSAHEDRO

The dodecahedron and the icosahedron are duals of each other - the vertices of one match the face midpoints of the other and vice versa.

The ten B lines of the thirty-one zone star go through the vertices of the dodecahedron or, equivalently, the face midpoints of the icosahedron while the six A lines go through the vertices of the icosahedron or, equivalently, the face midpoints of the dodecahedron.

The fifteen C lines go through the edge midpoints of either the icosahedron or the dodecahedron.

The middles of edges are commonly midway between vertices or face midpoints and C lines bisect all angles between A lines and three of the four kinds of angles formed between B lines.

In examining angles between lines we are also examining equators. There are six different equators and slicing through them we form the R, S, T, V, X and Y sections.

All pairs of lines lie in one of these six kinds of sections.

See Sections for a discussion of different angles and the polygons they form.

94

95

111

Framework for Robert Ford
residence.

A lines = $A\top^2$; base A = 40"
C lines = $C\top^2$ (see page 32)

162

Zome Climber

A lines = 40"
C lines = 42"

34

Figure 3.5 (above)
"Framework for Robert Ford Residence"
and "Zome Climber." Steve Baer, *Zome
Primer: Elements of Zonohedra Geometry*
(Albuquerque, NM: Zomeworks Corporation,
1970), 34. Courtesy of Steve Baer.

Figure 3.6 (opposite)
"Even a 'point' turns out to be 'nothing' until
it becomes a sphere. Thus the 'point' is the
bridge from metaphysics to physics." Keith B.
Critchlow, *Order in Space: A Design Source
Book* (New York: Viking Press, 1970), 8–9.
Courtesy of Keith B. Critchlow.

112

25 Keith Critchlow, *Order in Space*, 1969

It was the carpenter, designer-builder, and author Lloyd Kahn, coordinating in 1968 the building of seventeen domes at Pacific High School, an alternative school in the Santa Cruz mountains, who selected Keith Critchlow's *Order in Space* for inclusion in the *LWEC*: "A new book by an experimental mathematician on order in space...space defining, distribution patterns, space filling properties, packing & stacking, economy grids and communication linkages. There are exciting insights into structure in nature, and exploratory diagrams of the functions possible in space." As noted in the book's preface, "The book is designed for the visually oriented. It is a manual of space functions. It aims to show experimental ways in which related division is possible in space—by a process of economy or least effort between elements" (*Order in Space*, 3). The book contains a beautiful series of plates showing the subdividable possibilities of different space-filling solids, such as the rhombic dodecahedron or tetrahedron. At the time of its publication, Critchlow, an architect educated at the Royal College of Art, was lecturing at the Architectural Association School of Architecture in London. He studied geometry not only for its practical attributes but also for its metaphysical and mystical aspects as a key to understanding the universe. In 1976 he became, and was for many years, a professor of Islamic Art at the Royal College of Art in London. —CM

Source
Keith B. Critchlow, *Order in Space: A Design Source Book* (New York: Viking Press, 1970), 8. Courtesy of Keith B. Critchlow. (Referred to in *LWEC*, 91.)

Further Reading
Borrego, John. *Space Grid Structures: Skeletal Frameworks and Stressed-Skin Systems*. MIT Report. Cambridge, MA: MIT Press, 1968.

Critchlow, Keith. *Into the Hidden Environment: The Oceans*. New York: Viking Press, 1973.

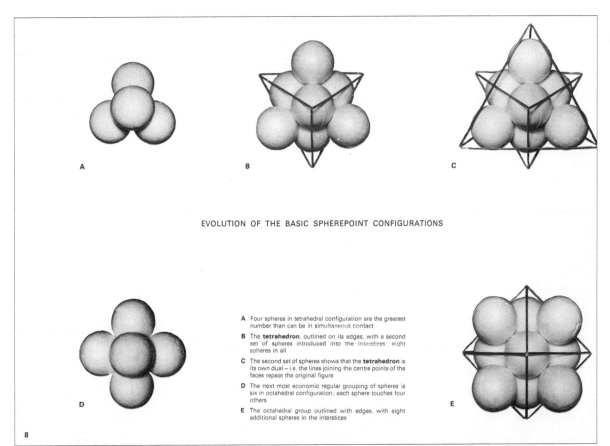

EVOLUTION OF THE BASIC SPHEREPOINT CONFIGURATIONS

A Four spheres in tetrahedral configuration are the greatest number than can be in simultaneous contact

B The **tetrahedron**, outlined on its edges, with a second set of spheres introduced into the interstices: eight spheres in all

C The second set of spheres shows that the **tetrahedron** is its own dual—i.e. the lines joining the centre points of the faces repeat the original figure

D The next most economic regular grouping of spheres is six in octahedral configuration; each sphere touches four others

E The octahedral group outlined with edges, with eight additional spheres in the interstices

8

26 Lloyd Kahn, *Domebook 2*, 1971

Lloyd Kahn, one of the editors of the "Shelter" section, wrote, in his review of his own book *Domebook 2*: "This is an instruction manual for builders and a story book of some new communities in America." He opened the text with a passage from the popular book *Black Elk Speaks*, in which the Lakota medicine man Black Elk asserted that "the power of the world," from the movement of the sun and moon to the shape of birds' nests, "always works in circles," and so the rectangular architecture imposed on Native American tribes by white men had devitalized his community (*Domebook 2*, 1). With this quote, Kahn intimated the righteousness and organicism of dome living, the philosophical underpinning of his book.

Domebook 2, an updated, extended version of *Domebook 1*, published the year before, provided the practical information, seen in the following image, that inexperienced dome builders would need to efficiently erect their own structures. Images of innovative built examples that employed a variety of materials, including aluminum, stone, and plastic, followed, often accompanied by the particular directions required for their construction. The enthusiastic guidance proffered by *Domebook 2* ensured its great success within countercultural circles in the years to come, despite Kahn's disillusionment with geodesics and espousal of vernacular architecture in his book *Shelter* (1973), only two years later. — MG

Source
Lloyd Kahn, *Domebook 2* (Bolinas, CA: Pacific Domes, 1971), 9. Courtesy of Lloyd Kahn. (Referred to in *LWEC*, 93.)

Further Reading
Goldstein, Barbara. "Lloyd Kahn Talks to Barbara Goldstein." *Architectural Design* 11 (November 1975): 653–654.

Kahn, Lloyd, and Edward Dorn. *Smart but Not Wise: Further Thoughts on Domebook 2, Plastics, and Whitman Technology*. Bolinas, CA: Shelter Publications, 1972.

Figure 3.7
"Geometric Geometry." Lloyd Kahn, *Domebook 2* (Bolinas, CA: Pacific Domes, 1971), 9. Courtesy of Lloyd Kahn.

GEODESIC GEOMETRY

Jonathan Kanter

A dome is a multifaceted polyhedron in which all the vertices lie on the surface of a sphere. Domes are developed from the tetrahedron, octahedron and icosahedron. Domes have the symmetry of the tetrahedron, octahedron and icosahedron, and also of their duals in the triacon breakdown.

The sphere encloses the most volume with the least surface, and is the strongest shape against internal and radial pressure.

The Platonic solids are the best shapes to start from because they have the most symmetry and regularity. A structure composed of triangles is best because a triangle is the simplest subdivision of a surface, and the only stable polygon with flexible connectors (see Elliptical Domes). Domes could also be developed from semi regular or other shapes. Domes can also be produced to form any other besides spherical shapes (see Elliptical Domes).

What follows explains domes developed from the icosahedron. The icosahedron is used because, of the Platonic solids with triangular faces (the tetrahedron, octahedron, and icosahedron) it most closely approximates the sphere. Domes can be developed in the same way from the tetrahedron and octahedron.

CHORD — A LINE BETWEEN 2 POINTS ON THE SPHERE ← SURFACE OF SPHERE

ICOSAHEDRON

ICOSAHEDRON WITHIN SPHERE ALL VERTEXES TOUCH SPHERE

ICOSAHEDRON PROJECTED ONTO SPHERE

Two breakdowns are possible when: the face is subdivided into triangles and the symmetry of the equilateral triangle face is retained. They are the alternate and triacon. This is a general geometrical explanation.

There are different sets of chord factors within the basic geometry. Clinton derives his from the coordinates of vertices on the sphere and uses analytical geometry. See Geodesic Math, p. 106 and Chord Factors, P. 108. Fuller derives his by spherical trig. calculations using arcs on a sphere, see Spherical Trig and 31 Great Circles. Fuller develops domes by projecting the icosahedron onto the sphere and then dividing the face of the spherical icosahedron with great circle arcs on the surface of the sphere, (a great circle arc cuts a sphere in half, it is the shortest distance between two points on the surface of a sphere) except for some arcs in the Alternate truncatable (Clinton's Class 1 method 4).

PLATONIC SOLIDS

The platonic solids are the most basic structures known. The cube is the most commonly used of these solids as an archetype for construction. The structures in this book represent an attempt to reexamine these basic solids and to find different channels and foundations for new forms of construction.

The platonic solids are defined as having equal faces (regular polygons), equal vertices, and equal dihedral angles between the faces. To generate the solids from equilateral polygons there are only 3 polygons which will fit together in three dimensional (3-d) space: the equilateral triangle, the square, and the pentagon. Hexagons when put together in threes lie in a plane and polygons larger than six sides won't fit together in threes(and will therefore not make a 3-d solid without a different polygon thrown in at each vertex). Combining the equilateral triangles results in the 3 stable solids, the tetrahedron, the octahedron, and the icosahedron, by having 3, 4, and 5 triangles around each vertex (6 becomes a plane and more will not fit).

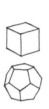

Squares and pentagons will fit together in threes at each vertex; this produces the cube and the dodecahedron.

Two solids are considered to be duals if the number of vertices of one equals the number of faces of the other and they have an equal number of edges. The dual of a solid can be generated by inscribing a sphere so that it is tangent to the solid at the midpoints of its edges and then by drawing lines perpendicular to the edges and tangent to the sphere at those points.

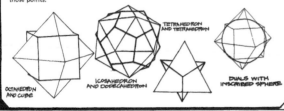

TETRAHEDRON AND TETRAHEDRON

ICOSAHEDRON AND DODECAHEDRON

DUALS WITH INSCRIBED SPHERE

OCTAHEDRON AND CUBE

9

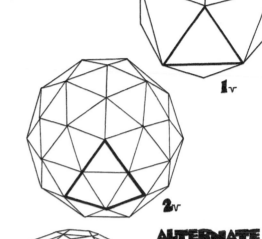

1v

2v

ALTERNATE
BREAKDOWN
• VERTEX UP •
ONE ICOSA FACE IS OUTLINED
CHECK SIMPLE GEODESICS

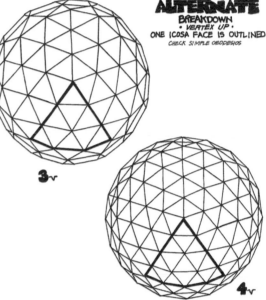

3v

4v

For any breakdown, as the frequency increases the number of chords and the number of facets increases so the dome becomes more spherical. Frequency is the same as the number of divisions of the edge.

| 1 FREQUENCY 1v | 2v | 3v | 4v |

The face is divided by lines (arcs on a sphere) parallel to each edge. The lines go from edge to edge.

All repeating patterns in alternate can be shown in one face (△). Henceforth, ∨ = frequency

27 Ken Kern, *The Owner-Built Home*, 1961

As Lloyd Kahn described *The Owner-Built Home* **in the** *LWEC,* **"Ken Kern makes a unique offer to anyone thinking about building his own home: for $10 he furnishes a preliminary house design as well as a copy of the** *Owner-Built Home* **which is about the most useful book on building available."**

Not licensed to practice as a registered architect because he did not complete his studies, Kern became a builder. He also set up a publishing house which he later called OwnerBuilder Publications, intended to promote building research. Highly critical of current architectural practices, he engaged in a dialogue with his readers, encouraging them to think for themselves about how to plan and build a home. Unlike an architect, Kern did not use measured drawings but illustrated his texts with informal sketches as though he were talking directly with the reader. Self-builders should make their own decisions, use local materials, and purchase only essential tools of good quality.

The book takes the reader through the whole process of building a house, step by step, from the first site sketches to constructions details. For example, the chapter dealing with climate control uses diagrams to show how to create cross-ventilation. The book aims to encourage the home builder to take complete responsibility for the construction.

Kern's introduction is both a severe critique of the building trades and an agenda in the form of seven axioms for empowering ordinary people to dare to build their own home.—CM

Source
Ken Kern, *The Owner-Built Home* (Oakhurst, CA: Ken Kern Drafting, 1961), 1–7. (Referred to in *LWEC*, 94.)

Further Reading
Kern, Ken, and Barbara Kern. *The Owner-Built Homestead*. New York: Scribner, 1977.

Introduction to *The Owner-Built Home*

The Owner-Built Home is intended to be a *how-to-think-it* book. Alternatives to the professionally executed, contractor-built home are presented in text and through sketches. None of the author's sketches are meant to be more than schematic representations of a thinking through process for one's house-building project.

The author has long had the compulsion to express feelings and thoughts in regard to the home-building industry and the wish to do something constructive for the people who suffer under it—both the construction worker and the home buyer alike. No critic as yet comprehends entirely why our houses are so poorly constructed, why they look so abominable, why they cost so much for construction and for maintenance, and why they are so uncomfortable. ...

Tracing these causes to their sources has helped me to view the problem in perspective—comprehensively. This process has also suggested some workable alternatives as solutions to personal housing needs. Here they are in the form of seven axioms, listed in order of importance for the prospective owner-builder.

When building your home, pay as you go. A building loan is a type of legalized robbery. More than any other agency, banks have been successful in reducing would-be democratic man to a state of perpetual serfdom. The banks have supported and helped to determine social and political conventions and have amassed phenomenal fortunes through unearned increment. If one expects any success at all with keeping costs of his new home down to a reasonable price one must be entirely free from interest rates.

Supply your own labor. Building trades unions have received—and not unjustly so—a notorious reputation as wasters of speed and efficiency in building work. ... The disinterest that the average journeyman has in his work, despite his high union-pay rate, is appalling. The lack of joy-in-work or of acceptance of responsibility among average workmen can be accounted for partly by the dehumanizing effect of the whole wage system. So long as the master-and-slave type of employer-employee relationship continues to exist in our society one can expect only the worst

performance from his "hired help." Until the dawn of the new era approaches one would do well, from an economic as well as from a self-satisfying standpoint, to supply his own labor for his own home insofar as he can.

Build according to your own best judgment. At the apex of the poor-building hierarchy—and perhaps the greatest single impediment to good housing—is convention. Building convention takes two forms: first, there is convention that is socially instilled, commonly called "style," which can be altered through education. The second type of convention is more vicious and politically enforced. Building codes, zoning restrictions, and ordinances all fall into this class of impediments. In urban jurisdictions politically controlled convention calls the shots for practically every segment of the building industry. ... If we are to be at liberty to build our own home at less cost, we must necessarily be free from building-code jurisdiction. This means we must locate outside urban control—in the country or in small-township districts.

Use native materials whenever possible. ... Basic materials, like common cement, have not appreciably advanced in price over the past dozen years, but some of the newer surfacing materials and interior fixtures have skyrocketed in price during this same period. By not using these high-cost materials one, of course, avoids this problem. Emphasis should, instead, be placed on the use of readily available natural resources—materials that come directly from the site or from a convenient hauling distance. Rock, earth, concrete, timber, and all such materials have excellent structural and heat-regulating qualities when properly used.

Design and plan your own home. One ten-percenter with whom we can well afford to do without when building a low-cost home is the architect-designer-craftsman-supervisor. Experience in this aspect of home building has led me to conclude that *anyone* can and everyone *should* design his own home. There is only one possible drawback to this: the owner-builder must know what he wants in a home, and he must be familiar with the building site and the regional climatic conditions. Without close acquaintance with the site and the regional climate and without a clear understanding of the family's living needs the project is doomed to failure no matter who designs the house. An architect—even a good architect—cannot interpret a client's building needs better than the client himself. Anyway, most contemporary architects design houses for themselves, not for their clients....

Use minimum but quality grade hand tools. If the house design is kept simple and the work program is well organized an expensive outlay in specialized construction equipment can be saved. The building industry has been mechanized to absurd dimensions. And even with more and better power tools labor costs rise.... If a prospective home owner is unprepared to accept the challenge of building his own home—and falls into the over-stocked power-tool trap—then he must be prepared to spend greater sums for a product which could very well prove to be inferior.

Assume responsibility for your building construction. The general contractor has become such a key functionary in practically every building operation that one soon loses sight of the fact that he is a relative newcomer to the housing scene. Not many years ago the contractor's job was performed by a supervising carpenter, a so-called "master builder," who had control of the whole project. Once people realize how little is involved in implementing a set of building plans they will better appreciate the fact that the contractor is the most expendable element on any job....

A positive philosophical outlook and way of life must necessarily precede the achievement of a quality owner-built home. This is to say that a truly satisfying home must *develop* from other and more subtle patterns. The mere technical problems of building a home are insignificant when compared to an understanding and an interpretation of one's innermost feelings and thoughts concerning his shelter needs.

28 Reginald and Gladys Laubin, *The Indian Tipi*, 1957

Although Reginald and Gladys Laubin were not of Native American descent, the couple devoted their lives to educating others about the history and culture of the Plains Indians through dance performances, lectures, and books, including *The Indian Tipi* (1957) and *American Indian Archery* (1980). While living with their adoptive Sioux family, the Laubins learned to pitch a tipi and soon became enamored with the comfort and beauty afforded by the structure. Reginald recalled a moment when Gladys exclaimed, "If there is any such place as Heaven, I hope it is a big tipi" (*The Indian Tipi*, 252–253), a statement that demonstrated the Laubins' particular enthusiasm for the subject of their first book. *The Indian Tipi* examined the historical, technical, and cultural dimensions of the vernacular building type, presenting the material in both visual and narrative form, and its expansive nature allowed the work to function as not only an educational text but also an instruction manual for readers seeking to build and reside in these traditional dwellings. —MG

Source
Reginald and Gladys Laubin, *The Indian Tipi: Its History, Construction, and Use* (Norman: University of Oklahoma Press, 1957), 19–24. (Referred to in *LWEC*, 100.)

Further Reading
Hungrywolf, Adolf. *Tipi Life.* Invermere, BC: Good Medicine Books, 1972.

Nabokov, Peter, and Robert Easton. *Native American Architecture.* New York: Oxford University Press, 1989.

Pearce, Roy Harvey. *Savagism and Civilization: A Study of the Indian and the American Mind.* Berkeley: University of California Press, 1953.

"Utility and Beauty," from *The Indian Tipi*

No dwelling in all the world stirs the imagination like the tipi of the Plains Indian. It is without doubt one of the most picturesque of all shelters and one of the most practical movable dwellings ever invented. Comfortable, roomy, and well ventilated, it was ideal for the roving life these people led in following the buffalo herds up and down the country. It also proved to be just as ideal in a more permanent camp during the long winters on the prairies.

One need not be an artist to appreciate its beauty of form and line, and no camper who has ever used a tipi would credit any other tent with such comfort and utility. Warm in winter, cool in summer, easy to pitch, and, because of its conical shape, able to withstand terrific winds or driving rain, the tipi is a shelter that should appeal to every outdoorsman. It is not only an all-weather tent but a home as well.

Often the tipi is confused with the wigwam. Actually both words mean the same thing—a dwelling—but "tipi" is a Sioux word, referring to the conical skin tent common to the prairie tribes, and "wigwam" is a word from a tribe far to the East, in Massachusetts, and refers to the dome-shaped round or oval shelter, thatched with bark or reed mats, used by the people of the Woodland area. Consequently, we prefer to use the term "tipi" to designate the Plains type of shelter and "wigwam" to designate the bark- or mat-covered Woodland dwelling even though both types are now covered with canvas. Indians, wherever they lived, made their dwellings of the best material at hand and to suit their conditions of life.

The word "tipi" has often been spelled "tepee" or "teepee," but since learning to write in their own language, the Sioux themselves spell it "tipi," which is simple and is accepted by students everywhere.

To one familiar with the charm and comfort of a real Indian tipi, it is little wonder that Kit Carson refused to accompany Lieutenant John Charles Fremont on his first expedition of exploration unless a tipi was taken along. The famous scout did not mind roughing it when necessary, but when the time came for rest and relaxation, he wanted the most comfort a traveling existence could afford. Now Kit had lived for years in tipis pitched by his Indian wives, but there were no women in Fremont's party, and Kit did not know how to erect his tipi. Fortunately a trader happened along with his Indian wife. She taught Kit how to do it. He found it quite a simple trick after all.

For all that, very few white men ever acquired the skill; and today hardly any Indians remember or understand the technique.

Differences in the outward appearance of various tribal styles of tipis are due primarily to the arrangement of the poles, since this determines the cut of the hide or canvas cover. Generally speaking, all the tribes about whom we have reliable information used tipis employing one of two possible types of pole arrangement. One of these types has a foundation of three poles, a tripod; the other has a foundation of four poles. This structural difference determines certain features that distinguish the two types....

Actually, the three-pole foundation provides the better solution, for it is possible to group more poles in the front crotch, and obviously it is more efficient to place the remainder in two crotches than in three. Consequently, the cover fits better, making it easier to close the smoke hole during a storm.

The three-pole tipi is also stronger than the four-pole, for two reasons. It usually has more poles and they are bound together at the top, which is not done in the four-pole type. The four-pole tipi has from one to four outside guys, whereas the three-pole tipi seldom uses even one. Also, a tripod is a more rigid foundation than (to coin a word) a quadripod. For service, therefore, the three-pole type is unequaled.

Tribes using the three-pole tipi included the Sioux, Cheyennes, Arapahoes, Assiniboines, Kiowa, Gros Ventres of the Prairie, Plains-Cree, Mandans, Arikaras, Pawnees, and Omahas.

The Teton Sioux, or Lakota, were located centrally on the Plains and many traits of Plains Indian culture seem to have been diffused from them. They were also the largest group on the Plains. Therefore, we have chosen the Sioux tipi as representative of the three-pole tents.

Chief One Bull and his wife, Scarlet Whirlwind, who adopted us into their family because of our interest in the old ways and our efforts to bring about a better understanding of the Indian people, showed us how to pitch a Sioux tipi. One Bull was a nephew of the famous Sitting Bull and lived with his uncle until the latter's death in 1890. These old people used to win all the tipi-pitching races at the tribal fairs. Such races were popular until about 1915, but since that time there have been few tipis on the Sioux reservations....

As the tipis wore out, they were replaced with wall tents. At one time the government issued new canvas for tipis, but a policy of discouraging everything Indian was inaugurated and ended this practice.

The tipi represented savagery, whereas the wall tent was the white man's, consequently civilized. Poles were also hard to get. The nearest place was the Black Hills, more than three hundred miles from Standing Rock, and in those days Indians had to get a pass from the agent to leave the reservation. Such permission was not granted for any such heathen mission as gathering new tipi poles.

For many summers we have been the only persons actually living in a tipi on the Standing Rock Sioux reservation in North and South Dakota and on the Crow reservation in Montana. The Indians tell us they have not seen a lodge like ours, furnished and arranged as in buffalo days, in more than seventy years. In writing this book therefore we hope we can bring to others something of the fun, fascination, and relaxation that tipi life has brought to us.

Poles – Hupa
Tripod – Caŋ'kazuntapi
Smoke Flaps – Wipibaha
Smoke Flaps straight up –
            Wipipaje
Smoke Hole – Wipage

Smoky part (top) – Wizi
Yellow lodge (from smoke) – Wihia
Base – Wihuta
Lacing Pins – Wiceškibasis'e
Pegs – Wihinpaspe
Anchor Rope – Tahukawikaŋ
Anchor Peg – Huŋpe

Lining – Unhnugaška
Back rest – Wakazuŋtapi
Travois – Hupah'an

R.K. Laubin

Figure 3.8
"Parts of the Sioux Tipi." Reginald and
Gladys Laubin, *The Indian Tipi: Its History,
Construction, and Use* (Norman: University
of Oklahoma Press), 55. Reprinted with
permission.

29 Frei Otto, *Tensile Structures*, 1967

The German architect and engineer Frei Otto (1925–2015) made an immediate impact with his West German Pavilion, designed for Expo 67 in Montreal. Jay Baldwin commented: "Every designer we know who's seen this book has commenced to giggle and point, jump up and down, and launch into enthusiastic endorsement of Otto, design, being a designer, and look at this here. The book is comprehensive in its field, technically thorough, beautifully presented" (Jay Baldwin, *LWEC*, 106). The formal photographic quality of the images capturing the results of Otto's experiments conducted from the 1950s—at the several institutions dedicated to lightweight structures that he would establish in Berlin, then Stuttgart—is a fascinating aspect of the books published by MIT Press, selected by Baldwin for the catalog.

The establishment of the Biology and Building research group at the Technical University of Berlin in 1961 marked the beginning of Otto's cooperative work between architects, engineers, and biologists. The group members applied their knowledge of tents, grid shells, and other lightweight structures to better understand the designs of biological structures and forms. In 1962 Otto published the first volume of his major opus *Tensile Structures: Design, Structure, and Calculation of Buildings of Cables, Nets, and Membranes* (the second volume was published in 1966).

His Institute for Lightweight Structures, a tentlike structure built in 1966 in a forested area of the new campus in Stuttgart, explored a wide range of innovative lightweight structures. Powerfully influenced by close observation of biological structures—animal and vegetable, such as soap bubbles, spiderwebs, and so on—Otto used physical test models to derive structural information that allowed him to develop tensile and inflatable structures by trial and error. One method of experimentation was the use of soap film to find optimal forms.—CM

Source
Frei Otto, ed., *Tensile Structures: Design, Structure, and Calculation of Buildings of Cables, Nets, and Membranes*, vol. 1 (Cambridge, MA: MIT Press, 1967), 82. Reprinted with the permission of MIT. (Referred to in *LWEC*, 106.)

Further Reading
Drew, Philip. *Frei Otto: Form and Structure*. London: Crosby Lockwood Staples, 1976.

Otto, Frei. *Frei Otto: Complete Works: Lightweight Construction, Natural Design*. Ed. Winfried Nerdinger. Basel: Birkhäuser, 2005.

Figure 3.9
Frei Otto, "Pneumatic Structures." Frei Otto, ed., *Tensile Structures*, vol. 1 (Cambridge, MA: MIT Press, 1967), 82. Reprinted with the permission of MIT.

Axisymmetrical Forms

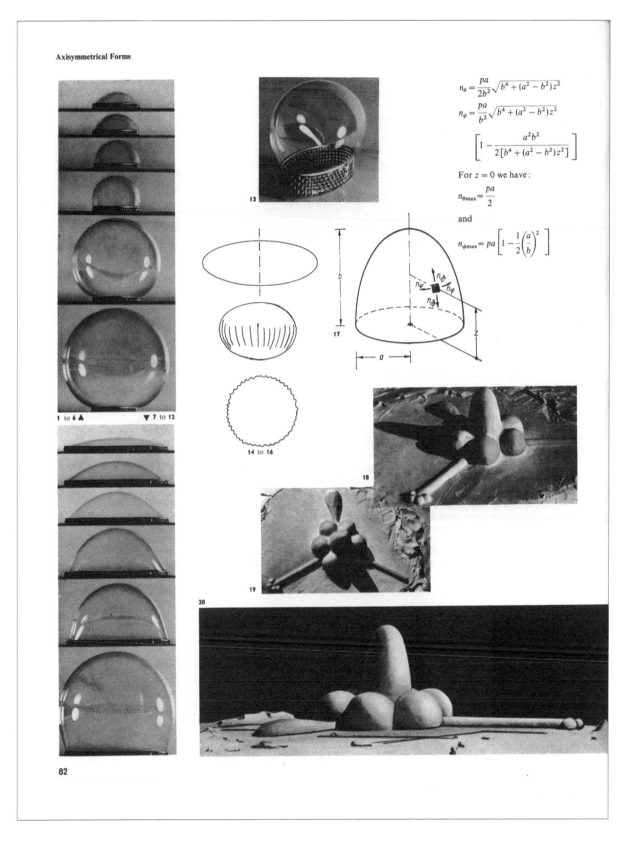

$$n_\theta = \frac{pa}{2b^2}\sqrt{b^4 + (a^2 - b^2)z^2}$$

$$n_\phi = \frac{pa}{b^2}\sqrt{b^4 + (a^2 - b^2)z^2}$$

$$\left[1 - \frac{a^2 b^2}{2\left[b^4 + (a^2 - b^2)z^2\right]}\right]$$

For $z = 0$ we have:

$$n_{\theta\max} = \frac{pa}{2}$$

and

$$n_{\phi\max} = pa\left[1 - \frac{1}{2}\left(\frac{a}{b}\right)^2\right]$$

13

17

1 to 6 ▲   ▼ 7 to 12

14 to 16

18

19

20

82

30 Ant Farm, *Inflatocookbook*, 1971

The inflatable shelter was an example of a structural technology imported from the military-industrial complex into the counterculture. Building and occupying a pneumatic structure were often the high point of alternative gatherings. As Brand noted: "An Inflatable is an event, like circus tents used to be. Dogs bark, kids gather, old ladies get up tight, cops drift by, youths take off their clothes…" However, he also noted, from considerable experience: "They're terrible to work in. The blazing redundant surfaces disorient; one wallows in space. When the sun goes behind a cloud you cease cooking and immediately start freezing (insulation schemes are in the works). Environmentally, what an inflatable is best at is protecting you from a gentle rain. Wind wants to take the structure with it across the country, so you get into heavy anchoring operations."

The radical architectural coalition Ant Farm, for instance, used inflatables in the Happenings it organized on West Coast college campuses. The authors—Chip Lord, Hudson Marquez, Doug Michels, and Curtis Schreier—were not named in the publication; only the group's name is mentioned.

Since Ant Farm valued feedback as a fundamental part of the learning process, readers were encouraged to share their own experiences, successes, and failures. *Inflatocookbook*, a how-to-do-it booklet, adopted a playful approach to construction technology, employing communication techniques that reflected that spirit. The tools needed for putting up the type of structures found in *Inflatocookbook* were easily acquired: big sheets of polyethylene plastic, bags of water or sand, fans, and so on. Both the amateur quality of the publication and the simplicity and humor of its language undermined the hierarchic quality of knowledge. The variety of graphic representations and the diversity of paper sizes used in *Inflatocookbook* was a way for Ant Farm to insist on the originality of each team member's contribution.—CM

Source
Ant Farm, *Inflatocookbook* (Sausalito, CA: Rip Off Press, 1971), plate titled "The World's Largest Snake." (Referred to in *LWEC*, 107.)

Further Reading
Burns, Jim. *Arthropods: New Design Futures*. New York: Praeger, 1972.

Maniaque, Caroline. "Searching for Energy." In *Ant Farm (1968–1978)*, ed. Constance Lewallen and Steve Seid. Berkeley: University of California Press, 2004, 14–21.

Scott, Felicity, ed. *Living Archive 7: Ant Farm; Allegorical Time Warp; The Media Fallout of July 21, 1969*. New York: Actar, 2008.

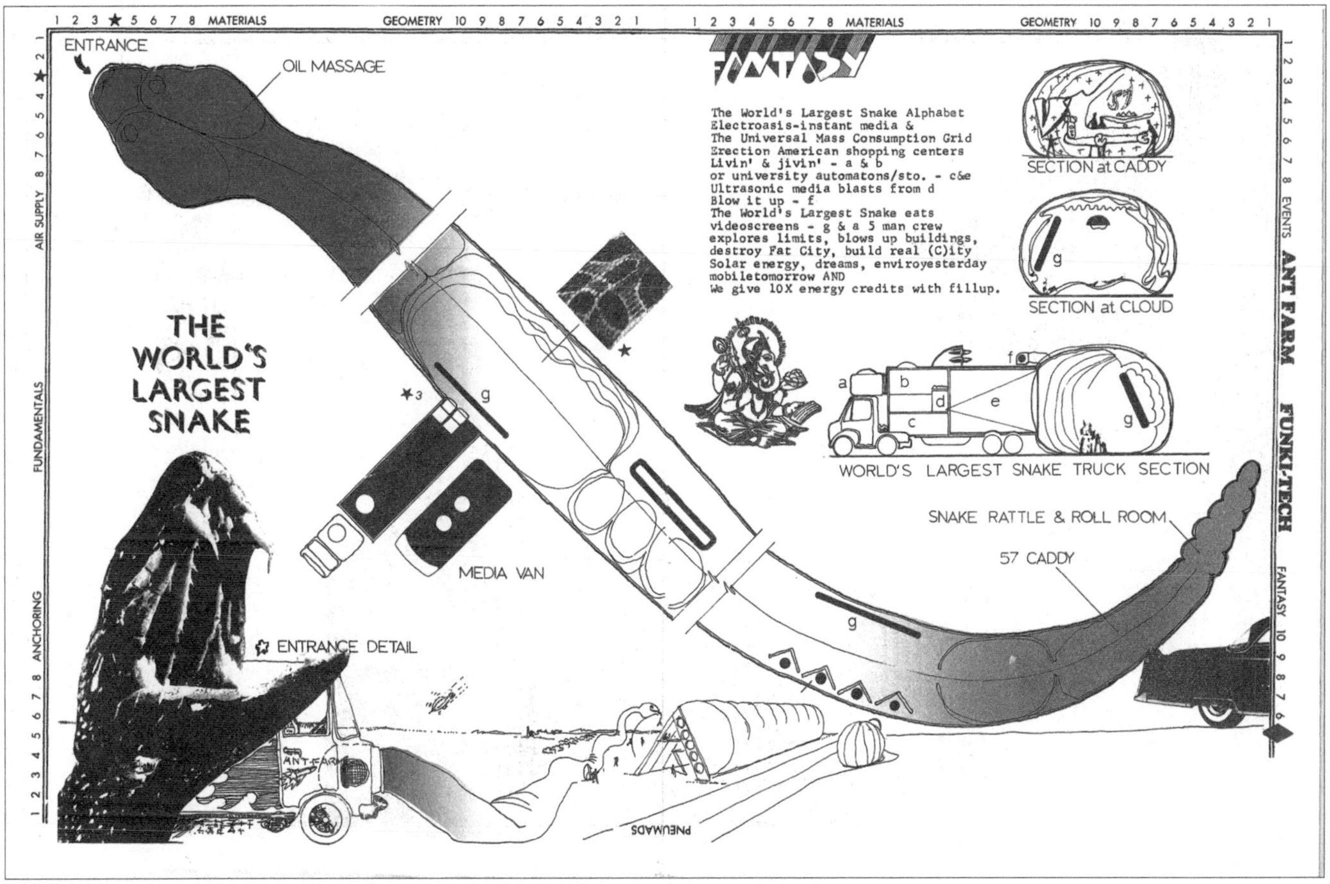

1 2 3 ★ 5 6 7 8 MATERIALS
GEOMETRY 10 9 8 7 6 5 4 3 2 1
1 2 3 4 5 6 7 8 MATERIALS
GEOMETRY 10 9 8 7 6 5 4 3 2 1
ENTRANCE
OIL MASSAGE
FANTASY
The World's Largest Snake Alphabet
Electroasis-instant media &
The Universal Mass Consumption Grid
Erection American shopping centers
Livin' & jivin' - a & b
or university automatons/sto. - c&e
Ultrasonic media blasts from d
Blow it up - f
The World's Largest Snake eats
videoscreens - g & a 5 man crew
explores limits, blows up buildings,
destroy Fat City, build real (C)ity
Solar energy, dreams, enviroyesterday
mobiletomorrow AND
We give 10X energy credits with fillup.
SECTION at CADDY
SECTION at CLOUD
THE WORLD'S LARGEST SNAKE
WORLD'S LARGEST SNAKE TRUCK SECTION
SNAKE RATTLE & ROLL ROOM
57 CADDY
MEDIA VAN
ENTRANCE DETAIL
AIR SUPPLY
FUNDAMENTALS
ANCHORING
EVENTS
ANT FARM
FUNKI-TECH
FANTASY
PNEUMADS

Figure 3.10
"The World's Largest Snake." Ant Farm, *Inflatocookbook* (Sausalito, CA: Rip Off Press, 1971). Courtesy of Ant Farm.

Industry

The contents of Section IV, "Industry," might appear quite surprising, ranging from criticisms of conspicuous consumption, to a discussion of education and design, to a selection of tools and materials available by mail order.

A number of texts referred to problems caused by too-rapid industrialization in the Western world. The section included books and items arranged in a number of subsections, such as "Design," "Chinese Technology," "Engineering," "Inventions," "Village Technology," "Knots," "Science," "Handbooks," "Plastics," "Data," "Modular Materials," "Appliances," "Lab Suppliers," "Welding and Winching," "Nifty Tools," "Tools," "Government Surplus," and "Tool Use." A whole page was devoted to the quirky and stimulating designer Don Gillert, who describes his approach to designing kites (*LWEC*, 131). Several important books were included, such as *Technology and Change*, by Donald Schön, or Victor Papanek's *Design for the Real World*, but "Industry" was one of the most pragmatic sections,

resembling more a conventional mail-order catalog than a guide to significant works of philosophy and lifestyle. Readers were also encouraged to subscribe to technical periodicals such as *Scientific American*, *New Scientist*, *Science*, and *Popular Science*, and there were also serious technical books on chemistry, engineering, and plastics. The moral, familiar to the *WEC* reader, was that expertise could be acquired in all domains. The catalog could not challenge big industry directly, but it could provide the theoretical groundwork, know-how, and tools for small-scale production. Many of the items were suggested and reviewed by readers at Brand's request.

In some ways, "Industry" was the least coherent of the sections. Many of the items could equally have appeared in "Craft" or "Understanding Whole Systems." Some tension was also evident between the implied or explicit criticism of industrial production and the catalogs and handwritten lists of industrial products. —CM

31 Stewart Brand, "A Report from Alloy," 1969

By some distance, the longest "text" in the *LWEC* was the seven-page montage of words and images recording the gathering of one hundred engineers, poets, and intellectuals for four days from March 20 to 23, 1969, on
an abandoned industrial site near Albuquerque, New Mexico. By opening the
"Industry" section with this feature, Brand made clear that new approaches
and people were needed to tackle the problems. Behind the idea of getting
a variety of heterogeneous thinkers together was the hope that some sort
of alchemical fusion would create a new kind of knowledge, based on both
experience and imagination. Their intention was to replace the thinking of
industrial consultants and the architectural profession by combining a range
of approaches—practical, scientific, spiritual, and traditional—to resolve the
major environmental problems of the day. The event was meant to create a
shock; the metaphor Brand used was that of rattling the cage. A feature of the
discussion was a deep sense of soul searching: "We want to change ourselves
to make things different," says Steve Durkee.

The event itself was characteristic of the alternative movement in architecture, since it was not only a talking venue; the participants also built domes.
The layout of the report published in the *LWEC*, found on pages 111–117, doc-
umented this double aspect, both talking and making. The text also, in small,
poetic, unattributed statements, reflected spontaneous dialogue rather than
expanded dissertation.—CM

Source
Stewart Brand, "Report from Alloy," *Difficult but Possible Supplement to the
Whole Earth Catalog*, March 1969,
reprinted in *LWEC*, 111–117.

Further Reading
Maniaque, Caroline. "We Want to Change
Ourselves to Make Things Different." In
*Investigating and Writing Architectural
History: Subjects, Methodologies and
Frontiers*, ed. Michela Rosso. Papers
from the third EAHN International Meeting,
Torino: Politecnico di Torino, 2014.

Figure 4.1
"A Report from Alloy Report." *LWEC*, 112.

ALLOY

New Mexico is the center of momentum this year, and maybe for the next several. More of the interesting intentional communities are there. More of the interesting outlaw designers are.

ALLOY was their first programmatic gathering. Conducted at an abandoned tile factory near La Luz, N.M. between the Trinity bomb-test site and the Mescalero Apache reservation, the conference lasted from Thursday to Sunday, March 20-23: the Vernal Equinox.

The event was initiated and organized by Steve Baer with Berry Hickman. Ray Graham donated the use of the land. Baer had in mind a meld of information on Materials, Structure, Energy, Man, Magic, Evolution, and Consciousness, so he invited individuals with amateur or professional interest in these areas to take responsibility for their coverage in the discourse. He also invited

performances by Libre, Drop City, Bill Pearlman's theatre group from Santa Fe, and the Whole Earth Truck Store.

150 people were there. They came from northern New Mexico, the Bay Area, New York, Washington, Carbondale, Canada, Big Sur, and elsewhere. They camped amid the tumbleweed in weather that baked, rained, greyed, snowed, and blew a fucking dust storm. Who were they? (Who are we?). Persons in their late twenties or early thirties mostly. Havers of families, many of them. Outlaws, dope fiends, and fanatics naturally. Doers, primarily, with a functional grimy grasp on the world. World-thinkers, drop outs from specialization. Hope freaks.

Among the voices here are J. Baldwin, Lloyd Kahn, Steve Baer, Dave Evans, Gandalf, Berry Hickman, Stewart Brand, Steve Durkee, Peter Rabbit, Dean Fleming, and brother anonymous.

How can you exchange with 150 people?

It can't get out of control because it already is out of control.

No one wants to be an audience. You can talk if you give them a dessert at the end. You need some way of recording it, otherwise it devolves into an endless circle of rap.

The mike is a ploy to create structure. It doesn't have to be plugged in.

Everybody's assuming the human potential is a static thing.

We're in some incredible bind that we can't even move around enough to know that we're in it. Maybe the only way you can get in the right place is by rattling. If this thing we're doing isn't a big rattle, I don't know what is.

The whole system is going the wrong goddamn way. We're not investigating what we should be investigating.

These relationships between people have not been based on some kind of appreciation of what the structure of the world is.

We want to change ourselves to make things different.

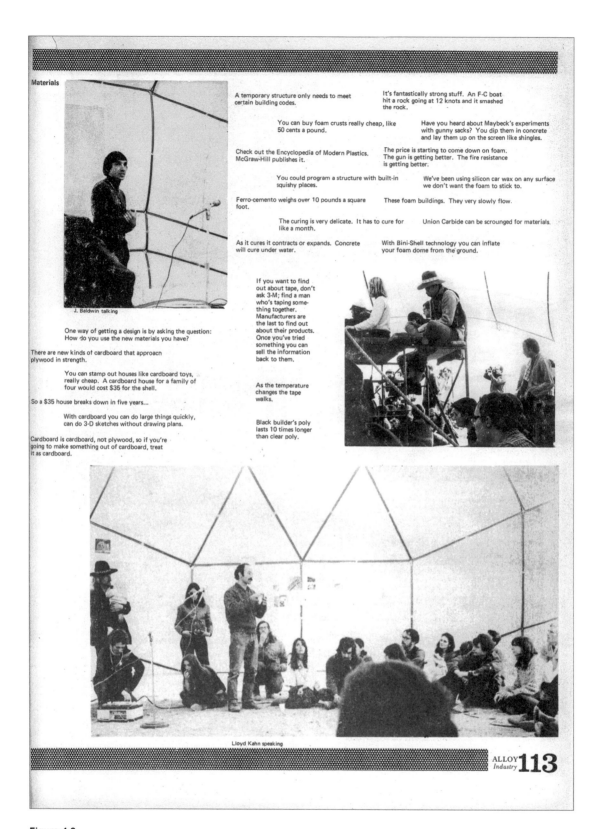

Figure 4.2
"A Report Alloy Report." *LWEC*, 113.

32 Paul Jacques Grillo, *What Is Design?* 1960

In the opening pages of *What Is Design?* Paul Jacques Grillo, a French architect and educator who received the Grand Prix de Rome in 1937, challenged young designers to "live dangerously … avoid brainwashing … [and] flout the rules as often as you can in a masterly way" (*What Is Design*, 4–5). It is no wonder, then, that the book appealed to Steve Baer, a so-called outlaw designer who sought to push the boundaries of building technologies, and captured the attention of Stewart Brand, who noted that he could find "no better introduction to all that is admirable in design."

For Grillo, a successful design project would provide the most efficient, purest solution to a given problem during a specific moment in time and space; yet, he admitted, this achievement would require constant reassessment as new technological or material possibilities developed. The tactic he described mirrored the process of biological evolution, and Grillo believed that the best architects would aspire to produce an "organic beauty" in their structures equal to that of living organisms. This proposed connection between the natural environment and architectural design was not solely conceptual, however; Grillo, the only architect invited to speak at the United Nations New Sources of Energy Conference in 1961,[1] also advocated the integration of eolian and solar energy systems into new buildings.

In one of his three speeches at the conference, Grillo envisioned a holistic new approach to architecture that could "join in spirit the superb anonymity of the highly successful solutions which force our admiration in farmhouses and peasant dwellings the world over."[2] The passage that follows expands on his perspective. —MG

1. Grillo, "The United Nations Conference on the New Sources of Energy in Rome: A Report in the Form of an Italian Concerto (Quasi una Fantasia)," *Architecture at Rice University* 2 (1961): 2.

2. Ibid.

Source
Paul Jacques Grillo, *What Is Design?* (Chicago: P. Theobald, 1960), 9–11. (Referred to in *LWEC*, 118.)

Further Reading
Grillo, Paul Jacques. "The United Nations Conference on the New Sources of Energy in Rome: A Report in the Form of an Italian Concerto (Quasi una Fantasia)." *Architecture at Rice University* 2 (1961): 1–24.

"Introduction to Design," from *What Is Design?*

Unless we gain a better understanding of design, we shall witness our environment getting steadily worse, in spite of the constant improvement of our machines and tools.

Take the instance of the airplane, a machine still pure, and the car, which is rapidly losing its self-respect as a machine since it has become a typical product of stylism:

In the making of an airplane, the fight is toward more simplicity. Everything not absolutely necessary, absolutely vital to the organism of the airplane, is shaved off, eradicated, cut out. Every rivet counts, as no one wants to add even one ounce of useless weight in the design of a machine planned for flying. Why such concern? Why has the airplane evolved so rapidly to become one of the most beautiful expressions of our time? The answer is simple: a pure necessity for survival. The airplane design is dictated by one criterion only: efficiency. Thus the airplane has become a model of elegant simplicity. No flourish of fashion has come to mar the honest purity of its design. The result is a design which achieves an organic beauty, a near-natural perfection that we can only find in living organisms. In this fantastic creation, man is slowly developing a new species of natural organism which becomes the first cousin of the other speed monsters designed by nature. The truly great designers arrived at their conclusions through the observation of nature: their masterpieces show close kinship of form and design with plants or animals which have a similar problem to overcome....

Design requires a constant remodeling of our ideas as it must adapt its language to new possibilities offered by new structural materials. We used to talk architecture in terms of columns, beams, and walls. Today, we talk of surfaces, of forms like egg shells, that we can build without yesterday's crutches of columns, beams, and walls.

However, we are so saturated with columns and walls that a thing built still means columns and walls to us. We are shocked by what the best designer puts up today. We are intrigued—yes—as we would be by some acrobatic circus act, and we keep asking ourselves: "Is this design?"

The world has not been aware of such radical possibilities in building since the invention of the brick. For the last five centuries, architecture has only been a series of variations on the Greco-Roman and Gothic themes. Our libraries, post offices and railroad stations, are enlarged carbon copies of antique temples. Churches have become photographically reduced cathedrals. We gradually have become such addicts of the habit-forming drug of monumental history that we cannot take the fresh air of the open spaces created by the athletic structures of today.

But when we analyze good contemporary design, we should acknowledge the fact that its language of form is much less foreign to us than the architectural wardrobe our ancestors used to wear. We see these "new" forms in leaves—every-day—in trees, flowers, or fruit. We treasure them on the beach as beautiful shells and stones rolled by the waves.

The designer today is offered more than ever before the opportunity of falling in step with the world around him.

Design is an end in itself. It is the achievement of man's logic in adapting his creations to his natural environment and way of life. Machines rapidly become obsolete because they represent only a small and temporary part of the solution to a problem, and as such they are all a somewhat awkward expression of man's ignorance. The more learned and experienced man becomes, the more new machines he builds to replace the old ones, but no one can ever attain complete replacement, as the perfect machine would annihilate all other machines and render them all useless and no machine can approximate nature's perfection, even in a small detail like the eye.

Each true work of design should be a complete achievement in itself. It should be a permanent solution that cannot be duplicated either in time or space. Each design is unique and deserves the same amount of respect regardless of size, cost, or location. Whether it is a rice bowl, a cobbler's bench, a wooden plough, the temples at Ise or Saint Sophia, it is in itself a total and admirable lesson.

33 Victor Papanek, *Design for the Real World*, 1971

Stewart Brand presented Victor J. Papanek's (1923–1998) *Design for the Real World* by emphasizing: "Industrial design is in deservedly bad odor lately, and this book will give you many of the reasons—most of them having to do with pandering to self-exaggerating marketplace fashion, the disposable environment. Victor Papanek is a Bush Designer. He prefers to hang out in Third-World places, where the problems are physical and serious, and where native design genius can be learned and applied to them using industrial materials. This book works on distinguishing between real and phony design problems, and their solutions" (*LWEC*, 119). Designer, teacher, and author, Papanek proposed practical solutions to real problems. For example, he designed a radio from ordinary metal food cans powered by a thermopile and a candle. He also developed simple designs such as cheap irrigation pumps, food-cooling units, organic packaging, and wheelchairs built from indigenous materials.

However, more than his actual design descriptions, the book was an uncompromising critique of contemporary design culture. In a fun, lively, and witty way, Papanek analyzed the macroeconomic and political forces that controlled production and consumption, and he predicted the environmental and social crises that today threaten the planet.

Papanek's writing was informed by his background. Born in Vienna in 1923, he left the country with his mother after the Anschluss of Austria to Nazi Germany and lived for a while in England before settling in the United States in 1939. Educated at Cooper Union in the late 1940s, he enrolled in an engineering course at the Massachusetts Institute of Technology. There he became a follower and ally of Buckminster Fuller, who wrote the preface to the first English-language edition of Papanek's seminal publication *Design for the Real World: Human Ecology and Social Change* (1971). The first edition was published in Swedish.

Despite, or perhaps because of, its polemical tone, the book was a huge success; translated into twenty-three languages, it remains one of the most widely read design books to date.—CM

Source
Victor Papanek, *Design for the Real World: Human Ecology and Social Change* (New York: Bantam Books, 1971), 1–4. (Referred to in *LWEC*, 119.)

Further Reading
Banham, Reyner. *The Aspen Papers: Twenty Years of Design Theory from the International Design Conference in Aspen*. New York: Praeger, 1974.

Hennessey, James, and Victor J. Papanek. 1973. *Nomadic Furniture: How to Build and Where to Buy Lightweight Furniture That Folds, Collapses, Stacks, Knocks-Down, Inflates or Can Be Thrown Away and Re-cycled; Being Both a Book of Instruction and a Catalog of Access for Easy Moving*. New York: Pantheon Books, 1973.

Papanek, Victor J. *The Green Imperative: Natural Design for the Real World*. New York: Thames and Hudson, 1995.

Selections from *Design for the Real World*

Preface to the First Edition

There are professions more harmful than industrial design, but only a very few of them. And possibly only one profession is phonier. Advertising design, in persuading people to buy things they don't need, with money they don't have, in order to impress others who don't care, is probably the phoniest field in existence today. Industrial design, by concocting the tawdry idiocies hawked by advertisers, comes a close second. Never before in history have grown men sat down and seriously designed electric hairbrushes, rhinestone-covered shoe horns, and mink carpeting for bathrooms, and then drawn up elaborate plans to make and sell these gadgets to millions of people. Before (in the "good old days"), if a person liked killing people, he had to become a general, purchase a coal mine, or else study nuclear physics. Today, industrial design has put murder on a mass production basis. By designing criminally unsafe automobiles that kill or maim nearly one million people around the world each year, by creating whole new species of permanent garbage to clutter up the landscape, and by choosing materials and processes that pollute the air we breathe, designers have become a dangerous breed. And the skills needed in these activities are carefully taught to young people.

In this age of mass production when everything must be planned and designed, design has become the most powerful tool with which man shapes his tools and environments (and, by extension, society and himself). This demands high social and moral responsibility from the designer. It also demands greater understanding of the people by those who practice design and more insight into the design process by the public. Not a single volume on the responsibility of the designer, no book on design that considers the public in this way, has ever been published anywhere....

Design must become an innovative, highly creative, cross-disciplinary tool responsive to the true needs of men. It must be more research oriented, and we must be more research oriented, and we must stop defiling the earth itself with poorly designed objects and structures.

... This book is also written from the viewpoint that there is something basically wrong with the whole concept of patents and copyrights. If I design a toy that provides therapeutic exercise for handicapped children, then I think it is unjust to delay the release of the design by a year and a half, going through a patent application. I feel that ideas are plentiful and cheap, and it is wrong to make money from the needs of others. I have been very lucky in persuading many of my students to accept this view. Much of what you will find as design examples throughout this book has never been patented. In fact, quite the opposite strategy prevails; in many cases students and I have measured drawings of, say, a play environment for blind children, written a description of how to build it simply, and then mimeographed drawings and all. If any agency, anywhere, will write in, my students will send them all the instructions free of charge. I try to do the same myself....

What Is Design? A Definition of the Function Complex

All men are designers. All that we do, almost all the time, is design, for design is basic to all human activity. The planning and patterning of any act toward a desired, foreseeable end constitutes the design process. Any attempt to separate design, to make it a thing-by-itself, works counter to the fact that design is the primary underlying matrix of life. Design is composing an epic poem, executing a mural, painting a masterpiece, writing a concerto. But design is also cleaning and reorganizing a desk drawer, pulling an impacted tooth, baking an apple pie, choosing sides for a backlot baseball game, and educating a child.

Design is the conscious and intuitive effort to impose meaningful order.

It is only in recent years that to add the phrase "and intuitive" seemed crucial to my definition of design. Consciousness implies intellectualization, celebration, research, and analysis. The sensing/feeling part of the creative process was missing from my original definition. Unfortunately intuition itself is difficult to define as a process or ability. Nonetheless it affects design in a profound way. For through intuitive insight we bring into play impressions, ideas, and thoughts we have unknowingly collected on a subconscious, unconscious, or preconscious level. The "how" of intuitive reasoning in the design doesn't readily yield to analysis but can be explained through example. Watson and Crick intuitively felt that the underlying structure of the DNA chain would express itself most elegantly through a spiral. Beginning with this intuition, they began their research. Their instinctive precognition paid off: a spiral it is!

Our delight in the order we find in frost flowers on a window pane, in the hexagonal perfection of a honeycomb, in leaves, or in the architecture of a rose, reflects man's preoccupation with pattern. We constantly try to understand our ever-changing highly complex existence by seeking order in it. And what we seek we find. There are underlying biological systems to which we respond on levels that are often conscious or unconscious. The reason we enjoy things in nature is that we see an economy of means, simplicity, elegance, and an essential rightness there. But all these natural templates, rich in pattern, order, and beauty, are not the result of decision making by mankind and therefore lie beyond our definition. We may call them "design," as if we were speaking of a tool or artifact created by humans. But this is to falsify the issue, since the beauty we see in nature is something we ascribe to processes we don't understand. ...

"Should I design it to be functional," the students say, "or to be aesthetically pleasing?" This is the most often heard, the most understandable, and yet the most mixed-up question in design today. "Do you want it to look good or to work?" Barricades are erected between what are really just two of the many aspects of function.

34 Donald A. Schön, *Technology and Change: The New Heraclitus*, 1967

"Recommended reading for changebringers," urged Jay Baldwin in his review of Donald A. Schön's *Technology and Change*. Schön, who held a doctorate in philosophy from Harvard University, worked for the large industrial research firm Arthur D. Little and the U.S. Department of Commerce while writing the book, in which he sought to "present the dynamics of industrial change as a metaphor for change in our society as a whole" (*Technology and Change*, xix). The book's subtitle, *The New Heraclitus*, referred to a model of stability and change, developed by the ancient Greek philosopher, which posited, "All things are in process and nothing stays still. ... You could not step twice into the same river" (xix); Schön regarded this Heraclitan concept as a useful approach to understanding the rapid pace of mid-twentieth-century technological progress and the social shifts it effected.

While this brief description may not, perhaps, suggest the "subversive stuff" Baldwin promised, Schön did, with a vigorous logic born of experience, broach many of the same topics countercultural thinkers regularly confronted, especially with regard to the societal and environmental tolls of technological innovation. His views on "Return" and "Revolt," included in the second passage, may have been particularly resonant with the catalog's readers. —MG

Source
Donald A. Schön, *Technology and Change: The New Heraclitus* (New York: Delacorte Press, 1967), 193–195, 202–203. Courtesy of Nancy Schön. (Referred to in *LWEC*, 122.)

Further Reading
Bronowski, Jacob. "Technology and Culture in Evolution." *American Scholar* 41, no. 2 (spring 1972): 197–211.

Schön, Donald A. *Beyond the Stable State*. New York: Norton, 1971.

Schön, Donald A. *The Reflective Practitioner: How Professionals Think in Action*. New York: Basic Books, 1983.

"An Ethic of Change," from *Technology and Change: The New Heraclitus*

To summarize, we believe, according to the conventional wisdom, in the theory of the stable state. We believe that change represents a series of transitions from one stable state to another. We believe that technology is a neutral instrument of change, and we believe in a Technological Program according to which we march steadily toward a good society whose objectives remain stable and clear throughout the march. Time is perceived under the concept of Progress, measured in terms of human satisfactions and individual freedom. Through work and the steady development of character the happiness and peace of all mankind are] to be attained.

This view of the Technological Program has become a part of our ideological equipment. When we speak in approving tones of a "changing society," what we mean is a society changing within this stable framework. If we talk of social change, we keep the model of technological change foremost in our minds: hence, we have "mechanisms" of social change as though we were dealing again with instruments deliberately used to achieve prior objectives within a stable framework. The question, What's technology for?, asked in the United States in 1966, will usually produce a pretty good facsimile of the Technological Program.

Nevertheless, we seem in recent years to have become increasingly aware that something about the program has gone wrong. This does not mean that we speak its language any less. On the contrary, we give it clearer and more frequent expression than ever before, but there is a growing sense of uneasiness about it. This uneasiness we sometimes attribute to the discovery that technology, as a neutral instrument, may be used for good or evil ends and that its by-products may outweigh in evil the good anticipated for its consequences. By now we are more than familiar with pronouncements about the violence of two world wars made more terrible by advanced technology, and we struggle with our sense of the potential for total annihilation of mankind by the Bomb. We are familiar with sermons against the pollution of air, water and land by industrial by-products and the destruction of the beauty and order of our cities and countryside by the automobile and its appendages. All of this, however, can be comprehended under the heading of "technology badly used." Its remedy appears to be a more intelligent use of technology: a more complete application of reason to the human goals of the eighteenth century program.

… But there is a deeper source of moral uneasiness associated with the Technological Program. It is the sense, the intuitive awareness, that technological change has been undermining the very objectives constitutive of the Good Society to which technology is supposed to lead.

In all areas and at all levels of society, technological change has been eroding established identities. It has been undercutting the sense of self on which commitment to the Technological Program depends.

The call, then, is to Return. The appeal of the call is that it offers comfort in the face of chaos. There is change, but it represents a relaxation of standards, a loss of the good old objectives, a deviation from the ways of our ancestors, to which we are urged to return. There is great appeal in the call. … The call to return gives rise to a form of sleepwalking, to use Arthur Koestler's term. Since the good society of the eighteenth-century program is, indeed, bankrupt, it cannot sustain us in the face of the problems and complexities we now face. The attempt to return to it, therefore, condemns us to a sleepwalking toward the future, of necessity leaving that future unexamined.

The call to Revolt has as much appeal as the call to Return. It is an alternate response to the destruction of the myth of stability. In effect, it says that the old objectives are hollow and inadequate. There is oppression, slavery, humiliation, constraint and injustice. There is no Promised Land in sight. Revolt against these things! The call to revolt need not produce its own image of a Promised Land. It takes its image from the present against which it reacts. It gets its

concreteness from the life it refuses to accept and is, in this sense, a form of conservatism.

If we cannot be inattentive to change of objectives in our society and cannot accept the silent calls of return and revolt, what is left? How do we ever tolerate and stand up to change in objectives? …

For a child becoming an adult, in revolt against the objectives of childhood and, characteristically, of family, there is a need to seek out the next stable state and somehow in the process to preserve a sense of self and self-worth. Typically there are individuals, and groups—peers, teachers, movements in society—with which a young person identifies. He is en route to their objectives. They are front-runners and he is catching up. As he learns the new objectives, he preserves his sense of self and self-worth through identification with these people. Front-running and catching-up make a pattern of change of objectives in society. But what happens, as at present, when at a fundamental level the society itself is in process of change in objectives and there are no front-runners at hand with whom to catch up?

When we are neither catching up nor in revolt, we can abandon objectives which were central for us only through an ethic governing the process of change itself. An ethic of change—which we may call a meta-ethic—provides discipline for the process of change, making possible abandonment of old positions without a loss of self. It permits us to tolerate the dizzying freedom and the surplus of information that come from abandoning old objectives.

35 Farrington Daniels, *Direct Use of the Sun's Energy*, 1964

Farrington Daniels's *Direct Use of the Sun's Energy* **was immensely influential on solar designers, including Steve Baer, who called it "the best book on Solar Energy that [he knew] of." A physical chemist who had worked as the director of the Manhattan Project's Metallurgical Laboratory but had become disillusioned when he discovered the government's almost exclusive interest in nuclear weapon design and not more peaceful applications of the technology, Daniels had committed himself to the research of solar power shortly after World War II, becoming one of the founding members of the American Solar Energy Society in 1954. He published** *Direct Use of the Sun's Energy* **ten years later, and it became a paradigmatic text for Americans interested in solar power, especially after the 1973 oil crisis. The book included a diversity of visual information—from scientific diagrams to line drawings explaining how one might make a simple solar water heater—and explained sophisticated concepts in a way even laypeople could understand. In the following passage and figures, Daniels described the ways in which geographical location affected solar radiation, and envisaged a solar-driven future world that surely captured the imagination of the catalog's readers and others seeking to minimize the use of nonrenewable resources and thwart further environmental deterioration.—MG**

Source

Farrington Daniels, *Direct Use of the Sun's Energy* (New Haven, CT: Yale University Press, 1964), 36–44. Reprinted with the permission of Yale University Press. (Referred to in *LWEC*, 124.)

Further Reading

Daniels, Farrington. "Atomic and Solar Energy." *American Scientist* 38, no. 4 (October 1950): 520–548.

Laird, Frank. "Constructing the Future: Advocating Energy Technologies in the Cold War." *Technology and Culture* 44, no. 1 (January 2003): 27–49.

Total Environmental Action. *Solar Energy Home Design in Four Climates.* Harrisville, NH: TAE, 1975.

"Geographical Distribution of Radiation," from *Direct Use of the Sun's Energy*

As indicated in the preceding section, the solar radiation varies greatly with the length of the day, with the angle of the sun's rays to the ground, with the length and quality of the atmosphere through which it passes, and particularly with the cloud coverage. Accordingly the radiation varies greatly with the geographical location, the altitude, and the weather.…

A wealth of information on weather in the United States including some data on solar radiation has been collected and summarized by Visher. Figure 12, adapted from Visher, gives the normal annual number of hours of sunlight visibility in the continental United States. The range is from 2,200 hr in the northeast to 4,000 in the southwest. Figure 13…also adapted from Visher, gives…the number of langleys/day^{-1} in the United States east of the Rocky Mountains for June 21 and December 21.…

Table [4.1] summarizes solar radiation in a few selected regions of the United States.

It must be emphasized that the data given in these maps and tables, and almost all data reported in langleys by weather stations throughout the world, apply to the radiation received on a horizontal plane. This is much less than can be obtained with a tilted surface particularly in winter, as already explained.

Caryl found that in Arizona the solar radiation between 9 A.M. and 3 P.M. on a surface facing south and inclined 45° from the horizontal varied only from 13,000 langleys in the month of January to 19,000 langleys in March and September.…

…In general the greatest amount of solar energy is found in two broad bands encircling the earth between 15° and 35° latitude north and south. In the best regions there is a minimum

Table 4.1

Solar Radiation in Langleys (1 cal cm^{-2}) per Day on a Horizontal Surface

	December	March*	June
New York and Chicago	125	325	550
Southern California and Arizona	250	400	700
Florida	250	400	500
Nevada	175	400	650

*September radiation is close to that of March.

monthly mean radiation of 500 langleys/day^{-1} and a monthly overall variation of less than 250 langleys/day^{-1}. These regions are on the equatorial side of the world's arid deserts. They have less than ten inches of rain in a year. In some of the countries more than two thirds of their area is arid land, and there is usually over 3,000 hr of sunshine a year, over 90 percent of which comes as direct radiation. These areas are well suited for applied solar energy.

The next most favorable location is in the equatorial belt between 15° N and 15° S. Here the humidity is high, the clouds frequent, and proportion of scattered radiation high. There are about 2,300 hr of sunshine per year and very little seasonal variation. Radiation is from 300 to 500 langleys/day^{-1} throughout the year, and there are few successive days of low radiation.

Between 35° and 45° at the edge of the desert areas the radiation can average 400 to 500 langleys/day^{-1} on a horizontal surface throughout the year, but there is a marked seasonal effect and the winter months have low solar radiation. This seasonal variation can be greatly minimized by tilting the receiving surfaces to face the sun.

The regions north of 45° N and south of 45° S are limited in their year-round direct use of solar energy. One might think of the Arctic and Antarctic regions as hopeless for solar energy utilization but in the summer season they may be quite important. Some of the Antarctic records show long days that receive 700 langleys of solar radiation, which is extraordinarily high. Of course in the winter time there is no opportunity to use solar devices. Where heat and power come only from gasoline brought in over thousands of miles by air transport at a cost of $5 per gallon, solar radiation might be particularly important if lightweight solar heaters could be developed.

One of the best locations in the world for solar applications is the desert of north Chile, which extends 100 miles in width from the Pacific Ocean to the Andes Mountains and for 280 miles along the coast from latitude 20° S to latitude 25° S. It has 1 mm of rainfall a year, 364 days of bright sunshine, and no dust. ... There is no water and no life. The land is valueless now except for a few rich nitrate deposits and copper mines. In the future, when our resources of fossil fuels diminish, this area and other similar deserts could supply tremendous amounts of energy. Theoretically this waste area in north Chile alone receives annual solar heat greater than all the heat produced in the world in a year by the burning of coal, oil, gas, and wood, which in 1957 amounted to 5×10^{13} hp hr or 3.2×10^{16} kcal^{-1}. The 28,000 square miles of this desert receive about 1.3×10^{17} kcal of solar heat in a year (28,000 mile^2 × 1.3×10^{10} kcal/day^{-1}/mile^{-2} × 365 days/year^{-1} = 1.3×10^{17} kcal/year^{-1}). It is clear that *theoretically* this desert could supply all the energy needs of the world. There are other sunny wastelands such as the Sahara desert in Africa, parts of the middle east, parts of the southwestern United States, parts of Australia, and calm areas of the oceans that could be used for collecting and storing solar energy. Also there are many less favorable wastelands, some of which are closer to populations where the energy is needed. The solar radiation could be used to produce hydrogen by the electrolysis of water and the hydrogen could be stored and transported through pipelines or combined with carbon dioxide to give methanol or other transportable fuels.

SOLAR RADIATION

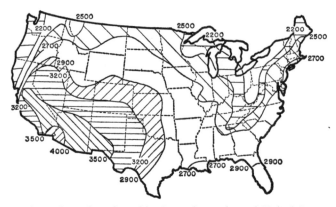

12. Annual number of sunshine hours in continental United States. [Adapted with permision of the Harvard University Press from Visher's *Climatic Atlas of the United States* (1954), p. 177.]

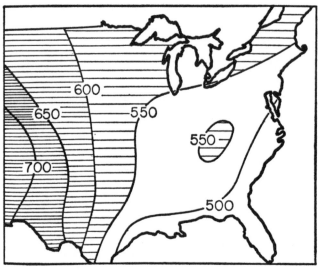

13. Solar radiation, in langleys, in the United States east of the Rocky Mountains on June 21. [Adapted with permission of the Harvard University Press from Visher's *Climatic Atlas of the United States* (1954), p. 183.]

Figure 4.3
"Figure 11. Worldwide distribution of solar energy in hundreds of hours per year"; "Figure 12. Annual number of sunshine hours in continental United States"; and "Figure 13. Solar radiation, in langleys, in the United States east of the Rocky Mountains on June 21." Farrington Daniels, *Direct Use of the Sun's Energy* (New Haven, CT: Yale University Press, 1964), 38–39. Reprinted with permission of Yale University Press.

36 "Snap-On Tools"

In the *WEC*, tools always combined associations of the body and the mind. As Jay Baldwin said: "I've remarked that tools are extensions of your hands. No mystery here; a hammer is just a hard fist; a screwdriver, a tough fingernail. But hands usually operate according to instructions from the head, so it can also be said that tools are an extension of your mind."[1]

Organizing tools was also a metaphor for the structures of memory and the translation of thought into action. Physical location has always played a role in memory, as Frances Yates demonstrated in her 1966 book *The Art of Memory*.[2] Baldwin organized the tools in his traveling workshop around the tasks they perform rather than by conventional categories. The illustration of the very expensive Snap-on tool cabinet was as much about ordering and classification as about the actual tools themselves. As Brand observed, these tools were extremely expensive and only valuable for the professional. One of the attractions of the *WEC*, as in other trade catalogs, was the appeal of things you could not afford, but which stimulated the imagination.—CM

1. JB [Jay Baldwin], *Next Whole Earth Catalog*, 1980, 139.

2. Frances Yates, *The Art of Memory* (Chicago: University of Chicago Press, 1966).

Source
"Snap-On Tools Corp., Kenosha, Wisconsin," in *LWEC*, 138.

Further Reading
Baldwin, Jay. "One Highly Evolved Toolbox." In *The Next Whole Earth Catalog: Access to Tools*, 138–141. Sausalito, CA: Point Foundation, 1980.

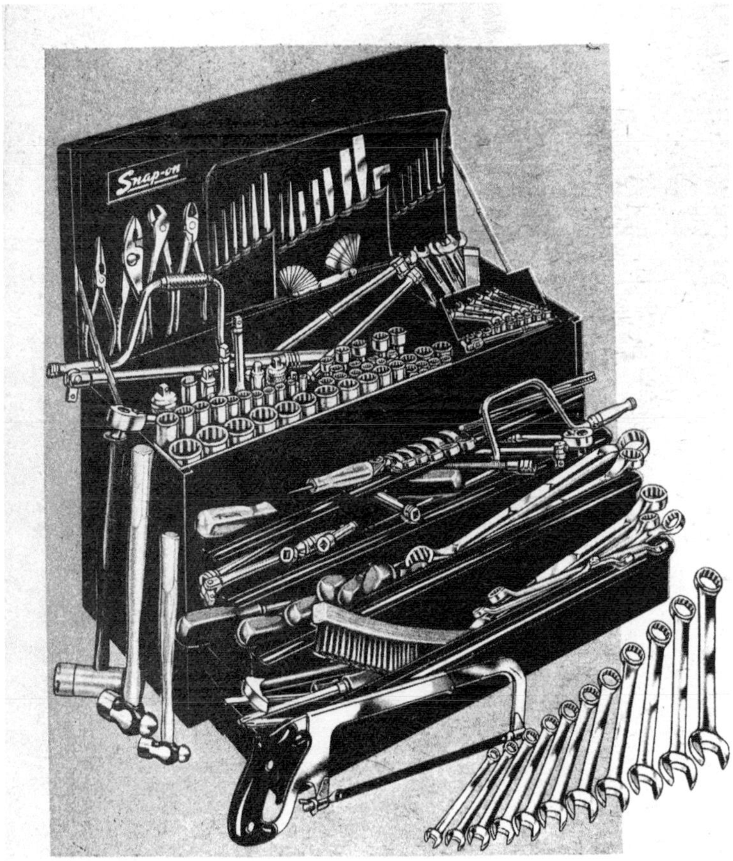

Figure 4.4
"Snap-On Tools Corp., Kenosha, Wisconsin," *LWEC*, 138.

Craft

On the first pages of the "Craft" section, J. D. Smith, a former Tufts- and Harvard-educated classics scholar, composed a short poem in free verse titled, aptly, "Craft." In the piece, Smith, who edited the fall 1970 edition of the catalog and managed the Whole Earth Truck Store from 1970 to 1971, aired his concerns regarding the adage "Do it right the first time."

wyatt frydenland,
a student of mine
once said
"do it right the first time."
now that's a little bit tightass
when applied to homedone things.
maybe it should read
"practice until you can
do it right the first time."
or
"a thing worth doing is worth doing."
or
"just go ahead on and do it,
whatever it is,
and learn from your mistakes."
or
"find somebody who knows something
and learn it from him."
or
"through industry we prosper,"
and all that craft.

Smith's "Craft" is an excellent encapsulation of the purpose of this diverse section, which provided information and resources for woodworking, basket weaving, macramé, and leather-tooling projects, among many others. The "Craft" section was particularly directed to readers concerned with the social and environmental ramifications of a mass-consumption culture; they adopted a do-it-yourself attitude, throwing their own cups and bowls, sewing and mending their own clothing, even spinning and dyeing their own wool. Smith's poem was both an inspiring and reassuring introduction to a section filled with learning aids and products to support a multitude of crafts.

The following passages and images demonstrate the catalog's approach toward craft. Recommended texts included both pictures and words to properly guide novice craftspeople in new projects. For the most part, as in the previous section, "Craft" contains far fewer historical or philosophical texts, focusing on the "how" rather than the "why." R. J. De Cristoforo's *How to Build Your Own Furniture* (1965), with clear instructions and ample illustrations, was deemed by the catalog to be the ultimate do-it-yourself furniture book, as it was not prescriptive of the design but instead supplied readers with the explanatory tools to create their own. *Foxfire*, a quarterly journal, which chronicled not only the crafts but also the cultural traditions of southern Appalachia, offered instructional guides but also connected the do-it-yourself

countercultural spirit to its American roots. On the single page titled "Philosophy and Craft Access," E. F. Schumacher's "Buddhist Economics" and Kakuzō Okakura's *Book of Tea* discussed the spirituality of craftsmanship, and the last two selections—*Your Handspinning* and *The Illustrated Hassle-Free Make Your Own Clothes Book*—clearly manifested the meaning of J. D. Smith's directive "A thing worth doing is worth doing." This section of the *Whole Earth Catalog* maintained the grassroots, ecologically responsible stance of the four previous chapters but applied it to a more intimate, individual set of everyday goods and products.—MG

37 R. J. De Cristoforo, *How to Build Your Own Furniture*, 1965

In the four parts of his book *How to Build Your Own Furniture*, R. J. De Cristoforo provided the information needed to craft furniture to one's own speci-
fications: from the first part, "Planning the Project," in which he discussed
aspects of style and composition; to the second, "Materials and Assembly
Tools," in which he introduced the proper resources with which to build; to the
last two sections, "Furniture Components" and "The Furniture," in which he
considered both the individual elements of the design and the final results. "If
you put a hundred craftsmen in a hundred different shops and asked each to
build a table," De Cristoforo estimated, "it's highly unlikely that you would find
duplicates despite the fact that a table is a table and that each consists of a
broad surface (slab), legs, and maybe rails." His book was premised on this
conviction.

De Cristoforo's approach distinguished *How to Build Your Own Furniture*
from the many other furniture-making guides of the 1950s and '60s, which
presented definitive plans and instructions for each piece of furniture, not
allowing for the adjustments that many do-it-yourself projects required. The
images that follow, both from part 3, "Furniture Components," demonstrate
the author's strategy and the possibilities it offered both amateur and more
experienced craftsmen. —MG

Source
R. J. De Cristoforo, *How to Build Your
Own Furniture* (New York: Popular
Science, 1965), 78, 88. (Referred to in
LWEC, 148.)

Further Reading
Atkinson, Paul. "Introduction: Do It Yourself: Democracy and Design." *Journal of
Design History* 19, no. 1 (spring 2006):
1–10.

Roland, Albert. "Do-It-Yourself: A Walden
for the Millions?" *American Quarterly* 10,
no. 2 (summer 1958): 154–164.

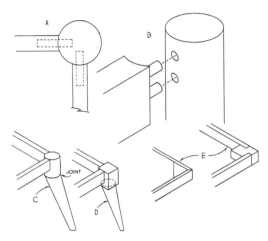

Round legs can be used in leg-and-rail assembly if end of rail is shaped on drum sander to conform to leg (A and B). Two-piece leg (C and D) can have either a round or a square top. Corner blocks (E) provide a seat for attaching the leg, thus eliminating the rail-to-leg joint.

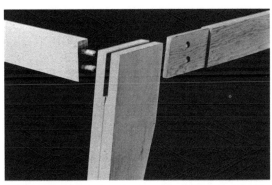

Here is a combination dowel-and-tenon joint that provides a locking device which will keep parts together even if the glue fails. The dowels on the one rail pass through half the leg and enter the holes drilled in the tenon on the other rail.

78

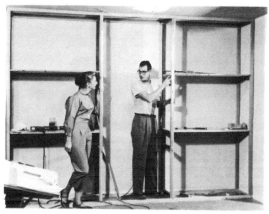

Skeleton frame can be used to construct a built-in which will cover the entire wall of a room. Note that some of the frame members are utilized as supports for shelves. Installing a face-frame and then hanging the doors will complete the project.

PLYWOOD OR GLUED-UP STOCK ALSO FORMS A FRAME

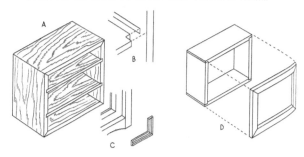

A. Frame of slabs can be joined at edges in many ways; dividers or shelves can be let into the sides in dadoes. B. Blind dadoes avoid cut at front of case. C. Front edges may be banded, decoratively framed, or hidden with moulding. D. Front frame is the means of hanging hinged doors, or it may be grooved for sliding doors.

88

Figure 5.1 (left)
"Leg-and-rail assemblies." R. J. De Cristoforo, *How to Build Your Own Furniture* (New York: Popular Science, 1965), 78. Reprinted with permission.

Figure 5.2 (right)
"Plywood or glued-up stock also forms a frame." R. J. De Cristoforo, *How to Build Your Own Furniture* (New York: Popular Science, 1965), 88. Reprinted with permission.

38 Mary Garth, "Making a Hamper of White Oak Splits," 1970

Eliot Wigginton and his English students at the Rabun Gap-Nacoochee School in the Appalachian Mountains of northern Georgia founded *Foxfire* magazine in 1966; the content of the student-run quarterly, named after a bioluminescent fungus found on decaying wood in the area's surrounding forests, addressed the traditions and culture of southern Appalachia, as expressed by the students' family, friends, and neighbors. Folktales, recipes, book reviews, and articles on traditional crafts such as quilting, furniture building, and soap making appeared on the pages of the magazine, evidence of what the Rabun Gap-Nacoochee students found intriguing and what their relatives deemed important to transmit to a wider audience. One significant result of this approach was the ability of the young authors to produce meaningful reportage on the culturally rich region while avoiding both "social/economic biases" and "pasty sentimentality"—characteristics that, according to Gurney Norman in his review of *Foxfire*, had infected much of the national coverage of Appalachia. "These kids in Georgia," Norman declared, "are living in a real world, studying real things, and in consequence they are creating a wonderfully real publication in *Foxfire*."

That a student publication from the mountains of Georgia could find a place in the internationally circulated *Whole Earth Catalog* revealed an overwhelming countercultural preoccupation with the folk traditions of a bygone, apparently simpler or more organic era in American history; and yet the inclusion of *Foxfire* was also a testament to the quality of research and writing found in the magazine and the unique voice it offered in a section of the catalog dominated by straightforward guides and manuals.—MG

Source
Mary Garth, "Making a Hamper of White Oak Splits," *Foxfire* 4, no. 4 (winter 1970): 210–213. (Referred to in *LWEC*, 152.)

Further Reading
Chappell, Fred. "The Ninety-ninth Foxfire Book." *Appalachian Journal* 11, no. 3 (spring 1984): 260–267.

Wigginton, Eliot, ed. *The Foxfire Book: Hog Dressing, Log Cabin Building, Mountain Crafts and Foods, Planting by the Signs, Snake Lore, Hunting Tales, Faith Healing, Moonshining, and Other Affairs of Plain Living.* Garden City, NY: Doubleday, 1972.

"Making a Hamper of White Oak Splits"

Sitting on her front porch hammering away at the heavy white oak ribs, Beulah Perry looked as if she had been making baskets for a long time. Actually, she had never made one before, but after years of watching her father, she knew just how to do it.

Even though I've known Mrs. Perry for over a year, she still amazes me with all her knowledge of the old times, and how she and her family lived before there were stores in which to buy canned foods, cloth and electric lamps.

Like many of the other people we have interviewed, she knows what it was like to have the closest neighbors five miles away; to have a cooked possum head as a reward for being good, and to maybe get a stick of peppermint when her father had a few extra pennies.

Her house is spotless. While she was showing us how to make the basket, she served us coffee and cake. Each of us had our own china cup and saucer—all different. When Jan and I helped her do the dishes afterwards, we were afraid that she would think us bad housekeepers if we left anything undone, so we scrubbed the sink and cabinets with Comet. We were sure she did it every time!

CAPTIONS FOR THE PLATES THAT FOLLOW: 211
    The splits (see Volume 4, Number 3 for directions for
making them) should be as long as possible. 24 heavy, 6
foot ribs are also needed. All should be used green.
    Crease each rib, making three 2-foot sections (below
and photo 2). Hammer out all knots and rough places.
    Place one rib on a flat surface, and a second perpen-
dicular to it with its corner inside and on top of the
first. The third rib should be placed parallel to the
second and <u>under</u> the first (3). Short tacks may be needed at first to keep them in place. As the others are
added, weave them through in an over/under pattern (4)
until the bottom of the basket is woven - 12 ribs going
in each direction and forming a square - and the sides
are sticking up (5).
    To form the sides, use thin, 1-inch-wide splits,
weaving them in and out of the side ribs (6). This will
take nearly all the splits you can make from two good
oaks. When the end of a split is reached, lap another
about two inches over the end of the last and continue.
    Pull the sides up straight as you go. When you reach
the top, you will probably find the ends of the ribs to
be uneven. These should be trimmed off even (7) before
adding the rim.
    ˜ For the rim, simply run a split around the inside and
outside top of the basket, and bind it there by wrapping
around it with thin, narrow splits (8,9).

Figure 5.3
"Making a Hamper of White Oak Splits."
Foxfire 4, no. 4 (winter 1970): 211. Reprinted
with permission of the Foxfire Fund.

In knowing Mrs. Perry, I've learned tremendous respect for her and all the others who
shared similar hardships, if you can call them
hardships at all. If kids my age didn't have TV,
cars and mothers that did all the work, maybe we
would be a lot more stable.

But it's too late to go back now. We have to
compensate by listening to what they have to say
and learning from it. That's one reason why we
asked Beulah Perry to show us how to make this
basket.

—Mary Garth

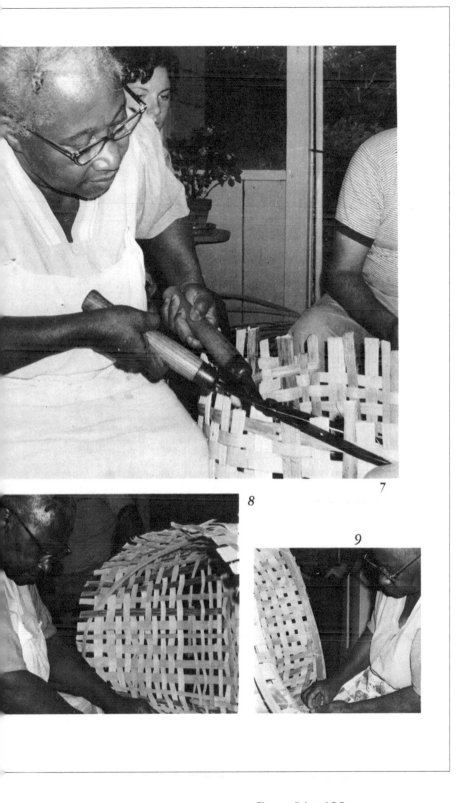

7

8

9

Figures 5.4 and 5.5
"Making a Hamper of White Oak Splits."
Foxfire 4, no. 4 (winter 1970): 212–213.
Reprinted with permission of the Foxfire Fund.

39 Kakuzō Okakura, *The Book of Tea*, 1906

Kakuzō Okakura (1862–1913) wrote *The Book of Tea* in English in 1906 as a protest against the modernization of Japan under the Meiji regime. The book was already in its sixteenth edition by 1932. Republished in 1964, it acquired a cult following. Okakura saw in the tea ceremony a means of maintaining traditional Japanese values threatened by Westernization. He sought to explain to Western readers the blend of spirituality, simplicity, and symbolic association contained in the tea ceremony. The first extract in the *WEC* explains that the tea ceremony not only was an aesthetic ritual but also had lessons to teach about hygiene, economics, and "moral geometry, inasmuch as it defines our sense of proportion to the universe." The second extract points up the differences between the Western view of Japan and the Japanese view of the West. The third extract reflects on man's relationship with nature: "We boast that we have conquered Matter and forget that it is Matter that has enslaved us. What atrocities do we not perpetrate in the name of culture and refinement."

Born in Japan, Okakura moved to Boston in 1904 at the invitation of the art collector William Sturgis Bigelow, who donated a large collection of Japanese art and artifacts to the Museum of Fine Arts. In the United States, Okakura found an environment sympathetic to Japanese art and culture. *The Book of Tea* was written for a cultured American audience, but it was also a kind of criticism of American values and behavior. Its great popularity in the 1960s was partly due to the anxiety among Americans about the dehumanizing effects of Western capitalism. In his introduction to the 1964 edition, E. F. Bleiler pointed out that there was also a deeply conservative aspect to Okakura's thought, encouraging Japanese isolationism and distrust of foreigners. Japanese tradition required, as Bleiler put it, "a submission of one's self to the ways of the fathers. It means that you must do things according to the old days." As paradoxical as it might seem for a generation trying to free itself from parental control, this book is another example of the catalog's search for an alternative to Western civilization, capitalism, and materialism. —CM

Source
Kakuzō Okakura, *The Book of Tea* (1906; reprint, Mineola, NY: Dover, 1964), 1–3, 22–23, 42–44, 52. (Referred to in *LWEC*, 153.)

Further Reading
Bleiler, Everett F. Introduction and afterword to *The Book of Tea*, by Kakuzō Okakura, vii–xxv, 85–99. Mineola, NY: Dover, 1964.

"The Philosophy of Tea," from *The Book of Tea*

The Philosophy of Tea is not mere aestheticism in the ordinary acceptance of the term, for it expresses conjointly with ethics and religion our whole point of view about man and nature. It is hygiene, for it enforces cleanliness; it is economics, for it shows comfort in simplicity rather than in complex and costly; it is moral geometry, inasmuch as it defines our sense of proportion to the universe. It represents the true spirit of Eastern democracy by making all its votaries aristocrats in taste.…

The long isolation of Japan from the rest of the world, so conducive to introspection, has been highly favorable to the development of Teaism. Our home and habits, costume and cuisine, porcelain, lacquer, painting,—our very literature,—all have been subject to its influence.…

Those who cannot feel the littleness of great things in themselves are apt to overlook the greatness of little things in others. The average Westerner, in his sleek complacency, will see in the tea ceremony but another instance of the thousand and one oddities which constitute the quaintness and childishness of the East to him.…

The Schools of Tea

…It is in the Japanese tea-ceremony that we see the culmination of tea-ideals. Our successful resistance of the Mongol invasion in 1281 had enabled us to carry on the Sung movement, so disastrously cut off in China itself through the nomadic inroad. Tea with us became more than an idealization of the form of drinking; it is a religion of the art of life.…

We have already said that it was the ritual instituted by the Zen monks of successively drinking tea out of a bowl before the image of Bodhidharma, which laid the foundations of the tea-ceremony. We might add here that the altar of the Zen chapel was the prototype of the Tokonoma,—the place of honor in a Japanese room where paintings and flowers are placed for the edification of the guests.

All our great tea-masters were students of Zen and attempted to introduce the spirit of Zennism into the actualities of life. Thus the room, like the other equipments of the tea-ceremony, reflects many of the Zen doctrines. The size of the orthodox tea-room, which is four mats and a half, or ten feet square, is determined by a passage in the Sutra of Vikramaditya. In that interesting work, Vikramaditya welcomes the Saint Manjusri and eighty-four thousand disciples of Buddha in a room of this size,—an allegory based on the theory of the non-existence of space to the truly enlightened. Again the rōji, the garden path which leads from the machiai to the tea-room, signified the first stage of meditation,—the passage into self-illumination. The rōji was intended to break connection with the outside world, and produce a fresh sensation conducive to the full enjoyment of aestheticism in the tea-room itself. One who has trodden this garden path cannot fail to remember how his spirit, as he walked in the twilight of evergreens over the regular irregularities of the stepping stones, beneath which lay dried pine needles, and passed beside the moss-covered granite lanterns, became uplifted above ordinary thoughts. One may be in the midst of a city, and yet feel as if he were in the forest far away from the dust and din of civilization.…

The simplicity of the tea-room and its freedom from vulgarity make it truly a sanctuary from the vexation of the outer world. There and there alone one can consecrate himself to undisturbed adoration of the beautiful. In the sixteenth century the tea-room afforded a welcome respite from labor to the fierce warriors and statesmen engaged in the unification and reconstruction of Japan. In the seventeenth century, after the strict formalism of the Tokugawa rule had been developed, it offered the only opportunity possible for the free communion of artistic spirits. Before a great work of art there was no distinction between daimyo, samurai, and commoner. Nowadays industrialism is making true refinement more and more difficult all the world over. Do we not need the tea-room more than ever?

40 E. F. Schumacher, "Buddhist Economics," 1968

E. F. Schumacher (1911–1977) published this version of his essay "Buddhist Economics" in the British magazine *Resurgence* in 1968.[1] The magazine specialized in decentralization, deurbanization, libertarian technology, and alternative lifestyles. Schumacher discussed the issue of labor, which was fundamental to the psychic equilibrium of the individual and essential to a sense of satisfaction and social integration. In short, Schumacher deconstructed the whole capitalist economy so as to focus on individual well-being rather than on a system of financial exchange.

Born in Germany, Schumacher came to England in 1930 as a Rhodes Scholar to read economics at New College, Oxford. Schumacher was also an expert on farming, active in the Soil Association, which promoted organic farming and challenged the orthodoxy of chemical-based agriculture. He became an economic adviser to the British Control Commission in Germany (1946–1950), and then had a long career in the National Coal Board in Britain. The turning point came in 1955, when he was sent as economic development adviser to the government of Burma. He was supposed to introduce there the Western model of economic growth, but he discovered that the Burmese did not need economic development along Western lines, as they themselves had an indigenous economic system well suited to their conditions, culture, and climate.

Schumacher's collection of essays, written in the 1950s and 1960s and published in 1973 under the title *Small Is Beautiful: Economics as If People Mattered*, became part of the shared consciousness of the 1970s. The "Buddhist Economics" essay was rewritten for this publication.

Opposing small to big was a recurrent theme of the catalog. In simple terms, Schumacher provided convincing arguments for replacing industrial production with hand labor. In fact, Schumacher's argument was extremely radical, substituting the emphasis on consumption with a value-based ideology, founded on satisfaction in production. —CM

1. The version in *Small Is Beautiful* is different. See http://www.centerforneweconomics.org/content/lindisfarne-tapes.

Source
E. F. Schumacher, "Buddhist Economics," *Resurgence* 1, no. 11 (January 1968). First published in *Asia: A Handbook*, ed. Guy Wint (London: Anthony Blond, 1966). (Referred to in *LWEC*, 153.)

Further Reading
Schumacher, E. F. *Small Is Beautiful: Economics as If People Mattered; 25 Years Later; With Commentaries*. Point Roberts, WA: Hartley & Marks, 1999.

"Buddhist Economics"

The Buddhist point of view takes the function of work to be at least threefold: to give a man a chance to utilize and develop his faculties; to enable him to overcome his ego-centeredness by joining with other people in a common task; and to bring forth the goods and services needed for a becoming existence. Again, the consequences that flow from this view are endless. To organize work in such a manner that it becomes meaningless, boring, stultifying, or nerve-racking for the worker would be little short of criminal; it would indicate a greater concern with goods than with people, an evil lack of compassion and a soul-destroying degree of attachment to the most primitive side of this worldly existence. Equally, to strive for leisure as an alternative to work would be considered a complete misunderstanding of one of the basic truths of human existence, namely that work and leisure are complementary parts of the same living process and cannot be separated without destroying the joy of work and the bliss of leisure.

From the Buddhist point of view, there are therefore two types of mechanization which must be clearly distinguished: one that enhances a man's skill and power and one that turns work over to a mechanical slave, leaving man in a position of having to serve the slave. How to tell the one from the other? The craftsman himself, says Ananda [a historian and philosopher of Indian art], a man equally competent to talk about the Modern West as the Ancient East, "the craftsman himself can always, if allowed to, draw the delicate distinction between the machine and the tool. The carpet loom is a tool, a contrivance for holding warp threads at a stretch for the pile to be woven round them by the craftsmen's fingers; but the power loom is a machine, and its significance as a destroyer of culture lies in the fact that it does the essentially human part of the work." It is clear, therefore, that Buddhist economics must be very different from the economics of modern materialism, since the Buddhist sees the essence of civilization not in a multiplication of wants but in the purification of human character. Character, at the same time, is formed primarily by a man's work. And work properly conducted in conditions of human dignity and freedom, blesses those who do it and equally their products. The Indian philosopher and economist J. C. Kumarappa sums the matter up as follows: "If the nature of the work is properly appreciated and applied, it will stand in the same relation to the higher faculties as food is to the physical body. It nourishes and enlivens the higher man and urges him to produce the best course and disciplines the animal in him into progressive channels. It furnishes an excellent background for man to display his scale of values and develop his personality."

If a man has no chance of obtaining work he is in a desperate position, not simply because he lacks an income but because he lacks this nourishing and enlivening factor of disciplined work which nothing can replace. A modern economist may engage in highly sophisticated calculations on whether full employment "pays" or whether it might be more "economic" to run an economy at less than full employment so as to ensure a greater mobility of labor, a better stability of wages, and so forth. His fundamental criterion of success is simply the total quantity of goods produced during a given period of time. "If the marginal urgency of goods is low," says Professor Galbraith in *The Affluent Society*, "then so is the urgency of employing the last man or the last million men in the labor force." And again: "If ... we can afford some unemployment in the interest of stability—a proposition, incidentally, of impeccably conservative antecedents—then we can afford to give those who are unemployed the goods that enable them to sustain their accustomed standard of living."

From a Buddhist point of view, this is standing the truth on its head by considering goods as more important than people and consumption as more important than creative activity. It means shifting the emphasis from the worker to the product of work, that is, from the human to the subhuman, a surrender to the forces of evil.

While the materialist is mainly interested in goods, the Buddhist is "The Middle Way" and therefore in no way antagonist to physical well-being. It is not wealth that stands in the way of liberation but the attachment to wealth; not the enjoyment of pleasurable things but the craving for them. The keynote of Buddhist economics, therefore, is simplicity and non-violence. From an economist's point of view, the marvel of the Buddhist way of life is the utter rationality of its pattern—amazingly small means leading to extraordinarily satisfactory results.

41 Elsie G. Davenport, *Your Handspinning*, 1964

According to Stewart Brand, *Your Handspinning*, by Elsie G. Davenport, was "THE book on handspinning"; in his review, he detailed the contents of each chapter, demonstrating the book's informational breadth. With line drawings, photographs, and precise instructions, Davenport's text contained a great deal of information for novice handspinners, from shearing a sheep to constructing a spinning wheel to preparing spun yarn. Countercultural craftsmen and women, especially those motivated to return to a less mechanized way of life, embraced the book for its educational value. —MG

Source
Elsie G. Davenport, *Your Handspinning* (Big Sur, CA: Craft & Hobby Book Service, 1964), 6, 35. (Referred to in *LWEC*, 166.)

Further Reading
Auther, Elissa. "Fiber Art, the Craft Revival of the 1960s and 1970s, and Popular Craft." In *String, Felt, Thread: The Hierarchy of Art and Craft in American Art*, 25–27. Minneapolis: University of Minnesota Press, 2010.

Fannin, Allen. *Handspinning: Art and Technique*. New York: Van Nostrand Reinhold, 1970.

PLATE I

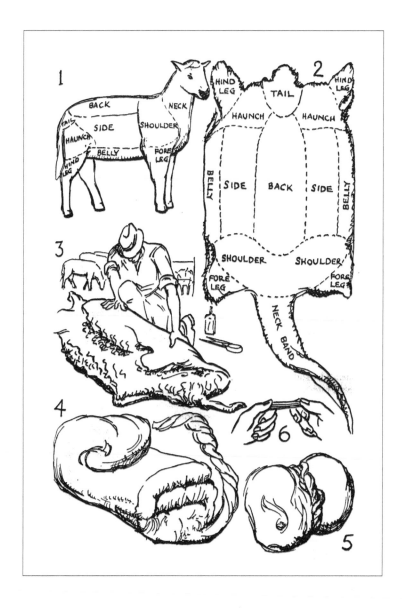

Figure 5.6 (left)
"Plate 1. Some typical wools. Shoulder locks from pedigree sheep." Elsie G. Davenport, *Your Handspinning* (Big Sur, CA: Craft & Hobby Book Service, 1964), 6.

Figure 5.7 (right)
"Opening Out." Elsie G. Davenport, *Your Handspinning* (Big Sur, CA: Craft & Hobby Book Service, 1964), 35.

42 Sharon Rosenberg and Joan Wiener, *The Illustrated Hassle-Free Make Your Own Clothes Book*, 1971

"Relax and have a good time" (2) were the words with which Sharon Rosenberg and Joan Wiener concluded the introduction to *The Illustrated Hassle-Free Make Your Own Clothes Book,* **a guide for readers interested in mending, remodeling, and hand-making their own clothing. Following this advice, Rosenberg and Wiener proceeded to write, as the passages that follow suggest, in a conversational manner, with anecdotes and casual yet informational instructions, and accompanied their text with playful hand-drawn sketches. Within this mellow prose, however, existed a broader commentary on the environmental, economic, and social effects of fashion consumption.**

Diana Shugart, in her catalog review, focused on the subversiveness of the work. "Revolution," she asserted, "takes all sorts of forms these days, & making your own clothes is one way to play fiddle-fuck-around with fashion— sexism, consumption-pushing, & related evils. Not to mention such personal benefits as being comfortable, feeling good, & looking however you want to." *The Illustrated Hassle-Free Make Your Own Clothes Book* **thus served a dual purpose in the "Craft" section: as both a practical manual for clothes making and a philosophical reminder of the significance of the handmade in fulfilling one of the most basic human needs.—MG**

Source
Sharon Rosenberg and Joan Wiener, *The Illustrated Hassle-Free Make Your Own Clothes Book* (San Francisco: Straight Arrow Books, 1971), 19, 24, 30. Reprinted with permission of Skyhorse Publishing. (Referred to in *LWEC*, 172.)

Further Reading
Duncan, Ida Riley. *The Complete Book of Progressive Knitting.* New York: Liveright, 1966.

Jacopetti, Alexandra, and Jerry Wainwright. *Native Funk and Flash: An Emerging Folk Art.* San Francisco: Scrimshaw Press, 1974.

Selections from *The Illustrated Hassle-Free Make Your Own Clothes Book*

Embroidery

Embroidery is so easy, you can probably do it with just an idea of what stitches are available and with a few pictures as hints. When something you love is ripped beyond repair, a big butterfly will generally save it. When you dribble spaghetti sauce on your nice white Indian shirt, a big yellow embroidered sun will make it wearable. Embroidering on your dull old sheets and pillows will help you sleep better. Embroidering on pieces of material in a smart fashion will turn them into wall hangings. There are about 20 million things you can use embroidery for, some of which are illustrated here.

Recycling Scraps and Waste Materials

At some time or other, everyone has had an indispensable pair of pants—jeans or chinos. And just like Roy Rogers when Trigger died, they are loath to part with them until the irreversible moment when these pants rip at the knee. Rather than stuffing them—as was Trigger's alleged fate—the pants can be turned into cut-offs by a mere snipping off at the knee. This, then, is re-cycling in its most common incarnation. Just as organic wastes are re-cycled, clothing scraps can be used again, old clothes can be magically transformed by a bit of imaginative remodeling, rips can be repaired and all sorts of things can be re-made into perfectly pleasurable, wearable garments without having to be thrown out.

Have a special place where you can save all fabric scraps of substantial size. They're good to use for the following things:

Bags	Pot Holders
Belt Loops	Purses
Bikinis	Ruffles
Bows	Scarves
Collars	Sleeves

Rips and What to Do about Them

You don't have to throw something out just because it's ripped. Once I threw out an old pair of tights, a *shmata* with a horrifying gash and Sharon said, "Ork. What are you doing?"

"I am throwing this rag away."

"It's still wearable."

Then I take it out and dutifully embroider over the rip and have something I can really dig.

Figure 5.8
"Embroidery." Sharon Rosenberg and Joan Wiener, *The Illustrated Hassle-Free Make Your Own Clothes Book* (San Francisco: Straight Arrow Books, 1971), 20. Reprinted with permission of Skyhorse Publishing.

Community

The "Community" section of the *Whole Earth Catalog* was one of its most diverse; it included subjects that directly addressed countercultural communities as such but also acknowledged the complexity of defining this idea by incorporating an assortment of other topics, from the mundane to the radical. "Community" included pages on Japanese communes and cooking, dogs and emergency medicine, "bargain living" and justice; the novels *Dune* and *Atlas Shrugged* appeared alongside an entire page devoted to grinders and juicers. And yet, in spite of the section's topical breadth, conceptual absences remained, particularly with regard to race and gender issues. To demonstrate the full range of the section, both its successes and its failures, a wide-ranging selection of excerpts and images follows, beginning with Paul and Percival Goodman's seminal text *Communitas* and concluding with James Baldwin's and Margaret Mead's thought-provoking conversation *A Rap on Race*.

"Community" established its essential theme in the first pages: *Communitas*, written well before the catalog's inception, nevertheless imagined a societal structure aligned with the countercultural values of freedom, creativity, and equality; and in texts such as *Drop City*, *Living on the Earth*, and the *Journal of the New Alchemists*, the catalog presented contemporary projects that sought to realize such aspirations. Analogously, Gary Snyder's essay "Why Tribe" from his book *Earth House Hold*

considered the historical lineage of communal living and the rationale behind contemporary "tribes," offering context to the previous subject sections.

Beyond *Earth House Hold*, however, the catalog began to interpret the meaning of community more creatively, deconstructing the conceptual framework to address more specific, substantive aspects of the term. An instance of this shift emerged from the four-page "Cooking" subsection; Ita Jones's *The Grubbag* and Michel Abehsera's *Zen Macrobiotic Cooking* signified a narrower, yet equally relevant, approach to community life by way of food and nutritional philosophies. Alternatively, Saul Alinsky's *Rules for Radicals* and the OM Collective's *Organizer's Manual* discussed the ideological and concrete methods of organizing communities at a grassroots level to initiate locally oriented change. And although the relative paucity of material on race and women's issues is problematic, particularly given the prominence of the Black Power and women's liberation movements during this period, the inclusion of *Women and Their Bodies* and *A Rap on Race* suggests an attempt to represent gendered and racial minority experiences in the catalog.

That the editors selected two excellent texts—edifying lessons in feminism and racial politics, respectively—indicates an awareness of these timely issues, yet their failure to directly advocate for meaningful change, coupled with polarizing book reviews and essays, may have overshadowed any positive intentions from the perspective of the catalog's minority readership. Brand's opinion of race as a "deadly subject" and Diana Shugart's description of gendered domestic work as "luxurious" demonstrated the uninformed, ambivalent, or even dismissive stance of the catalog toward the problems faced by African Americans and women at the time, and this subjective inequity was perhaps thrown into further relief against the generally diverse texture of the section as a whole.

The expansiveness of the "Community" section, as well as its disparities, provides an intriguing portrait of how the catalog's countercultural readership may have understood "community," from the broad topics of societal structure and communal living, to everyday tasks such as cooking and bargain shopping, to urgent political issues.—MG

43 Percival and Paul Goodman, *Communitas*, 1947

Brand selected this book for its insight into how communities are formed and how they function. Percival Goodman (1904–1989) was an American architect and urbanist trained at the École des Beaux-Arts in Paris. Professor at Columbia University for twenty-five years, he also had a successful practice. His brother, Paul Goodman (1911–1972), was a social theorist. In *Communitas*, the Goodmans considered a range of urban-planning theories from Patrick Geddes to Le Corbusier. They used cartoons imaginatively and employed a lively and controversial writing style. They analyzed different forms of social organization, comparing, for example, one based on consumption, another oriented toward creative work, and a third intended to provide the greatest possible measure of human liberty. The ideal was a system where individuals contributed the skills and products required by the community without needing to earn substantial sums through wage labor. Young adults would be required to contribute two years of conscripted work to the community. This utopian project was exemplified in the extract selected by Brand, which discusses education as a model of social leadership. The Goodmans saw a progressive educational system as a potentially revolutionary project. In their introduction, they reflected on the problem of making planning decisions within a surplus economy where understanding the choices had become too difficult for most people to grasp. The witty cartoon illustrated here is the frontispiece of *Communitas*. Three calligrams representing three social ideals contain the names of eminent architects and thinkers. Whereas Le Corbusier and skyscrapers are dismissed as part of the capitalist enterprise, Frank Lloyd Wright and Robert Owen stand for a more planned form of organization, and, on the right, Gandhi and Tolstoi suggest a world free of administrative control. The cartoon, like the three models, was intended to provoke debate. — CM

Source
Percival and Paul Goodman, *Communitas: Means of Livelihood and Ways of Life* (Chicago: University of Chicago Press, 1947; 2nd ed., rev. 1960), 10–14. © 2015 Columbia University Press. Reprinted with permission of Columbia University Press. (Referred to in *LWEC*, 177.)

Further Reading
Goodman, Paul. *Growing Up Absurd: Problems of Youth in the Organized Society*. New York: Random House, 1960.

Selections from *Communitas*

Introduction

Any community plan involves a formidable choice and fixing of living standards and attitudes, of schedule, of personal and cultural tone. Generally people move in the existing plan unconsciously, as if it were nature (and they will continue to do so, until suddenly the automobiles don't move at all). But let a new proposal be made and it is astonishing how people rally to the old arrangement. Even a powerful park commissioner found the housewives and their perambulators blocking his way when he tried to rent out a bit of the green as a parking lot for a private restaurant he favored; and wild painters and cat-keeping spinsters united to keep him from forcing a driveway through lovely Washington Square. These many years now since 1945, the citizens of New York City have refused to say "Avenue of the Americas" when they plainly mean Sixth Avenue.

The trouble with this good instinct—not to be regimented in one's intimate affairs by architects, engineers, and international public-relations experts—is that "no plan" always means in fact some inherited and frequently bad plan. For our cities are far from nature, that has a most excellent plan, and the "unplanned" tends to mean a gridiron laid out for speculation a century ago, or a dilapidated downtown when the actual downtown has moved uptown. People are right to be conservative, but what is conservative? In planning, as elsewhere in our society, we can observe the paradox that the wildest anarchists are generally affirming the most ancient values, of space, sun, and trees, and beauty, human dignity, and forthright means, as if they lived in Neolithic times or the Middle Ages, whereas the so-called conservatives are generally arguing for policies and prejudices that date back only four administrations.

The best defense against planning—and people do need a defense against planners—is to become informed about the plan that is indeed existent and operating in our lives; and to learn to take the initiative in proposing or supporting reasoned changes. Such action is not only a defense but good in itself, for to make positive decisions for one's community, rather than being regimented by others' decisions, is one of the noble acts of man.

Technology of Choice and Economy of Abundance

The most curious anomaly, however, is that modern technology baffles people and makes them timid of innovations in community planning. It is an anomaly because for the first time in history we have, spectacularly in the United States, a surplus technology, a technology of free choice, that allows for the most widely various community-arrangements and ways of life.... And with this technology of choice, we have an economy of abundance, a standard of living that is in many ways *too* high—goods and money that are literally thrown away or given away—that could underwrite sweeping reforms and pilot experiments....

Think about a scarcity economy and a technology of necessity.... For almost every item that men have invented or nature has bestowed, there are alternative choices. What used to be made of steel (iron ore and coal) may now often be made of aluminum (bauxite and waterpower) or even of plastic (soybeans and sunlight). Raw materials have proliferated, sources of power have become more ubiquitous, and there are more means of transportation and lighter loads to carry. With the machine-analysis of manufacture, the tasks of labor become simpler, and as the machines have become automatic our problem has become, astoundingly, not where to get labor but how

to use leisure. Skill is no longer the arduously learned craftsmanship of hundreds of trades and crafts—for its chief habits (styling, accuracy, speed) are built into the machine; skill has come to mean skill in a few operations, like turning, grinding, stamping, welding, spraying and half a dozen others, that intelligent people can learn in a short time. Even inspection is progressively mechanized. The old craft-operations of building could be revolutionized overnight if there were worthwhile enterprises to warrant the change, that is, if there were a social impetus and enthusiasm to build what everybody agrees is useful and necessary.

Consider what this means for community planning on any scale. We could centralize or decentralize, concentrate population or scatter it. If we want to continue the trend away from the country, we can do that; but if we want to combine town and country values in an agrindustrial way of life, we can do that. In large areas of our operation, we could go back to old-fashioned domestic industry with perhaps even a gain in efficiency, for small power is everywhere available, small machines are cheap and ingenious, and there are easy means to collect machined parts and centrally assemble them. If we want to lay our emphasis on providing still more mass-produced goods, and raising the standard of living still higher, we can do that; or if we want to increase leisure and the artistic culture of the individual, we can do that. We can have solar machines for hermits in the desert like Aldous Huxley or central heating provided for millions by New York Steam. All this is commonplace; everybody knows it....

Technology is a sacred cow left strictly to (unknown) experts, as if the form of the industrial machine did not profoundly affect every person; and people are remarkably superstitious about it....

Indeed, they are outraged by the good-humored demonstrations of Borsodi that, in hours and minutes of labor, it is probably cheaper to grow and can your own tomatoes than to buy them at the supermarket, not to speak of the quality. Here once again we have the inevitable irony of history: industry, invention, scientific method have opened new opportunities, but just at the moment of opportunity, people have become ignorant by specialization and superstitious of science and technology, so that they no longer know what they want, nor do they dare to command it. The facts are exactly like the world of Kafka: a person has every kind of electrical appliance in his home, but he is balked, cold-fed, and even plunged into darkness because he no longer knows how to fix a faulty connection.

Certainly this abdication of practical competence is one important reason for the absurdity of the American Standard of Living. Where the user understands nothing and cannot evaluate his tools, you can sell him anything.

Figure 6.1
"Bibliography for Three Ways of Life Today." Percival and Paul Goodman, *Communitas: Means of Livelihood and Ways of Life* (Chicago: University of Chicago Press, 1947), frontispiece. © 2015 Columbia University Press. Reprinted with permission of the publisher.

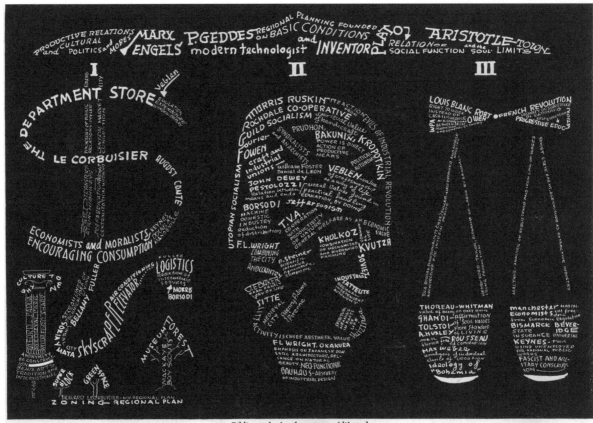

Bibliography for three ways of life today

44 Peter Rabbit, *Drop City*, 1971

Peter Rabbit, the pseudonym of Peter Douthit (1936–2012), was a close observer of the rise and decline of Drop City, near Trinidad, Colorado. In its heyday, from 1965 to 1967, Drop City, a vibrant laboratory for an alternative lifestyle, centered on artistic and theatrical performance, music, free love, and drugs. Drop City was a victim of its own sensational publicity, as hundreds of curious dropouts came to share the experience. Peter Rabbit thoughtfully analyzed the causes and decline, which he saw as essentially philosophical. After few months there, he moved in 1966 from Drop City to Libre (see text 22), near Gardner, Colorado, a community that is still active today. Brand commented on the extract: "Peter can write fine. Drop City was the first of [the] current crop of weed patch communes. The first dome-funk display in the U.S. The first nose-tickle of America's country sneeze. How to risk your life repeatedly and keep laughing. It's a good story." The selected extract emphasizes the use of recycled garbage: "An incredible number of people can live off the garbage heap of the U.S. Drop City is built on the garbage dump of a dying town of ten thousand strung-out coalminers." The selection also gives Rabbit's account of how Gene and JoAnn Bernofsky (Curley Bensen and Drop Lady) and Clark Richert (Clard Svensen) founded Drop City and focuses on the inspirational role of Steve Baer (called Luke Cool) and the excitement of constructing the dome.—CM

Source

Peter Rabbit, *Drop City* (New York: Olympia Press, 1971), 26–33. (Referred to in *LWEC*, 178.)

Further Reading

Sadler, Simon. "Drop City Revisited." *Journal of Architectural Education* 59, no. 3 (2006): 5–14.

Voyd, Bill. "Funk Architecture." In *Shelter and Society*, ed. Paul Oliver, 156–158. New York: F. A. Praeger, 1969.

"Car Tops," from *Drop City*

Drop City is a tribe-family, 20 people living together in domes, man-woman-child, living, determined to live.... Drop City is a fucked-up mess, Drop City is completely open, completely free; I own it, you own it, because we know that all energy comes from the same place. Ten domes under the skydome, overshadowed by the Rockies; silverdomes, domes that are paintings, multicolored cartopdomes and one black dome.

Droppers live out of the garbage dump. We know about garbage. The people of Trinidad call us Dump City—they know where it's at. They call Droppers when they want garbage picked up. It delights them to watch us rummaging around in the schmutz; it delights us too. An incredible number of people can live off the garbage heap of the U.S. Drop City is built on the garbage dump of a dying town of ten thousand strung-out coalminers. Think what can be done with L.A.—N.Y.—San Francisco—Des Moines—Dallas! The

most beautiful place in Drop City is the dump, the garbage dump of the garbage pickers. It's our greatest resource.

Droppers make things—paintings, films, novels, poems, domes, solar cookers and heaters, chairs, turkey cages, mandalas, junkyards, droppings, noise. We work at our work out of joy, from our energy centers, ramming it all through to lovemaking life. We feel few compulsions, we make things, we are dedicated to not being employed but rather being employed truly by ourselves in the making of things. We have found freedom in action, freedom in the making of things. Creation has as its source the fullness of joy. We want freedom for our bodies, we want health, we want freedom for our social selves, we want goodness, we want freedom for our total selves, we want love, love is joy, joy is love, we are all one.

Droppers have learned how to build beautiful houses out of cartops for less than 200 dollars, less than 100 dollars. We know how to use solar heating; we're hip to windpower. Droppers know how to best use the government doles and poverty programs. Each dollar we use is one less that goes into the making of napalm. Every cent that Drop City uses is one less that goes to those insane retarded creeps in Washington.

Luke Cool [Steve Baer] taught Droppers to design domes, build solar heaters, make dwellings in which a family can live in comfort and beauty. Luke Cool is the king of the junkheap, teaching Droppers and the world how to USE things. Jump up on top of the car with an axe and chop 'um out, all around the edges just like a can opener, stomp, pop out the back glass, use it for a window, slip off the mirrors, use them for a solar cooker, pull out the insulation, use it, USE IT; make a honeycomb sandwich out of beer cans and plywood—fantastic strength—and USE IT....

Domes use materials in the most reasonable efficient way. Bucky Fuller gave Drop City the 1966 Dymaxion Award for poetically economic structural accomplishments. Soon domed cities will spread across the world, anywhere land is cheap—on the deserts, in the swamps, on mountains, tundras, ice caps. The tribes are moving, building completely free and open way-stations, each a warm and beautiful conscious environment. We are winning.

Droppers make movies—black and white snow—wind poems, flickering TV beauties with all the subliminal delights of pulsing coke ads. The crystal-molecular good sense of a dome going up, the grunting goodness of eating and sex. We want videotape recorders and cameras and strobes—hundreds of them—and tape decks and amps and echo chambers and everything. We want millions of green energy flakes; we want to use everything, good junk and bad, to make limitless things; we want an atomic reactor.

Droppers absolutely put down power; Droppers absolutely put down money, hoarded energy; we absolutely put down ownership—everything we have is yours, everything you have is ours. GET RID OF YOUR HANGUPS, HELP US GET RID OF OUR HANGUPS.

There is no political structure in Drop City. Things work out; the cosmic forces mesh with people in a strange complex intuitive interaction. Sometimes it's slow, but when it comes out it's good and nobody's head is bent. Only three things have ever come to a vote: in all three cases Droppers went around with their heads bent for days. When things are done the slow intuitive way the tribe makes sense. The tribe is human, the tribe is ready to die for love and sanity.

Droppers are medicine men. I go to the mountain, he meditates, she fucks, he fasts, she eats, he prays, she serves, he chants, she has terrifying dreams. Droppers are looking for the truth. The truth is the truth.

Drop City is firmly established. We are alive. The cosmic forces have helped us keep going for five years, making a plea for sanity, trying one thing after another. FOR GOD'S SWEET SAKE BE SANE AND GENTLE AND LOVING AND STRAIGHT AND GOOD. No two Droppers think the same way about anything. One kicks the goat in the ribs to get him out of the garden, another coaxes him and feeds him flowers. It's a strange place, an incredible webbing of circumstances and chance, planning and accidents, thumbmashes and cartopchops. We are not responsible for who, what and where we are; we have only taken our places in space and time and light. We are straight people who want love, food, warmth and the chance to raise our families with a degree of integrity. We have no integrity: we borrow, copy, steal any and all ideas and things, we use things, we are responsible, we are not responsible, we take things, we give things. We are elements of the essential paradox....

Luke Cool is an authentic genius of our time. He has more energy than anyone I know and he uses it well, creating, writing, building, inventing. His being flashes with brilliance almost constantly.... The first structure he put up at Drop City was the Cartop Dome, a dome about 27' by 14' made entirely of cartops. It was an incredible material breakthrough for the Droppers. The entire Cartop Dome cost $15, mostly for sheet metal screws.

Luke had all the panels ready to be put together when he arrived at Drop City. He had cut the tops from junk cars with an axe. If you cut carefully around the edge, you can get a 3 1/2' by 7' sheet of 27-gauge steel with a baked-on enamel surface from each car, 3 1/2' by 9' from station wagons. Droppers loved station wagons.

After the tops were chopped out, Luke put a template of a given module on the top and outlined it with chalk. There was a 1" lip on each side of each panel. The module was then cut out with heavy-duty electric sheet metal shears. It takes 10 to 15 minutes to cut out each panel. After the modules are out, the lips are bent over at 90° with a sheet metal break, holes are drilled in the bent-over lips and the panels are fastened together with sheet metal screws. Fast, efficient, cheap. We also used the insulation from the cars. The inside skin was made from old paintings of Clard and Lard and the Wop....

In any feed store in the country you can get corn chop, soy meal, whole wheat, wheat shorts, dry molasses, pinto beans; good stuff for 5¢ a pound or less, to make your body super-healthy. In any junkyard in the country you can get cartops from 15¢ to 40¢ each. Sometimes the junkmen even give them to you free just to watch the crazy hippies bust their asses choppin' 'em out. You can build yourself the farthest-out environment you can imagine livin' in the USA.

45 Alicia Bay Laurel, *Living on the Earth*, 1970

This book, handwritten with delightful illustrations, is a guide to how to fend for yourself in a commune. It became a classic, selling 350,000 copies. Alicia Kaufman was born in Hollywood, California, in 1949, and after a stay on a houseboat at Sausalito in 1967, she joined the Hog Farm commune at Wheeler Ranch, where she adopted the name Alicia Bay Laurel. She wrote and illustrated *Living on the Earth* in 1971, at the age of nineteen.

Bay Laurel brought together what she had learned about how to survive in a commune, including the essential skills of sewing, making clothes, and growing and cooking food. She also included tips on building shelters and heating them. The book was originally published by Bookworks in 1970 and was eventually republished by Random House as a Vintage paperback, with a first printing of 100,000 copies. It even made it onto the *New York Times* bestseller list in 1971. The affection the book generated among its readers was captured by the *LWEC* reviewer: "This may well be the best book in this catalog. This is a book for people, so, if you are a person, it is for you. If you are a dog, for instance, and you can't read very well, it just might be for you too, because of the drawings" (JD, *LWEC*, 178). The reviewer also mentioned the vein of "how to" authors, proponents of back-to-the-earth living such as Bradford Angier (1910–1997), a wilderness survivalist who wrote more than thirty-five books on how to survive in the wild and live in a minimalistic way off the land. —CM

Source
Alicia Bay Laurel, *Living on the Earth: Celebrations, Storm Warnings, Formulas, Recipes, Rumors, and Country Dances* (New York: Vintage Books, 1970), 93. (Referred to in *LWEC*, 178.)

Further Reading
Binkley, Sam. *Getting Loose: Lifestyle Consumption in the 1970s*. Durham, NC: Duke University Press, 2007.

the book of tao says that every day the Scholar must know more & more, but the follower of tao must know less & 'less. Eventually I must say 'no' to this unceasing tide of information. This book is already too thick. But, if the tide bends me again, this book will have a sequel. Besides, it was fun drawing all these pictures. ...A.B.L. OG Dawn

193

winter heat: use 2 55 gallon drums. The lower one with door & grate is for the fire. The upper one simply holds the heat (has a damper to shut the pipe above). where the pipe goes through the roof, use a metal collar to protect from fire. elevate stove legs on bricks.

34

Figure 6.2 (left)
"The Book of Tao Says That Every Day the Scholar Must Know More and More…" Alicia Bay Laurel, *Living on the Earth: Celebrations, Storm Warnings, Formulas, Recipes, Rumors, and Country Dances* (New York: Vintage Books, 1970), 93. © 2015 by Alicia Bay Laurel. Reprinted with permission.

Figure 6.3 (right)
"Winter Heat." Alicia Bay Laurel, *Living on the Earth: Celebrations, Storm Warnings, Formulas, Recipes, Rumors, and Country Dances* (New York: Vintage Books, 1970), 94. © 2015 by Alicia Bay Laurel. Reprinted with permission.

46 John Todd, "A Modest Proposal," 1970

In 1969 the environmental scientist John Todd, aquaculturist William McLarney, and writer Nancy Jack Todd cofounded the New Alchemy Institute (NAI) to research the biology of self-sustaining food production systems. Their aim was to demonstrate how a small farm could be made almost self-sufficient without damaging the environment, at Woods Hole in Massachusetts. As with many other groups in the alternative culture of the 1960s and 1970s, the NAI was created as a response to the damage caused to the land, the food chain, and people's lives by the use of chemicals in agriculture, and the consequences of industrialization and patterns of consumption leading to monocultures and the depletion of the soil. The institute's founders believed that it was essential to bridge the gap between science in the laboratory and practice in the field; they wanted to provide an "alternative" lab, where biological research—to restore the soil, for example, or develop aquaculture or build an efficient greenhouse—could be carried out far from university labs. They emphasized small-scale, decentralized organization and ecological modeling, an attitude that contrasted with the dominant cultural scientific and technological practices in the industrial world.

Unlike many other groups, the institute outlived the countercultural period and created a number of satellites, some of which are still active. The *Journal of the New Alchemists*, published in seven issues from 1973 to 1981, was beautifully produced with full-page photographs of groups of children and their vegetable gardens, giving a vivid account of the diversity of the experiments conducted. The *WEC* review mentioned only a small booklet, "proposing to integrate solar heating, aquaculture, and agriculture in projects in New Mexico and the tropics."—CM

Source
John Todd, "A Modest Proposal," 1970, in *The Book of the New Alchemists*, ed. Nancy Jack Todd (New York: E. P. Dutton, 1977), xiii–xviii. Courtesy of John Todd. (Referred to in *LWEC*, 180.)

Further Reading
Todd, John. "Designing a New Science." *Organic Gardening and Farming*, October 1971, 83–84.

Todd, John. "Financial Status of the Institute." *Journal of the New Alchemists* 1 (1973): 5.

Todd, Nancy Jack. *The Book of the New Alchemists*. New York, NY: Dutton, 1977.

Todd, Nancy Jack, and John Todd. *Bioshelters, Ocean Arks, City Farming: Ecology as the Basis of Design*. San Francisco: Sierra Club Books, 1984.

"A Modest Proposal"

If the present trend continues, the world community will be shaped into a series of highly planned megalopolises, which are regulated by an advanced technology and fed by a mechanized and chemically sanitized agriculture. This future course is countered largely by the tenacity of many peoples throughout the world, including some of the Indians of North America, marginal and peasant farmers, traditional craftsmen, and many of the young seeking alternatives to the modern industrial state. ...

It is my contention that we are in danger of losing an important amount of social variability in the human community at the same time as we are losing the required amount of biological variability in our life-support bases. If we continue on our present path at the present rate, then our chances of maintaining healthy communities and environments will be dramatically reduced before the year 2000, perhaps beyond a point where society as we know it will be incapable of functioning. ...

A few years ago the New Alchemists began their search for ways in which science and the individual could come to the aid of humanity and the stressed planet. ... It was clear from the outset that social and biological diversity needed to be protected and, if at all possible, extended.

We felt that a plan for the future should create alternatives and help counter the trend toward uniformity. It should provide immediately applicable solutions for small farmers, homesteaders, native peoples everywhere, and the young seeking ecologically sane lives, enabling them to extend their uniqueness and vitality. Our ideas could also have a beneficial impact on a wider scale if some of the concepts were incorporated into society-at-large. Perhaps they could save millions of lives during crisis periods in the highly developed states. This modest and very tentative proposal suggests a direction that society might well consider.

At the foundation of the proposal is the creation of a biotechnology, which by its very nature would:

function most effectively at the lowest levels of society

be comprehensible and utilizable by the poorest of peoples

be based upon ecological rather than economic or efficiency considerations, although this would not negate the development of local eco-economies

permit the evolution of small decentralist communities, which in turn might act as beacons for a wiser future for many of the world's population

be created at local levels and require relatively small amounts of financial support (this would enable poorer regions or nations to embark upon the creation of indigenous biotechnologies.)

Unnatural Selection: Loss of Diversity

It is necessary before describing a way of reviving diversity at all levels to evaluate how its loss threatens our future....

For hundreds of years prior to the industrial revolution a wide variety of energy resources was used. Besides animal power and human toil there was a subtle integration of resources such as the wind and water power and a variety of fuels including peat, wood, coal, dung, vegetable starches, and animal fats. This approach of integration through diversity in providing the energy for society has been replaced by an almost exclusive reliance on fossil fuels and nuclear power....Modern society, by reducing the variety of its basic energy sources while increasing its per capita energy needs, is now vulnerable to disruption on an unprecedented scale....

A depression of the magnitude of the one that befell the country in 1929 could well take place, but if one should happen in the 1970s the social consequences would be much more severe. In 1929, a large percentage of Americans had friends or relatives on farms that could operate on a self-sufficient basis during lean periods. Today the situation is alarmingly different as the rural buffer is largely gone and far fewer people have access to the land....

The Green Revolution: Unnatural Selection

Over the past several decades the agricultural sciences have created a number of major advances in food raising and the widely acclaimed green revolution has come to symbolize the power of applied science and technology working on behalf of humanity. Our confidence has been renewed that the mushrooming populations can be fed if only western agriculture can be spread rapidly enough throughout the world....

The modernization of agriculture has resulted in the large scale use of chemical fertilizers upon which many of the new high-yielding strains of grains depend. Coupled with this is a basic emphasis on single cash crops, which are grown on increasingly large tracts of land. The dependency on fertilizers for successful crops has created depressed soil faunas and an alarming increase in nitrates in the ground waters in some areas....

The trend away from cultivating local varieties to a few higher yielding forms is placing much of the world's population out on a limb. If the new varieties are attacked by pathogens the consequences could be world-wide, rather than local, and plant breeders may not be able to create new strains before it is too late. Such events are not without precedent. An earlier counterpart of the green revolution occurred in Ireland in the eighteenth century with the introduction of the Irish potato from the western hemisphere. Production of food dramatically increased and by 1835 a population explosion had taken place as a result of the land's increased carrying capacity. During the 1840s a new fungal plant disease appeared, destroying several potato crops, and one-quarter of the Irish people died of starvation....A modern agriculture, racing one step ahead of the apocalypse, is not ecologically sane, no matter how productive, efficient, or economically sound it may seem.

47 Edward T. Hall, *The Hidden Dimension*, 1966

In his review of *The Hidden Dimension*, Stewart Brand wrote: "'Don't try to reform,' says Fuller, 'reform the environment.' This book tells you to do it. Here is fascinating information on how we relate to space and each other through space. If you know how you want to feel and act, you can design a place to ensure that, or, equally, to make it impossible" (*LWEC*, 182).

Like many of the authors whom Brand admired, the anthropologist Edward T. Hall drew on a range of disciplines in his work—anthropology, psychology, biology, ethology—and went on to make practical use of this knowledge. He explained the object of his research as "the structure of experience as it is molded by culture, these deep, common, unstated experiences which members of a given culture share, which they communicate without knowing, and which form the backdrops against which all other events are judged" (*The Hidden Dimension*, x).

Calling on scientific observations of animal behavior and anthropological studies, Hall attempted to develop the science of proxemics, or the study of spatial awareness. He asked what effect close proximity had on people in cities, how people perceived the space around them, and how architects and designers could use this knowledge to improve the living environment. He noted, for example, how chimpanzees are constantly looking around to check their position in the group. The book is organized into chapters with titles such as "Culture as Communication," "Distance Regulation in Animals," "Crowding and Social Behavior in Animals," "Perception of Space: Distance Receptors—Eyes, Ears, and Nose," "Perception of Space: Skin and Muscles," "Visual Space," and so on. The urban theorist Françoise Choay chose *The Hidden Dimension* as one of the essential American texts to translate in 1971 for the French readership. The book has become a standard text in the study of personal space, and Hall's findings have been applied in office management and psychological studies.—CM

Source
Edward T. Hall, *The Hidden Dimension* (Garden City, NY: Doubleday, 1966), 50–55. Excerpts from *The Hidden Dimension*, by Edward T. Hall, copyright © 1966, 1982, by Edward T. Hall. Used by permission of Doubleday, an imprint of the Knopf Doubleday Publishing Group, a division of Penguin Random House LLC. All rights reserved. (Referred to in *LWEC*, 182.)

Further Reading
Hall, Edward T. *The Hidden Dimension*. New York: Anchor Books, 1990.

Selections from *The Hidden Dimension*

Hidden Zones in American Offices

...Kinesthetic space is an important factor in day-to-day living in the buildings that architects and designers create. Consider for a moment American hotels. I find most hotel rooms too small because I can't move around in them without bumping into things. If Americans are asked to compare two identical rooms, the one that permits the greater variety of free movement will usually be experienced as larger. There is certainly great need for improvement in the layout of our interior spaces, so that people are not always bumping into each other. One woman (non-contact) in my sample, a normally cheerful, outgoing person, who had been thrown into a temper for the umpteenth time by her modern but badly designed kitchen, said:

"I hate being touched or bumped, even by people who are close to me. That's why this kitchen makes me so mad when I'm trying to get dinner and someone is always in my way." ...

Thermal Space

The information received from the distance receptors (the eyes, ears, and nose) plays such an important part in our daily life that few of us would even think of the skin as a major sense organ.

However, without the ability to perceive heat and cold, organisms including man would soon perish. People would freeze in winter and get overheated in summer. Some of the more subtle sensing (and communicating) qualities of the skin are commonly overlooked. These are the qualities which also relate to man's perception of space.

Nerves called the proprioceptors keep man informed of what is taking place as he works his muscles. Providing the feedback which enables man to move his body smoothly, these nerves occupy a key position in kinesthetic space perception. Another set of nerves, the exterioceptors, located in the skin, convey the sensations of heat, cold, touch, and pain to the central nervous system. One would expect that since two different systems of nerves were employed, kinesthetic space would be qualitatively different from thermal space. This is precisely the case even though the two systems work together and are mutually reinforcing most of the time. ...

Temperature has a great deal to do with how a person experiences crowding. A chain reaction of sorts is set in motion when there is not enough space to dissipate the heat of a crowd and the heat begins to build up. In order to maintain the same degree of comfort and lack of involvement, a hot crowd requires more room than a cool one. ...

When thermal spheres overlap and people can also smell each other, they are not only much more involved but, if the Bruce effect mentioned in Chapter III has meaning for humans, they may even be under the chemical influence of each other's emotions. Several of my subjects voiced the sentiments of many non-contact peoples (the ones who avoid touching strangers) when they said that they hated to sit in upholstered chairs immediately after they had been vacated by someone else. ...

Tactile Space

Touch and visual spatial experiences are so interwoven that the two cannot be separated. Think for a moment how young children and infants reach, grasp, fondle, and mouth everything, and how many years are required to train children to subordinate the world of touch to the visual world. ...

Two groups with which I have had some experience—the Japanese and the Arabs—have much higher tolerance for crowding in public spaces and in conveyances than do Americans and northern Europeans. However, Arabs and Japanese are apparently more concerned about their own requirements for the spaces they live in than are Americans. The Japanese, in particular, devote much time and attention to the proper organization of their living space for perception by all their senses. ...

Man's relationship to his environment is a function of his sensory apparatus plus how this apparatus is conditioned to respond. ... Our urban spaces provide little excitement or visual variation and virtually no opportunity to build a kinesthetic repertoire of spatial experiences. It would appear that many people are kinesthetically deprived and even cramped. ...

Man's sense of space is closely related to his sense of self, which is in an intimate transaction with his environment. Man can be viewed as having visual, kinesthetic, tactile, and thermal aspects of his self which may be either inhibited or encouraged to develop by his environment.

48 Gary Snyder, *Earth House Hold*, 1969

Typically, each book recommended by the *WEC* received its own review, with its title as the subsection heading, but, in the case of the three books by Gary Snyder included on the "Considerations" page of the "Community" section, the celebrated author's name became the heading for the entire group, while excerpts from the texts stood alone, with no editorial description. *Earth House Hold*, from which a passage follows, was accompanied by two of Snyder's other texts: *The Back Country* (1968), a collection of poems that reflected his experiences in Japan, India, and the American West; and *Regarding Wave* (1970), another poetry anthology, which explored man's relationship to nature, informed by Snyder's own spirituality and ecological philosophy.

Snyder, who would win the 1975 Pulitzer Prize in Poetry for *Turtle Island*, offered a series of prose works in *Earth House Hold*—including travelogues and essays regarding nature, religion, and contemporary society, among other topics. "Why Tribe" addressed the proclivity of countercultural groups, from Quakers to New Communalists, to form tribal units that quietly transgressed historically acceptable social structures. For readers of the catalog who aspired to such a lifestyle, the "timeless path of love and wisdom," Snyder's text would have been truly inspirational.—MG

Source
Gary Snyder, "Why Tribe," in *Earth House Hold: Technical Quotes and Queries to Fellow Dharma Revolutionaries* (New York: New Directions, 1969), 113–116. © 1969 by Gary Snyder. Reprinted by permission of New Directions Publishing Corp. (Referred to in *LWEC*, 182.)

Further Reading
Jung, Hwa Yol, and Petee Jung. "Gary Snyder's Ecopiety." *Environmental History Review* 14, no. 3 (autumn 1990): 74–87.

Snyder, Gary. *Turtle Island*. New York: New Directions, 1974.

"Why Tribe," from *Earth House Hold*

We use the term Tribe because it suggests the type of new society now emerging within the industrial nations. In America of course the word has associations with the American Indians which we like. This new subculture is in fact more similar to that ancient and successful tribe, the European Gypsies—a group without nation or territory which maintains its own values, its language and religion, no matter what country it may be in.

The Tribe proposes a totally different style: based on community houses, villages and ashrams; tribe-run farms or workshops or companies; large open families; pilgrimages and wanderings from center to center. ... The Tribe proposes personal responsibilities rather than abstract centralized government, taxes and advertising-agency-plus-Mafia type international brainwashing corporations.

In the United States and Europe the Tribe has evolved gradually over the last fifty years—since the end of World War I—in response to the increasing insanity of the modern nations.

As the number of alienated intellectuals, creative types, and general social misfits grew, they came to recognize each other by various minute signals. Much of this energy was channeled into Communism in the thirties and early forties. All the anarchists and left-deviationists—and many Trotskyites—were tribesmen at heart. After World War II, another generation looked at Communist rhetoric with a fresh eye and saw that within the Communist governments (and states of mind) there are too many of the same things as are wrong with "capitalism"—too much anger and murder. The suspicion grew that perhaps the whole Western Tradition, of which Marxism is but a (Millennial Protestant) part, is off the track. This led many people to study other major civilizations—India and China—to see what they could learn.

It's an easy step from the dialectic of Marx and Hegel to an interest in the dialectic of early Taoism, the *I Ching*, and the yin-yang theories. From Taoism it is another easy step to the philosophies and mythologies of India—vast, touching the deepest areas of the mind, and with a view of the ultimate nature of the universe which is almost identical with the most sophisticated thought in modern physics—that truth, whatever it is, which is called "The Dharma."

Next comes a concern with deepening one's understanding in an experiential way: abstract philosophical understanding is simply not enough. At this point many, myself included, found in the Buddha-Dharma a practical method for clearing one's mind of the trivia, prejudices and false values that our conditioning had laid on us—and more important, an approach to the basic problem of how to penetrate to the deepest non-self Self. Today we have many who are exploring the Ways of Zen, Vajrayana, Yoga, Shamanism, Psychedelics. The Buddha-Dharma is a long, gentle, human dialog—2,500 years of quiet conversation—on the nature of human nature and the eternal Dharma—and practical methods of realization.

In the course of these studies it became evident that the "truth" in Buddhism and Hinduism is not dependent in any sense on Indian or Chinese culture; and that "India" and "China"—as societies—are as burdensome to human beings as any others; perhaps more so. It became clear that "Hinduism" and "Buddhism" as social institutions had long been accomplices of the State in burdening and binding people, rather than serving to liberate them. Just like the other Great Religions.

At this point, looking once more quite closely at history both East and West, some of us noticed the similarities in certain small but influential heretical and esoteric movements. These schools of thought and practice were usually suppressed, or diluted and made harmless, in whatever society they appeared. Peasant witchcraft in Europe, Tantrism in Bengal, Quakers in England, Tachikawa-ryu in Japan, Ch'an in China. These are all outcroppings of the Great Subculture which runs underground all through history. This is the tradition that runs without break from Paleo-Siberian Shamanism and Magdalenian cave-painting; through megaliths and Mysteries, astronomers, ritualists, alchemists and Albigensians; gnoslics and vagantes, right down to Golden Gate Park.

The Great Subculture has been attached in part to the official religions but is different in that it transmits a community style of life, with an ecstatically positive vision of spiritual and physical love; and is opposed for very fundamental reasons to the Civilization Establishment.

It has taught that man's natural being is to be trusted and followed; that we need not look to a model or rule imposed from outside in searching for the center; and that in following the grain, one is being truly "moral." It has recognized that for one to "follow the grain" it is necessary to look exhaustively into the negative and demonic potentials of the Unconscious, and by recognizing these powers—symbolically acting them out—one releases himself from these forces. By this profound exorcism and ritual drama, the Great Subculture destroys the one credible claim of Church and State to a necessary function.

All this is subversive to civilization: for civilization is built on hierarchy and specialization. A ruling class, to survive, must propose a Law: a law to work must have a hook into the social psyche—and the most effective way to achieve this is to make people doubt their natural worth and instincts, especially sexual. To make "human nature" suspect is also to make Nature—the wilderness—the adversary. Hence the ecological crisis of today.

We came, therefore, (and with many Western thinkers before us) to suspect that civilization may be overvalued. Before anyone says "This is ridiculous, we all know civilization is a necessary thing," let him read some cultural anthropology. Take a look at the lives of South African Bushmen, Micronesian navigators, the Indians of California; the researches of Claude Lévi-Strauss. Everything we have thought about man's welfare needs to be rethought. The tribe, it seems, is the newest development in the Great Subculture. We have almost unintentionally linked ourselves to a transmission of gnosis, a potential social order, and techniques of enlightenment, surviving from prehistoric times.

The most advanced developments of modern science and technology have come to support some of these views. … The next great step of mankind is to step into the nature of his own mind—the real question is "just what is consciousness?" and we must make the most intelligent and creative use of science in exploring these questions. The man of wide international experience, much learning and leisure—luxurious product of our long and sophisticated history—may with good reason wish to live simply, with few tools and minimal clothes, close to nature.

The Revolution has ceased to be an ideological concern. Instead, people are trying it out right now—communism in small communities, new family organization. A million people in America and another million in England and Europe. A vast underground in Russia, which will come out in the open four or five years hence, is now biding. How do they recognize each other? Not always by beards, long hair, bare feet or beads. The signal is a bright and tender look; calmness and gentleness, freshness and ease of manner. Men, women and children—all of whom together hope to follow the timeless path of love and wisdom, in affectionate company with the sky, winds, clouds, trees, waters, animals and grasses—this is the tribe.

49 Ita Jones, *The Grubbag*, 1971

Published in 1972, *The Grubbag* was an anthology of columns Ita Jones had written from the late 1960s through 1970. Jones's commentaries, which focused on the societal, political, and environmental implications of food consumption, originally appeared in the *Rag*, an underground countercultural newspaper based in Austin, Texas, and then reached a wider audience through the Liberation News Service, a New Leftist press news service that distributed articles and photographs to radical newspapers worldwide. Her articles made for an unconventional cookbook: straightforward, practical recipes were fluently juxtaposed with philosophical expositions on a multitude of topics including "Capitalism's 'high holiday'" (i.e., Christmas; *The Grubbag*, 21), the Hong Kong flu epidemic, pesticide contamination, and communal living, creating an intriguing portrait of the alimentary counterculture. —MG

Source

Ita Jones, *The Grubbag* (New York: Random House, 1971), 21–24. Courtesy of Ita Willen. (Referred to in *LWEC*, 193.)

Further Reading

Belasco, Warren James. *Appetite for Change: How the Counterculture Took on the Food Industry*. Ithaca, NY: Cornell University Press, 1993.

Hartman, Stephanie. "The Political Palate: Reading Commune Cookbooks." *Gastronomica* 3, no. 2 (spring 2003): 29–40.

"6. The Plague," from *The Grubbag*

My Christmas gift to you is multifold in that it magically prevents the plague because it nourishes the flesh and the mind by unfolding the art of eating like a tapestry of taste and color, and if you like to smoke while cooking, you'll see what I mean. It's a Javanese omelet (really) and uses just about what most people have on hand.

Javanese Omelet

1. Lightly fry 4 strips of bacon in an ordinary frying pan, then drain on paper toweling. Pour off all the drippings (but save for other uses). Wipe the pan, then melt 3 tbsp. butter in it.

2. While the butter is melting, slice a large onion and a large ripe tomato.

3. Arrange the onion slices in the pan to cover the bottom and let them sauté a minute or two until they have softened. Over them arrange the tomato slices. Over the tomatoes sprinkle a handful of freshly snipped parsley, then the bacon, cut into small pieces, building carefully the omelet layer by layer and never stirring.

4. Over the top sprinkle a bit of crushed red pepper, black pepper, and salt.

5. Beat 4 eggs lightly with a fork, add a little milk, about 2 tbsp., if you have it, and then slowly pour this over the fantastic-smelling food in the pan. Turn the heat very, very low and cover.

6. When the eggs are just set (this takes about 20 minutes), sprinkle some soy sauce over the top and cut the omelet in half or thirds or quarters, depending on how many you want to feed. Serve with toast, soup, or tea for any meal.

There is no doubt in my mind that the plague has crept into America. When a population panics, you can be sure they are at least half sane. But when people continue to daydream while the rats of disaster eat their fingers, we can know that not only the physical but the metaphysical plague is among us.

According to the New York commissioner of health, the Hong Kong flu (Oran fever) is "different" in that it didn't spread from coast to coast but broke out all over the country simultaneously. I have always believed that colds are more often a result of depression than of getting cold, and so it may be that the flu could also have something to do with this type of reaction—the postelection slump, the meaninglessness of urban and

suburban life, an acute case of collective alienation, lack of hope, and bad diet. The population
lumbers slowly over the rippling American landscape like a pregnant cow.

I don't want to underestimate the mass psychology which produces plagues of this kind, but
neither should we ignore the fact that the population of the richest country in the world does
not know how to eat. I've mentioned before the
tragedy of the loss of the arts of living, loving,
and eating in this country. While this is the direct
result of the system's anesthetizing function, on
a deeper level this may stem from the long-
standing mind/body problem. Americans eat as
though the object of eating is to fill a pink box
called Stomach. If they thought for one moment
that every single cell of flesh and brain depends
on food, they wouldn't consume the mountains
of garbage they do. The mind and body are one in
that the body's chemistry affects the mind. And
we all know how the mind's disposition affects
the flesh.

A lousy diet is mainly responsible for the list-
lessness, fatigue, disinterest, and lack of joy most
Americans are at a loss to explain. The sensitive,
intelligent girls I work with don't understand
that they are their bodies and that the reason
half of them are out with the flu has something
to do with the breakfast of coffee and donuts I've
watched them consume for four months now.
Fruit and vegetable juices, grains, honey, and eggs
are valuable to the body.

I hope I've scratched the surface of the physical and metaphysical causes of plagues. There
are some who think the tragedy of plagues lies in
the meaningless and random way in which the
innocent are struck. As I think back over the first
plagues I encountered, they numbered ten upon
the Pharaoh's head, and seemed neither irrational nor random. Maybe we should reexamine this
recent message from Asia.

50 Michel Abehsera, *Zen Macrobiotic Cooking*, 1968

Five years after he began the dietary program of **Zen macrobiotics**, established by George Ohsawa in the 1930s, Michel Abehsera compiled a beginner's cookbook on the subject. Based on the nutritional habits of **Zen monks**, predominantly their reliance on brown rice for sustenance and, to a lesser extent, vegetarianism, **Zen macrobiotics** was presented by both Ohsawa and Abehsera not simply as a way of eating but as a holistic philosophy. According to Abehsera, Ohsawa "told us that from foods we are given life, and in the skillful joining and balancing of foods are we given the secret of life" (*Zen Macrobiotic Cooking*, 6). In many ways, the cookbook was a paean to the founder's teachings. Abehsera's reverence for the **Zen macrobiotic** lifestyle emerged in parables such as "The Children of Grain," and even straightforward explanations of basic macrobiotic concepts, such as "the law of opposites," were infused with a sincerity born of true devotion.

When Abehsera embarked on his journey toward the "secret of life," he confessed, he could barely cook the simplest of foods, but through trial and error, he became a skilled macrobiotic cook and greatly improved his and his family's health and well-being. This success encouraged him to compose *Zen Macrobiotic Cooking* to share the power of macrobiotics with newcomers to its tenets. Although the initial "Number 7 diet" Abehsera espoused, in which one ate only brown rice for ten full days before gradually moving into the "secondary" food groups, may have proved too ascetic for many readers, his essential message of harmonious and thoughtful eating appealed to readers seeking a more encompassing countercultural lifestyle, making his book an excellent contribution to the catalog. —MG

Source
Michel Abehsera, *Zen Macrobiotic Cooking: Book of Oriental and Traditional Recipes* (Secaucus, NJ: Citadel Press, 1968), 9–11, 56. (Referred to in *LWEC*, 193.)

Further Reading
Crowley, Karlyn. "Gender on a Plate: The Calibration of Identity in American Macrobiotics." *Gastronomica* 2, no. 3 (summer 2002): 37–48.

Ohsawa, George. *Zen Macrobiotics: The Art of Rejuvenation and Longevity*. Los Angeles: Ohsawa Foundation, 1965.

Selections from *Zen Macrobiotic Cooking*

A Simple Introduction to the Law of Opposites Before We Get to the Recipes

You don't have to understand this chapter to enjoy and profit from this recipe book. Just, please, read it and get from it what you can before going on to the recipes.

In Japan it is known as yin and yang. Here, in America, it goes by the rather chemical terms of acidity and alkalinity. For instance, you eat something acid and have heartburn; common knowledge tells you to take something alkaline, Alka-Seltzer, perhaps. Acid foods are rich in potassium; alkaline, in sodium.

We can equate it thus:

Yin = Acidity = Potassium = Sugar = Fruits, etc.

Yang = Alkalinity = Sodium = Salt = Cereals, etc.

Yin expands: sugar is yin. Sugar, when placed on the tongue, tends to make it expand.

Yang contracts: salt is yang. Salt, when placed on the tongue, tends to make it contract.

If any element affects the mouth, it invariably affects the body, which in turn affects the mind. What are commonly known as drugs can be taken generally as an extreme example of yin. Drugs, extremely acid, expand the mind, diffuse concentration, cloud sensibility, etc. Excessive eating and drinking by a longer yet equally efficient process bring about the same results. Yet do not be misled into thinking that acidity is bad! Acidity and alkalinity are necessary complementary opposites, good or bad in direct relation to quantity and balance.

In food, the perfect balance of alkalinity and acidity is found in only one grain—brown rice. Experts such as Dr. Rene Dubos, the Nobel prizewinner, tell us that we can possibly live on this cereal alone. It contains a balance of 5 parts potassium (yin) to 1 part sodium (yang). Thus, in eating other foods, it seems that if we adhere to this ratio of 5 to 1, we are in good hands.

If we eat rice exclusively, we need not worry about balancing it with other foods. Rice contains all the elements our bodies need. However, if we eat only meat, which is yang, we are not able to live in good health; a balance is necessary—some vegetables or some fruits. In a word, something yin. Even buckwheat, one of the yang cereals, must also be balanced by something yin, such as vegetables. We must keep balancing our meals or at least try to be as close as possible to the proper ratio. This kept in mind, there is yet another factor that enters the scene: our own individual need.

The 5-to-1 ratio may work for one person and not for another. A person who has much more potassium in his body surely needs more sodium; he would have to carefully introduce more sodium into his food, remembering, of course, that extreme action leads to extreme reaction.

Our cells, our body and mind keep changing all the time. Necessarily, our meals should follow this same law of change. What we are today we will not be tomorrow. Consequently: our meal of today should not be the same as the one of tomorrow. A very banal image of balance is the scale of justice. But there is another much more exciting image, that of the dynamic. We must modify our diet to fit with season and geography, time and place. Seasonal change is a drama in which we and our diet change also.

There is a proverb that says: Tell me with whom you associate and I will tell you who you are. This saying applies, not only to people, but also to our food. A man who eats only pork comes to resemble a pig and to behave like a pig in any circumstance. A meat eater is not as quiet as a vegetable eater. The meat eater gets angry easily. He is explosive, while the vegetable eater lives in the diurnal tranquility of a flower, opening to light, then closing with the day's end. A woman, if it is her wish to do so, can change the character of her husband by feeding him a certain kind of food. Choose your life! Wild or wise! Both perhaps, why not? Your destiny is on your tongue. . . .

The Children of Grain

"Master!" called another disciple, "is there any difference between a child who eats grain and a child who eats animal food?"

The master lit a cigarette.

"Aha! Is there any difference?" The Master was silent for a long moment. Those who were at his feet that day were accustomed to his silences. They waited, anticipating either facetiousness or simple, straightforward wisdom.

"The difference is very great. A child who eats grains and vegetables has a better memory and better understanding than the child who eats mainly meat. His mind is clearer, more agile. His answers in school are more precise. His decisions, in anything he undertakes, are very quick. I would be very curious to know the IQ of the one and the other.

"No child should be fed with exactly the same foods as its parents. The child, for instance, needs less salt and more water than an adult. Parents are the sculptors of their child until its adolescence; after adolescence, the boy is his own artist."

51 Boston Women's Health Collective, *Women and Their Bodies*, 1970

In "Diana's Rap," a piece on the women's liberation movement of the 1960s, Diana Shugart expressed an opinion that perhaps represented the catalog's wider perspective on women's issues: "that 'this is bigger than the bot'n us'—that men suffer from their cultural roles at least as much as women do, and that the *real* problem lies deeper—lies, in both cases, in a society and family structure which makes it very difficult for anyone to realize his full potential, or even any very close approximation."[1]

Shugart, who admittedly worked in what she called a "men's world" and lived in an open-minded communal setting, also opined on the freedoms allowed by domestic tasks—a flexibility of schedule, "doing different things on different days," having the opportunity to nap or dance during a man's work-day—and even called traditional women's work an indulgence in her own life. Her essay completely ignored the history of female subjugation in American society and the pervasiveness of such discrimination even in her own time; just seven months later, however, Shugart printed a retraction, beginning, "I take it all back! I didn't know what I was talking about!" And yet, in spite of this enlightenment on the part of Shugart, women's issues were still underrepresented by the catalog's editors: Shugart's retraction was printed in smaller font, underneath the still-present "Diana's Rap"; significant texts on women's lib and feminism were completely absent; and sections near the single "Women" page included "Birth," "Baby Stuff," and "Sex."

On the positive side, however, was the inclusion of *Women and Their Bodies*, published by the Boston Women's Health Collective, an organization that remains active today. Although passages on masturbation and birth control, as well as a strongly worded warning to male readers that the pamphlet "was not written for [them]," were included in the review (by Shugart), Lucy Candib's essay "Women, Medicine, and Capitalism" from the text was arguably the most illuminating feminist work suggested by the catalog, and its message still resonates today. —MG

1. Diana Shugart, "Diana's Rap," in *LWEC*, 222.

Source

Lucy Candib, "Women, Medicine, and Capitalism," in Boston Women's Health Collective, *Women and Their Bodies: A Course* (Boston: Boston Women's Health Collective, 1970), 6–8. (Referred to in *LWEC*, 221.)

Further Reading

Davis, Kathy. *The Making of Our Bodies, Ourselves: How Feminism Travels across Borders*. Durham, NC: Duke University Press, 2007.

Lemke-Santangelo, Gretchen. *Daughters of Aquarius: Women of the Sixties Counterculture*. Lawrence: University Press of Kansas, 2009.

Willis, Ellen. "Radical Feminism and Feminist Radicalism." *Social Text* 9–10 (spring–summer 1984): 91–118.

"Women, Medicine, and Capitalism," from *Women and Their Bodies*

Marcuse says that "health is a state defined by an elite." A year ago few of us understood that statement. What does he mean? We believed that all people want to be healthy and that some of us are more fortunate than others because we have more competent doctors. "Now you should go to Dr. A. Man. He's my doctor and he's just great!"

Today we understand the stark truth of Marcuse's statement. We have not only started to look at health differently, but have found that health is one more example of the many problems we as people, especially as women, face in this society. We have not had power to determine medical priorities; they are determined by the corporate medical industry (including drug companies, Blue Cross, the AMA and other profit-making groups) and academic research. We have learned that we are not to blame for choosing a bad doctor or not having the money to even choose. Certainly, some doctors have learned medical skills better than others, but how good are technical skills if they are not practiced in a human way?

We as women are redefining competence: a doctor who behaves in a male chauvinist way is not competent, even if he has medical skills. We have decided that health can no longer be defined by an elite group of white, upper middle class men. It must be defined by us, the women who need the most health care, in a way that meets the needs of all our sisters and brothers—poor, black, brown, red, yellow and pink.

I. The Ideology of Control and Submission

Perhaps the most obvious indication of this ideology is the way that doctors treat us as women patients. We are considered stupid, mindless creatures, unable to follow instructions (known as orders). While men patients may also be treated this way, we fare worse because women are thought to be incapable of understanding, or dealing with our own situation. Health is not something which belongs to a person, but is rather a precious item that the doctor doles out from his stores. Thus, the doctor preserves his expertise and powers for himself. He controls the knowledge and thereby controls the patient. He maintains his status in a number of ways: First,

he and his colleagues make it very difficult for more people to become doctors.... Second, he sets himself off from other people in a number of ways, including dressing in whites.... Another much more important way doctors set themselves off from other people is through their language. Pseudoscientific jargon is the immense wall which doctors have built around their feudal (private) property, i.e., around that body of information, experience, etc. which they consider as medical knowledge. (epistaxis = nosebleed, thrombosis = blood clot, scleral icterus = yellow eyeballs, etc.)

Thirdly, doctors insulate themselves from the rest of society by making the education process (indoctrination) so long, tedious, and grueling that the public has come to believe that one must be superhuman to survive it.... Thus, a small medical elite preserves its own position through mystification, buttressed by symbolic dress, language, and education.

It is important for us to understand that mystification is the primary process here. It is mystification that makes us postpone going to the doctor for "that little pain," since he's such a "busy man." It is mystification that prevents us from demanding a precise explanation of what is the matter and how exactly he is going to treat it. It is mystification that causes us to become passive objects who submit to his control and supposed expertise.

II. Objectification

We know that we as women are objectified as sex objects in our society. Any woman who has walked alone at night knows the feeling of vulnerability and helplessness that accompanies our awareness that we are being perceived as pure sex objects. The medical setting further objectifies a person. The patient is assumed to be an object on which one can "objectively" and "scientifically" perform certain operations. The patient is merely the vehicle which brings the disease to the interventionist (instrumentalist). The outgrowth of these assumptions is that the best place for a doctor to act on a patient is in the hospital, i.e., when the patient is horizontal, passive, most like an object. Finally, that part of a person which is

Figure 6.4
Cover image from Boston Women's Health
Collective, *Women and Their Bodies: A
Course* (Boston: Boston Women's Health
Collective, 1970). Reprinted with permission.

considered sick is further separated and removed. ("The ulcer in 417." or "We did a gall bladder today.") For us as women, the treatment of any gynecological or obstetrical problem thereby results in the alienation of us from our own body, from our own genitals.

III. Alienation

Naomi Weisstein, in her essay on women, *Kinder, Kirche, Kuche*, has outlined very well how the society has caused the alienation of a woman from her body. Freud's impact cannot be overestimated; we have internalized the notion that woman is incomplete, that something is missing. This alienation leads to a condition which is epitomized by the middle class woman, who, whenever she feels ill, goes to see her gynecologist. The implication: whatever is the matter with her has to do with her sexuality.

Alienation is also what makes it hard for us to talk about sex. Our sexual experience is so privatized that we never find out that other women have the same problems we do. We come to accept not having orgasm as our natural condition. We remain ignorant about our own sexuality and chalk it up to our own inadequacies. And if we should be so bold as to go to a doctor—and if we should summon up the courage to ask him about our common problem—chances are he will know nothing about it, although he will never or rarely admit this and will probably laughingly dismiss our questions. Doctors in general are as ignorant about sexuality as the rest of the men in society.

Doctors' blatant ignorance about sex stands in stark contradiction to the fact that they are considered the only legitimate person to consult about any sexual problem. Thus, we bring all our awkwardness and ignorance about sex to a doctor who cannot understand that his own ignorance and arrogance are the epitome of male chauvinism. ...

Which brings us to preventative medicine. We as women are made to feel uncomfortable about going to a doctor in the first place. If we cannot feel comfortable going to our doctors normally, then to go for preventative reasons will be all the more difficult. Thus, while the medical profession has come out in favor of massive screening of women for cancer of the breast and cervix (the cervix is the neck of the uterus, or womb), their practice, their approach, their manner—that is to say, their ideology—all works in the opposite direction. ... And when the visit involves a pelvic examination, it is even less likely a woman will go through with it. Small wonder that only 12% of the women in this country who ought to have "Pap" tests (short for Papanicolaou, the *guy* who invented it) for cervical cancer get them. This is one of the very concrete ways that male chauvinist medicine means poorer health care and health protection for us.

We cannot begin to write here about capitalist forms of medicine per se; that is to say, the prohibitive cost of medical care, the racist and inferior treatment of poor people and black people, the profit and prestige-making institutions of the "health industry" (hospitals, medical schools, drug companies, etc.), the total neglect of the public or preventative protection, or the fee-for-service, pay-as-you-die economic base upon which most medical practice is based. This is an important and extensive issue which must be dealt with elsewhere. Suffice it to say that capitalism is incapable of providing good health care, both curative and preventive, for all the people. ... The capitalist medical care system can be no more dedicated to improving the people's health than can General Motors become dedicated to improving the people's public transportation. Our difficulty in perceiving the similarity between the health care system and any other corporate capitalist enterprise in the society results from our acceptance of the rhetoric that medicine *helps* people.

52 Saul D. Alinsky, *Rules for Radicals*, 1971

At the beginning of the section titled "The Ideology of Change" in *Rules for Radicals*, Saul Alinsky asked, "What kind of ideology, if any, can an organizer have who is working in and for a free society?" (10). Part of his answer appears in the passage that follows, in which Alinsky suggested that organizers grasp the duality of political phenomena but be mindful that this is only one aspect of the "ideology of change." The organizer should also understand the eternal flux of society and respond to this change by maintaining a "loose, resilient, fluid" approach to his work; he should, in addition, trust that empowered people would typically "reach the right decisions" for their own lives (11).

Alinsky, an internationally renowned political activist and organizer who worked with communities across the United States from the 1930s until his death in 1972, shared many such organizational concepts, as well as personal reflections and anecdotes, in his second book, *Rules for Radicals*, which he intended as a training manual for aspiring organizers. In his review of the text, Brand noted, "Much radical literature is aimed at fighting. This book is aimed, by an expert, at winning"—a pithy recapitulation of Alinsky's own stated objectives.—MG

Source

Saul D. Alinsky, *Rules for Radicals: A Practical Primer for Realistic Radicals* (New York: Random House, 1971), 15–19. Used by permission of Random House, an imprint and division of Penguin Random House LCC. All rights reserved. (Referred to in *LWEC*, 236.)

Further Reading

Boyte, Harry. *The Backyard Revolution: Understanding the New Citizen Movement*. Philadelphia: Temple University Press, 1980.

Chalmers, David. *And the Crooked Places Made Straight: The Struggle for Social Change in the 1960s*. Baltimore, MD: The Johns Hopkins University Press, 1991.

Cohen, Mitchell, and Dennis Hale, eds. *The New Student Left: An Anthology*. Boston: Beacon Press, 1966.

"The Ideology of Change," from *Rules for Radicals*

Once we have moved into the world as it is then we begin to shed fallacy after fallacy. The prime illusion we must rid ourselves of is the conventional view in which things are seen separate from their inevitable counterparts. We know intellectually that everything is functionally interrelated, but in our operations we segment and isolate all values and issues. Everything about us must be seen as the indivisible partner of its converse, light and darkness, good and evil, life and death. From the moment we are born we begin to die. Happiness and misery are inseparable. So are peace and war. The threat of destruction from nuclear energy conversely carries the opportunity of peace and plenty, and so with every component of this universe, all is paired in this enormous Noah's Ark of life.

Life seems to lack rhyme or reason or even a shadow of order unless we approach it with the key of converses. Seeing everything in its duality, we begin to get some dim clues to direction and what it's all about. It is in these contradictions and their incessant interacting tensions that creativity begins. As we begin to accept the concept of contradictions we see every problem or issue in its whole, interrelated sense. We then recognize that for every positive there is a negative, and that there is nothing positive without its concomitant negative, nor any political paradise without its negative side.

Niels Bohr pointed out that the appearance of contradictions was a signal that the experiment was on the right track: "There is not much hope if we have only one difficulty, but when we have two, we can match them off against each other." Bohr called this "complementarity," meaning that the interplay of seemingly conflicting forces or opposites is the actual harmony of nature. Whitehead similarly observed, "In formal logic, a contradiction is the signal of defeat; but in the evolution of real knowledge it marks the first step in progress towards a victory."

Everywhere you look all change shows this complementarity. In Chicago the people of Upton

Sinclair's Jungle, then the worst slum in America, crushed by starvation wages when they worked, demoralized, diseased, living in rotting shacks, were organized. Their banners proclaimed equality for all races, job security, and a decent life for all. With their power they fought and won. Today, as part of the middle class, they are also part of our racist, discriminatory culture.

The Tennessee Valley Authority was one of the prize jewels in the democratic crown. Visitors came from every part of the world to see, admire, and study this physical and social achievement of a free society. Today it is the scourge of the Cumberland Mountains, strip mining for coal and wreaking havoc on the countryside.

The C.I.O. was the militant champion of America's workers. In its ranks, directly and indirectly, were all of America's radicals; they fought the corporate structure of the nation and won. Today, merged with the A.F. of L., it is an entrenched member of the establishment and its leader supports the war in Vietnam.

Another example is today's high-rise public housing projects. Originally conceived and carried through as major advances in ridding cities of slums, they involved the tearing down of rotting, rat-infested tenements, and the erection of modern apartment buildings. They were acclaimed as America's refusal to permit its people to live in the dirty shambles of the slums. It is common knowledge that they have turned into jungles of horror and now confront us with the problem of how we can either convert or get rid of them. They have become compounds of double segregation on the bases of both economy and race and a danger for anyone compelled to live in these projects. A beautiful positive dream has grown into a negative nightmare.

It is the universal tale of revolution and reaction. It is the constant struggle between the positive and its converse negative, which includes the reversal of roles so that the positive of today is the negative of tomorrow and vice versa.

This view of nature recognizes that reality is dual. The principles of quantum mechanics in physics apply even more dramatically to the mechanics of mass movements. This is true not only in "complementarity" but in the repudiation of the hitherto universal concept of causality, whereby matter and physics were understood in terms of cause and effect, where for every effect there had to be a cause and one always produced the other. In quantum mechanics, causality was largely replaced by probability: an electron or atom did not have to do anything specific in response to a particular force; there was just a set of probabilities that it would react in this or that way. This is fundamental in the observations and propositions which follow. At no time in any discussion or analysis of mass movements, tactics, or any other phase of the problem, can it be said that if this is done then that will result. The most we can hope to achieve is an understanding of the probabilities consequent to certain actions.

This grasp of the duality of all phenomena is vital in our understanding of politics. It frees one from the myth that one approach is positive and another negative. There is no such thing in life. One man's positive is another man's negative. The description of any procedure as "positive" or "negative" is the mark of a political illiterate.

Once the nature of revolution is understood from the dualistic outlook we lose our monoview of a revolution and see it coupled with its inevitable counterrevolution. Once we accept and learn to anticipate the inevitable counter-revolution we may then alter the historical pattern of revolution and counterrevolution from the traditional slow advance of two steps forward and one step backward to minimizing the latter. Each element with its positive and converse sides is fused to other related elements in an endless series of everything, so that the converse of revolution on one side is counterrevolution and on the other side, reformation, and so on in an endless chain of connected converses.

53 The OM Collective, *The Organizer's Manual*, 1971

If *Rules for Radicals* conceptualized what Brand called the "science of revolution," the OM Collective's *Organizer's Manual*, dedicated "to all those who will make a better revolution than we can prescribe," specified the myriad ways to actualize it. Although the book briefly described the role of an organizer, it began, in the first chapter, to detail the practical steps he or she might take to create a community group, raise funds, and expand. From there the manual provided instructions ranging from using tactics such as guerrilla theater, pamphlets, and canvassing to promote an activist group's message to preparing for a tear gas attack and treating wounds. It concluded by cataloging the various approaches that organizers should adopt when working with distinct constituencies including high school students, Native Americans, veterans, and the elderly. J. D. Smith, in his review of the text, mentioned its benefit for those "who know they should be doing something, but don't feel right about generating unstructured change." *The Organizer's Manual* afforded such structure, educating and potentially galvanizing the "polite revolutionaries" invoked by Smith. —MG

Source
The OM Collective, *The Organizer's Manual* (New York: Bantam Books, 1971), 50–52, 76–77. (Referred to in *LWEC*, 236.)

Further Reading
Horowitz, Irving Louis. *The Struggle Is the Message: The Organization and Ideology of the Anti-war Movement*. Berkeley, CA: Glendessary Press, 1970.

Myerhoff, Barbara G. "The Revolution as a Trip: Symbol and Paradox." *Annals of the American Academy of Political and Social Science* 395 (May 1971): 105–116.

Sahlins, Marshall. "The Teach-Ins: Anti-war Protests in the Old Stoned Age." *Anthropology Today* 25, no. 1 (February 2009): 3–5.

Selections from *The Organizer's Manual*

d. Posters

Posters and signs serve two broad purposes in the movement: to announce or advertise actions, meetings, and other events; to keep certain slogans, pictures, quotations, or other political reminders constantly before the eyes of the people and of the group itself.

Design. The sign to advertise a demonstration, meeting, or benefit performance will be posted, presumably, in public places—streets, storefronts, transport stations, walls, etc., where many people pass. It should be designed to catch the eye and deliver its message at a glance.

Include only the essential information: a brief slogan indicating the purpose of the event; the event itself (including speakers, etc.); the place; the time; the sponsorship. (If the place is unfamiliar, directions may be added in small type.) Decide in advance what is most important and make it stand out by size of letters or change of colors or placement on page. In general, one or two colors are sufficient. Many bright colors tend to compete for attention. Any illustration should be relevant to the action, simple and eye-catching. Abstract designs should call attention to the words, not compete with them. Leave lots of empty space around your copy—always.

Production. The simplest and probably cheapest method of poster printing is by hand with magic markers or broad grease pastels. Once the design and copy is set, you can produce a large number quite fast by an assembly-line technique. Printing block letters is harder for untrained people than italics. A large, clear handwriting is usually more attractive than inexpert printing. Attractive small posters can be made by mimeograph, if you learn the technique of hand printing by stylus on a stencil.

When the copy is large and brief, or when you want to include a simple design (like an explosion flash, a flower, a star or circle, or the silhouette of a building), cut a stencil out of stiff paper with a razor blade or knife. Place the stencil over your poster board and spray with a can of quick drying paint (available at art stores). When your design is dry, you can print over it, if you wish, with another color.

When you need a quantity of posters and a professional looking job, silk screen is your best bet. Unless you are blessed with an activist art school nearby where you need only supply the paper, ink, and copy, it will pay you to make the modest outlay required for silk-screening and to have several members train themselves to do it....

Distribution. There are good poster spots and bad, for visibility and for your relations with the public. There are no prescriptions for these save observation, experimentation, and consideration. Make a list of the centers most frequented by the people you want to reach, and the best spots in those centers.

Whether you observe "post no bills" warnings will depend on whom you are trying to reach and what your relationships with the community are. Indiscriminate sign posting can create hostility as well as support, so evaluate the local situation. Ask a storekeeper's permission to use his window and avoid covering up his own display. Don't deface nature or hammer nails into trees. Remove obsolete posters to make way for your new ones....

If you plan to paint on walls, a cardboard stencil and a spray can will add dash to a well-phrased slogan—but think twice about using hard-to-remove paints or adhesives for signs which will soon be out of date. Oil-based paints soak into stone or cement and have to be sanded off. Scotch tape will mar interior paint....

11. Zaps

A zap is symbolic shock therapy aimed against the status quo. A zap pits irrationality against hardened rationality. Zaps undermine the confidence and composure of the self-satisfied and complacent. They break old patterns of predictability and offer new avenues of communication.

A zap is contradictory. It can be humorous in content but deadly serious in intent. Anyone who comes into viewing, hearing, smelling, tasting, or feeling distance can be zapped.

Spread ZAP graffiti around. Carry an ink marker and gummed graffiti labels with you everywhere. Use every available space to your advantage. Put political quotes on popcorn boxes and in right-wing books.

Experiment with political definitions and timely slogans. Print up "wanted" posters of your most dangerous opponents.

Mailed Zaps: Send roaches to slumlords.—Send dirt back along with bills to major pollutants.—Don't put stamps on bills at all.—Postage will be due from receiver.

War Zaps: In your spare time, call the Pentagon, collect, and ask to speak to someone in the Intercommunication Center. Ask how the war's going.—Process draft forms using your dog's name.

Money-Economy Zaps: Give away free: money, posters, beer, etc. Burn money.—Design dollar-bill toilet paper and distribute it.

Ecological Zaps: Wear gas masks in the streets.—If you can guarantee shipping and adequate grazing area, the government will supply you with your own buffalo (keeps the herds at a controllable level). Park it in front of the Department of Interior in D.C.—Call Hickel and ask for a personal, guided tour of his newly interior-decorated office (well, that is his job, no?). You'd like to see his $55-per-yard carpet.—Call your local polluter and ask where you can purchase packaged, government-inspected air.

Miscellaneous Zaps: When filling out a job application (or other form), in the blank that asks for Race-----, write in Human!; where it asks for Sex-----, write Yes.—Call your TV station around 3:00 A.M. and tell them you were watching the test pattern on TV and ask them why they took it off. Tell them it works! Ask them who or what they are testing.—Sneak into a convention of liberal politicians. Cut the cord from the mike and insert a jack into the tape deck. Insert a recording of farts; or a recording of Country Joe and the Fish singing "Be the first in your block to have your son come home in a box."

54 Margaret Mead and James Baldwin, *A Rap on Race*, 1971

One of the catalog's greatest deficiencies, perhaps, was its coverage of African American subjects, racism, and civil rights, and perhaps the best example of this myopia may be found on the "Community" section's "Politics" page, where just two pieces—*A Rap on Race* and *Black Reading List*—appeared, accompanied by texts for conscientious objectors, radical magazines, and even *Robert's Rules of Order*. Relegating the black experience to a single shared page under the heading "Politics" is deeply problematic, especially when considering the expressed purpose of the catalog, to empower individuals to educate themselves via "tools" promoted by the magazine. By focusing on the broader structural dilemmas facing Western society, the predominantly white, male editors of the catalog may have failed to recognize the urgent, concrete struggles of minority groups: an inability to see the trees for the forest, to reverse the phrase.

Their neglect, however, should not overshadow the virtue of the selected text, *A Rap on Race*, which Brand reviewed in the catalog: "Some life and some light on deadly subjects like America, Race, The World, The Future, Where We Came From, What's Happening, by … is it a black and a white, or bright young man and a bright old lady? Foxy sweethearts." The book is a transcript of three consecutive discussions between the famous anthropologist Margaret Mead and the greatly esteemed African American author James Baldwin from the morning of August 26 through the evening of August 27, 1970. The unrehearsed, unscripted discussion did, as Brand noted, address a number of topics, but Mead and Baldwin, for the most part, concentrated on race and Western society, from both a historical and a contemporary perspective. Whether intentional or inadvertent on the part of the speakers, the discussion emerged as a surprisingly coherent, engaging, and thought-provoking text, a passage of which follows.—MG

Source

Margaret Mead and James Baldwin, *A Rap on Race* (Philadelphia: Lippincott, 1971), 64–70. Reprinted by arrangement with the James Baldwin Estate. (Referred to in *LWEC*, 237.)

Further Reading

Baldwin, James. *The Fire Next Time.* New York: Dial Press, 1963.

Bloom, Joshua, and Waldo E. Martin. *Black against Empire: The History and Politics of the Black Panther Party.* Berkeley: University of California Press, 2013.

Carson, Clayborne. *In Struggle: SNCC and the Black Awakening of the 1960s.* Cambridge, MA: Harvard University Press, 1981.

Selections from *A Rap on Race*

Baldwin: I agree with Malcolm X, who said that if you're south of the Canadian border you're South. You're born in Harlem, but you don't grow up in Harlem saying: "Thank God I wasn't born in Atlanta." You've never seen Atlanta, but you know what's happening to you in Harlem, and Harlem is a dreadful place. It's a kind of concentration camp, and not many people survive it. I think the black people in this country are quite extraordinary. They must obviously have been the strongest people. To have survived that mid-passage, as you said much earlier; to have survived those boats at all.

Mead: And then to continue to survive.

Baldwin: And to change, without the nation even knowing it, to actually change something in the very deepest part of the psychology of the nation. Because white people cannot really live without black people here. Not anymore. Not at all. And regardless of whether they act on it, they know and are in many ways unconsciously controlled by the fact that we are their brothers—not in the sense that all men are brothers, but in the literal sense of being related by blood.

Mead: You see, I think that's a different point. It's absolutely true of the South, of course. I remember meeting a very prominent Negro educator who had been sent abroad by UNESCO. He'd been in some part of Europe for a couple of years, and he said, "It's so wonderful to get back to my country. I got so tired of all those white people over there. It's wonderful to come back where there are both black and white." And certainly for the Southern consciousness, black or white, both are there. But the whole spirit of the North has been to keep other people out. It's not only been about keeping out black people, it's been about keeping out everybody. The Mayflower pact was itself a pact under which we were going to build our nice little society and keep everybody out. So the first thing you know you have Rhode Island, made up of refugees. The North has always tried to establish identity by cutting other people out and off.

Baldwin: I agree. But if you're black it takes a very long time to realize that. It's only beginning. It's only now that black people in the main are beginning to recognize that many white people, if not most white people, are excluded from what Lyndon Johnson would have called the Great Society. The Great Society hasn't room for the poor-white West Virginian miner, for example. So obviously there's no room for me. This is part of the crisis, in fact: that it takes a long time before one can see that; that for a long time I simply knew that I was black and I was excluded because I was black. It didn't make any difference to me that the Irish were excluded, or that the Jews and the Poles were excluded. I was excluded and that was all that mattered to me. And furthermore, from a black American's vantage point—

Mead: Also, remember that you're a Northerner.

Baldwin: Well, in a way that doesn't mean anything to me.

Mead: You are, though. You see you were shocked when you went South.

Baldwin: Yes, I was shocked by the extent of the society's paranoia.

Mead: You felt something strange and different. You are a Northerner. And one of the crises stems from the fact that there are many Northern black people who don't want other blacks around, especially black people from the South: you see, they're behaving like Northerners. So that in building any kind of an American society where we have to take into account the Southeast and take into account California and Old New England and all of this, we really have to take into account the way in which the Northern identity is dependent upon whom you can keep out.

Baldwin: The American identity has always, from my point of view, depended on keeping me out....

* * *

Baldwin: We are going to have to [build a country] according to premises which at this moment in this country are considered absolutely treasonable. For a radical example, I agree with the Black Panthers' position about black prisoners. I think that one can make the absolutely blanket statement that no black man has ever been tried by a jury of his peers in America. And if that is so, and I know that is so, no black man has ever received a fair trial in this country. Therefore, I'm under no illusions about the reason why many black people are in prison. I'm not saying there are no black criminals. Still, I believe that all black prisoners should be released and then retried according to principles more honorable and more just. Do you see what I mean?

Mead: Yes.

Baldwin: But, of course, that is not about to happen next Wednesday, is it?[1] ...

Mead: You treat this country as if it had one problem. It has a lot more than one problem.

Baldwin: Yes, but that one problem is a problem which has obsessed my life. And I have the feeling that that one problem, the problem of color in this country, has always contained the key to all the other problems. It is not an isolated particular, peculiar problem. It is a symptom of all the problems in this country. And, in a way, this is indicated by the reaction in the country, and finally in the world, to the civil rights movement. Even the Women's Liberation Movement, you know, owes a lot to those people who started marching in Montgomery in 1956. And the whole black power movement in London comes from the United States. And the whole ferment which began in Montgomery when Mrs. Rosa Parks refused to stand up and give her seat to a white man, to give it an arbitrary date, has begun to overturn the entire country. And even the language in which people express their discontent, the language in which they fashion their petitions, comes out of the whole black confrontation in this country in the last fifteen to twenty years. If that is so, then you know that the problem—and I think the American Indian agrees with me—that the problem of the black and the Indian is related. In fact, the Puerto Ricans in the cities have taken techniques from the black rebellion. The Young Lords are a direct result of the Black Panthers.

1. It is likely that Baldwin is referring to the Black Panther prosecutions in 1970 in New Haven, Connecticut.

Nomadics

Deeply embedded in American culture is the idea that "moving on" or "hitting the road" is a good way to change your life or stimulate the imagination. For the restless generation of the 1960s, this was particularly true. Being "on the road" was a paradigmatic site of poetic engagement. As Stewart Brand said of the Airstream Travel Trailer, "It frees us from owning land and encourages us to live in wilder places" (*LWEC*, 247). The "Nomadics" section was really about different forms of transportation, from cars, bicycles, canoes, and horses to airplanes. Nonetheless, several texts, which are letters from readers, were about people who spend their lives on the road. In the "Shelter" section (1973), Lloyd Kahn included a number of accounts of the lives and vehicles of nomads.

Survival in extreme climates or exotic locations was one theme of the section. You can find helpful books for when "you get stuck out there with nothing but some knowledge, no matches, food, compass, toothpaste or magazines, scrounge the tools food and shelter" (*LWEC*, 275). Different environments, including the desert, mountains, and Arctic were considered. However, texts on walking were also included: "Some of them old boys are into making houses out of chicken wire and condoms. Some of them can gather you a salad right off the forest floor. Some can make you a computer out of old Stomberg-Carlson radio parts and have enough wire and tubes left over for two laser death rays and a UFO. But Colin is into walking." This is how Colin Fletcher's book *The Complete Walker* was introduced. The book "gives a little walk philosophy, and then proceeds to discuss…how to put a nice little well equipped house and its fittings on your back" (reviewed by Roland Jacopetti, *LWEC*, 255).

A theme linking this section to the underlying ideas of the catalog is the attitude toward repair and maintenance. A recycled school bus is a perfect item for life on the road, as the bus is described as "cheap, dependable, adaptable, and impressive" (SB, *LWEC*, 246). The specialized brochure *Leisure Camping* was devoted to bus conversion: "The school bus is a very well insulated vehicle—made of double wall construction, which also makes them very easy to remodel into a motor home. They have large radiators which give excellent cooling capacity in travelling hilly or mountainous country. They are all equipped with excellent heating plants. Most of the 1955 or newer buses have 6 foot ceilings so that you do not have to stoop while walking inside and are very comfortable to ride in" (246). Central are comfort and price. To be able to replace seats "with good quality bucket seats out of cars" is the kind of attention to detail that the counterculture promoted particularly when discussing recycling: "Most junk yards charge outrageous prices for buckets, but one of the very best seats is from a Renault R8 or 10" (246).

Nomadic experiences would oblige the reader to think about outdoor equipment: sleeping bags, tents, and backpacks. Many of the reviews of this kind of equipment were written by Roland Jacopetti, former member of Morningstar Ranch, a commune created in 1966 by Lou Gottlieb in Occidental, California.

In his book *Counterculture Green: The Whole Earth Catalog and Environmentalism*, **Andrew Kirk** makes many interesting connections between environmentalism, the outdoors, and consumerism, demonstrating the link between the readers of the catalog who "headed to the mountains in search of 'authentic experiences' and their quest for excellence for the material they used." This search led some of them to develop their own enterprises to fill the gap, and their technical innovations and contributions would contribute to a major economic revolution in outdoor equipment and apparel.—CM

55 Stewart Brand, "The Great Bus Race"

The "Nomadics" section of *The Last Whole Earth Catalog* began surprisingly with an account of "The Great Bus Race,"[1] a gathering of nomadic groups in adapted school and hospital buses. The buses included "Ken Kesey's renowned 'Further,' with a skeleton crew of Farmsters aboard, and the Hog Farm's fleet: 'Road Hog,' 'Hospital Bus,' 'Kitchen Bus,' etc." (*LWEC*, 245). The event was organized by Ken Kesey and Hugh Romney for the 1969 summer solstice in Aspen Meadows, New Mexico. The text reads like an account of a pop festival, where everyone knows the participants, who are given star treatment in the photographs. What comes across from the description is the sense of a tribal culture, like a meeting of travelers or, more recently, a New Age festival. Presumably the intention was to communicate a sense of community and lighthearted entertainment, tinged with wild excess: "The mood is pure, fatalistic optimism." The mythical quality of the event was further emphasized by the addition of contradictory, trippy accounts of the event. There was an explicit sense of trying to immortalize memories of a particular kind of anarchic solidarity. —CM

1. In 1964, Ken Kesey and the Merry Pranksters drove from California to New York in a stripped-down, psychedelic school bus, an LSD-fueled adventure recounted in Tom Wolfe's *The Electric Kool-Aid Acid Test*.

Source
Stewart Brand, "The Great Bus Race," *LWEC*, 245.

Further Reading
Perry, Paul, Ken Babbs, Michael Schwartz, and Neil Ortenberg. *On the Bus: The Complete Guide to the Legendary Trip of Ken Kesey and the Merry Pranksters and the Birth of the Counterculture*. New York: Thunder's Mouth Press, 1990.

Figure 7.1
"The Great Bus Race." *LWEC*, 245.

Difficult But Possible

SUPPLEMENT to the

WHOLE EARTH CATALOG

$1   July 1969

At the starting line Hugh Romney mounts his bus.

Feverish last minute preparations aboard Further.

THE GREAT BUS RACE

Winning Time 2:32. Who Won Is Debated.

Aspen Meadows, New Mexico. Summer Solstice, 1969

Hugh Romney, who mustered this 1000 in this mountain pasture, is announcing the start of the Bus Race and that a couple named John and Mary are somewhere around and must be found because they have the bubonic plague and require innoculation immediately. Everybody's pretty innoculated already; it's the spaciest part of the afternoon. The race was going to be one bus at a time against the clock, but Ken Kesey and others are maintaining that's a chickenshit race. It's got to be all at once.

The buses are Kesey's renowned Further, with a skeleton crew of Farmsters aboard, and the Hog Farm's fleet: Road Hog, Hospital Bus, Kitchen Bus, etc. The Motherfuckers' tight little bus is around but not in the race.

Everyone will go one trial run around the course: ¼ mile up the meadow, around the flag, and back downhill to the starting line. Majestically the line of buses staggers up the course. "I don't know about this," says Mike Hagen, "Further ain't got much brakes." The Hospital Bus is stalled half-way up the hill right where it's almost impossible to get around. Reportedly it also ran over a pup tent (unoccupied).

The survivors of the trial run line up. They're immense, these faded colorful wrecks filled with folly. People are climbing off, others are climbing on. It's a test of passengers as much as drivers. The mood is pure: fatalistic optimism.

And it's begun. Road Hog gets off first. Further is dragging. The air is filled with buses and exhaust and din. They wallow up the pasture like berserk pigs. As Further sways past the grounded Hospital Bus Kesey disappears over the windshield shouting orders to driver Bucavich. Over there in the bushes the Foot Swami from India is watching it all go by.

Further approaches the flag from the left. The Kitchen Bus with Romney on top is coming around the other way. They intersect, pass narrowly, and ram into opposite hillsides, then rev and back blindly toward each other. Frantic shouting gets them forward again. For a moment the entire solstice is tied in a sweet mad knot at the top of the pasture.

Bucavich gets Further around. The Kitchen Bus is still hung with a rear bumper gouged in the creek bed. Rocket, who's driving, surges again and gets it free. The buses swarm downhill. Hugh Romney's mouth is wide open, his toothless warning wailing down the pasture, "BEEE CARRREFUULLLL." At the finish line a thousand longhairs cheer and dodge.

Afterward, at the award-giving argument, Romney is saying, "A miracle. It's a pure miracle nobody was killed." Kesey is saying, "Two minutes thirty-two seconds. It's something to think about for next year."

Further ain't got much brakes.

Barrelling toward the first turn Kesey has an excellent view of the rear of the Hog Farm fleet.

Tied up at the flag, Romney and Further pass in opposite directions. Fierce bus action is visible on the track below.

56 Stewart Brand, "Airstream Travel Trailer"

I've lived in a 22 foot Airstream with wife and two cats for most of three years now. It's the only high-tech home I've found at all lovable, indeed comparable to the way some ocean-going boat owners feel about living on board. The Airstream is an elegant honest design-job. It makes us parsimonious (and conscious) about using water, gas, power. It frees us from owning land, and encourages us to live in wilder places. It is proof from fire, earthquake, floods (drive away) and mice (except what the cats bring in and lose). It's one of the few domes I know that doesn't leak. When we travel with it, wherever we go we're coming home. One more month of production and we're hauling ass and house out of here to the desert. (SB, *LWEC*, 247).

The Airstream Clipper caravan, one of the first to sport the streamlined, aluminum-clad shape, was designed in 1936 by Hawley Bowlus, who had also worked on Charles Lindbergh's *Spirit of St. Louis*. Although expensive, the Airstream became extremely popular in the 1960s and '70s, especially among design-conscious Americans, and has remained in production ever since. Architects considered the Airstream a modern design classic, and Brand's choice of it for his mobile home demonstrates his willingness to blend modernity with an alternative lifestyle. —CM

Source
Stewart Brand, "Commentary on Airstream Travel Trailer," *LWEC*, 247.

Further Reading
Burkhart, Bryan, and David Hunt. *Airstream: The History of the Land Yacht*. San Francisco: Chronicle Books, 2000.

Kirk, Andrew G. *Counterculture Green: The Whole Earth Catalog and American Environmentalism*. Lawrence: University Press of Kansas, 2007.

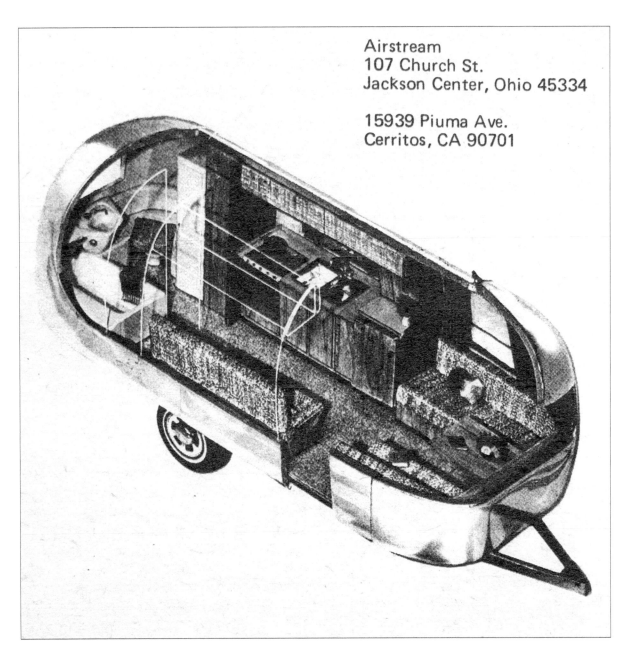

Airstream
107 Church St.
Jackson Center, Ohio 45334

15939 Piuma Ave.
Cerritos, CA 90701

Figure 7.2
"Airstream Travel Trailer." *LWEC*, 247.

57 John Muir, *How to Keep Your Volkswagen Alive,* 1969

The Volkswagen "Bug" or "Beetle" was chosen by many young intellectuals, students, and activists for its robust engineering and complete contrast with the traditional American automobile. The Volkswagen camper van was another icon of the counterculture. Page 248 of the *LWEC* included reviews of no fewer than four books on VW maintenance, one of them supplied by a reader. *How to Keep Your Volkswagen Alive* **was written by John Muir (1918–1977), a former aerospace engineer who opened a mechanical garage in Taos, New Mexico, in 1969. With illustrations by Juniperus Scopularum (Peter Aschwanden), the book provides a readable and comprehensible guide to the restoration and maintenance of the Volkswagen.**

The principle of keeping well-designed vehicles on the road rather than feeding the junkyard in the traditional American cycles of obsolescence and consumption was an article of faith. Muir considered his task not only to explain what to do but also to engender a culture of maintenance. "Read it all the way through like a novel, skipping the detailed steps in the procedures, but scan all the comments and notes. This will give you a feel for the operational viewpoint I took when I wrote it. The intent is to give you some sort of answer for every situation you'll run into in the life of the car" (*How to Keep Your Volkswagen Alive,* **1969). The book is extremely well structured, with procedures organized according to different situations and a detailed guide on how to approach each task. Above all, the tone is intimate: "You must do this work with love, or you will fail. You don't have to think, but you must love. This is one of the reasons I have nice tools. If I get hung up with maybe a busted knuckle or a busted stud, I feel my tools, like art objects or lovely feelies until the rage subsides and sense and love return. Try it, it works."—CM**

Source
John Muir and Tosh Gregg, *How to Keep Your Volkswagen Alive: A Manual of Step by Step Procedures for the Compleat Idiot* (Santa Fe, NM: John Muir Publications, 1969). (Referred to in *LWEC,* 248.)

Further Reading
McQuiston, Liz. *Graphic Agitation: Social and Political Graphics since the Sixties.* London: Phaidon, 1993.

Pirsig, Robert. *Zen and the Art of Motorcycle Maintenance: An Inquiry into Values.* New York: Morrow, 1974.

Introduction to *How to Keep Your Volkswagen Alive*

Your Volkswagen is not a donkey, but the communication considerations are similar. Your car is constantly telling your senses where it's at: what it's doing and what it needs. I don't speak "donkey," but am fairly conversant in "Volkswagen" and will help you learn the basic vocabulary of this language so your Bus, Bug or Ghia can become an extension of your sensory equipment. Perhaps the idea of feeling about your car is a little strange, but herein lies a type of rapport which will bridge the communication gap between you and your transportation.

I am a man, engineer, mechanic, lover-feeler who has worked and felt with cars of all descriptions for many years. This book contains the product of these years: clear and accurate procedures to heal and keep well your Volkswagen. I don't expect you to become a mechanic—I have done that! My understanding and knowledge will be yours as you work. You supply the labor, the book will supply the direction, so we work as a team, you and I.

While the levels of logic of the human entity are many and varied, your car operates on one

Figure 7.3
Front cover of *How to Keep Your Volkswagen Alive*. John Muir and Tosh Gregg, *How to Keep Your Volkswagen Alive* (Santa Fe, NM: John Muir Publications, 1969).

simple level and it's up to you to understand its trip. Talk to the car, then shut up and listen. Feel with your car, seek the operation out and perform it with love. The type of life your car contains differs from yours by time scale, logic level and conceptual anomalies, but is "Life" nonetheless. Its Karma depends on your desire to make and keep it—ALIVE!

58 Tom Cuthbertson, *Anybody's Bike Book*, 1971

Countercultural bicycle owners enjoyed a variety of benefits associated with the machine: bikes allowed freedom of travel at significantly less cost than an automobile, were an ecologically responsible form of transit, and could even be easily individualized. Tom Cuthbertson, the "John Muir of bicycles," according to Stewart Brand, provided an inexpensive, thorough manual in *Anybody's Bike Book*, sociably instructing both new and more experienced riders in the process of purchasing, maintaining, and repairing a bicycle. Illustrations by Rick Morrall, such as the one shown here, contributed to the text's affable yet educational character. Although *Anybody's Bike Book* was less comprehensive than *The Complete Book of Bicycling*, included below the former book's review on the "Bicycles" subject page, its approachable character and sound advice endeared it to the catalog's editors and presumably the readership as well. —MG

Source

Tom Cuthbertson and Rick Morrall, *Anybody's Bike Book* (Berkeley, CA: Ten Speed Press, 1971). (Referred to in *LWEC*, 253.)

Further Reading

How to Improve Your Cycling. Chicago: Athletic Institute, 1966.

Sloan, Eugene A. *The Complete Book of Bicycling*. New York: Trident Press, 1970.

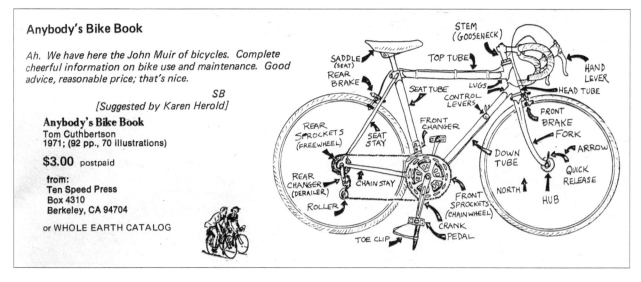

Anybody's Bike Book

Ah. We have here the John Muir of bicycles. Complete cheerful information on bike use and maintenance. Good advice, reasonable price; that's nice.

SB
[Suggested by Karen Herold]

Anybody's Bike Book
Tom Cuthbertson
1971; (92 pp., 70 illustrations)

$3.00 postpaid

from:
Ten Speed Press
Box 4310
Berkeley, CA 94704

or WHOLE EARTH CATALOG

Figure 7.4
"Bike diagram." *LWEC*, 253. Tom Cuthbertson and Rick Morrall, *Anybody's Bike Book* (Berkeley, CA: Ten Speed Press, 1971).

59 William Hillcourt, *The Golden Book of Camping*, 1971

Although only two catalog pages were specifically dedicated to camping, the subject—which encompassed many of the "Nomadics" section's dominant themes, including transience, human enterprise, and environmental connection—pervaded the entire section, from the Luger Camper Kits on the first page through Herb and Judy Klinger's *Knapsacking Abroad* (1967) on the last. The books on the "Camping" and "Camp" pages, such as *Camping and Woodcraft* (1917), *Wildwood Wisdom* (1945), and *The Sierra Club Wilderness Handbook* (1951), all offered comprehensive accounts of both the concept and act of camping and provided necessary context for other texts and products featured in the chapter, many of which adopted a narrower scope with regard to the topic.

To the field of definitive texts just mentioned, Brand introduced William Hillcourt's *The Golden Book of Camping*, which, despite its intended audience of young campers and scouts, contained, according to Brand, "better camping information than the authoritative tomes [on the adjacent pages]" and even "some tips [he had] never heard anywhere before." Hillcourt, or "Green Bar Bill," as he was known to readers of *Boys' Life* magazine, developed a passion for camping and scouting as a youth in early twentieth-century Denmark. By the time he arrived in the United States in the mid-1920s, he had amassed an astonishing knowledge of camping and scouting, which he shared in many texts for the Boy Scouts of America, including the *Handbook for Patrol Leaders* (1929), the *Scout Fieldbook* (1948), and three editions of the *Boy Scout Handbook* (1959, 1965, 1979). *The Golden Book of Camping*, first published in 1959 and once again in 1971, was a slight departure from this oeuvre inasmuch as it was not oriented toward Boy Scouts alone; the author instead sought to inspire a broader readership with a passion for camping, offering a wealth of straightforward educational material, vividly illustrated by Ernest Kurt Barth, as demonstrated in the passages that follow.—MG

Source
William Hillcourt, *The Golden Book of Camping: For Family Groups, Scouts, and Trailblazers—Everything You Need to Know* (New York: Golden Press, 1971), 29. (Referred to in *LWEC*, 273.)

Further Reading
Bear, Sun. *At Home in the Wilderness.* Healdsburg, CA: Naturegraph, 1968.

Carlson, Reynold E. "Organized Camping." *Annals of the American Academy of Political and Social Science* 313 (September 1957): 83–86.

Jaeger, Ellsworth. *Wildwood Wisdom.* New York: Macmillan, 1945.

Camp Expeditions

MAP SYMBOLS

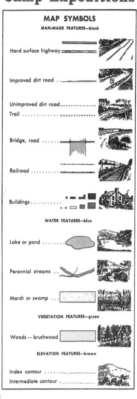

MAN-MADE FEATURES—black

Hard surface highway

Improved dirt road

Unimproved dirt road

Trail

Bridge, road

Railroad

Buildings

WATER FEATURES—blue

Lake or pond

Perennial streams

Marsh or swamp

VEGETATION FEATURES—green

Woods — brushwood

ELEVATION FEATURES—brown

Index contour

Intermediate contour

For expeditions out of camp, you should know how to use map and compass.

The best map is a topographic map in a scale of 1:24000. To get such a map, drop a postcard to Map Information Office, U.S. Geological Survey, Washington, D.C. 20242, and ask for Topographic Map Index Circular for the state in which you expect to travel. From this circular, order the maps you'll need.

Pick a good compass. Among the best is the official Boy Scout Pathfinder compass. Learn to use it by following the instructions on the opposite page. For wilderness traveling, study the Orienteering Handbook, *Be Expert with Map and Compass* ($2.50, American Orienteering Service, Silva, Inc., La Porte, Indiana 46350).

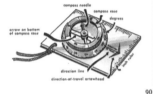

compass needle
compass case
degrees
arrow on bottom of compass case
direction line
direction-of-travel arrowhead

90

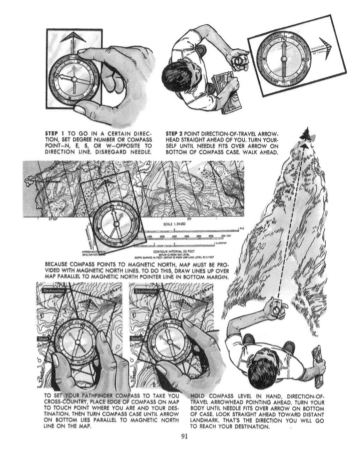

STEP 1 TO GO IN A CERTAIN DIRECTION, SET DEGREE NUMBER OR COMPASS POINT—N, E, S, OR W—OPPOSITE TO DIRECTION LINE. DISREGARD NEEDLE.

STEP 2 POINT DIRECTION-OF-TRAVEL ARROWHEAD STRAIGHT AHEAD OF YOU. TURN YOURSELF UNTIL NEEDLE FITS OVER ARROW ON BOTTOM OF COMPASS CASE. WALK AHEAD.

SCALE 1:24000

CONTOUR INTERVAL 20 FEET

BECAUSE COMPASS POINTS TO MAGNETIC NORTH, MAP MUST BE PROVIDED WITH MAGNETIC NORTH LINES. TO DO THIS, DRAW LINES UP OVER MAP PARALLEL TO MAGNETIC NORTH POINTER LINE IN BOTTOM MARGIN.

TO SET YOUR PATHFINDER COMPASS TO TAKE YOU CROSS-COUNTRY, PLACE EDGE OF COMPASS ON MAP TO TOUCH POINT WHERE YOU ARE AND YOUR DESTINATION, THEN TURN COMPASS CASE UNTIL ARROW ON BOTTOM LIES PARALLEL TO MAGNETIC NORTH LINE ON THE MAP.

HOLD COMPASS LEVEL IN HAND, DIRECTION-OF-TRAVEL ARROWHEAD POINTING AHEAD. TURN YOUR BODY UNTIL NEEDLE FITS OVER ARROW ON BOTTOM OF CASE. LOOK STRAIGHT AHEAD TOWARD DISTANT LANDMARK. THAT'S THE DIRECTION YOU WILL GO TO REACH YOUR DESTINATION.

91

Figure 7.5
"Comfortable Camp Beds." William Hillcourt and Ernest Kurt Barth, *The Golden Book of Camping* (New York: Golden Press, 1971), 28. Reprinted with permission of the "Green Bar Bin" Hillcourt Trust.

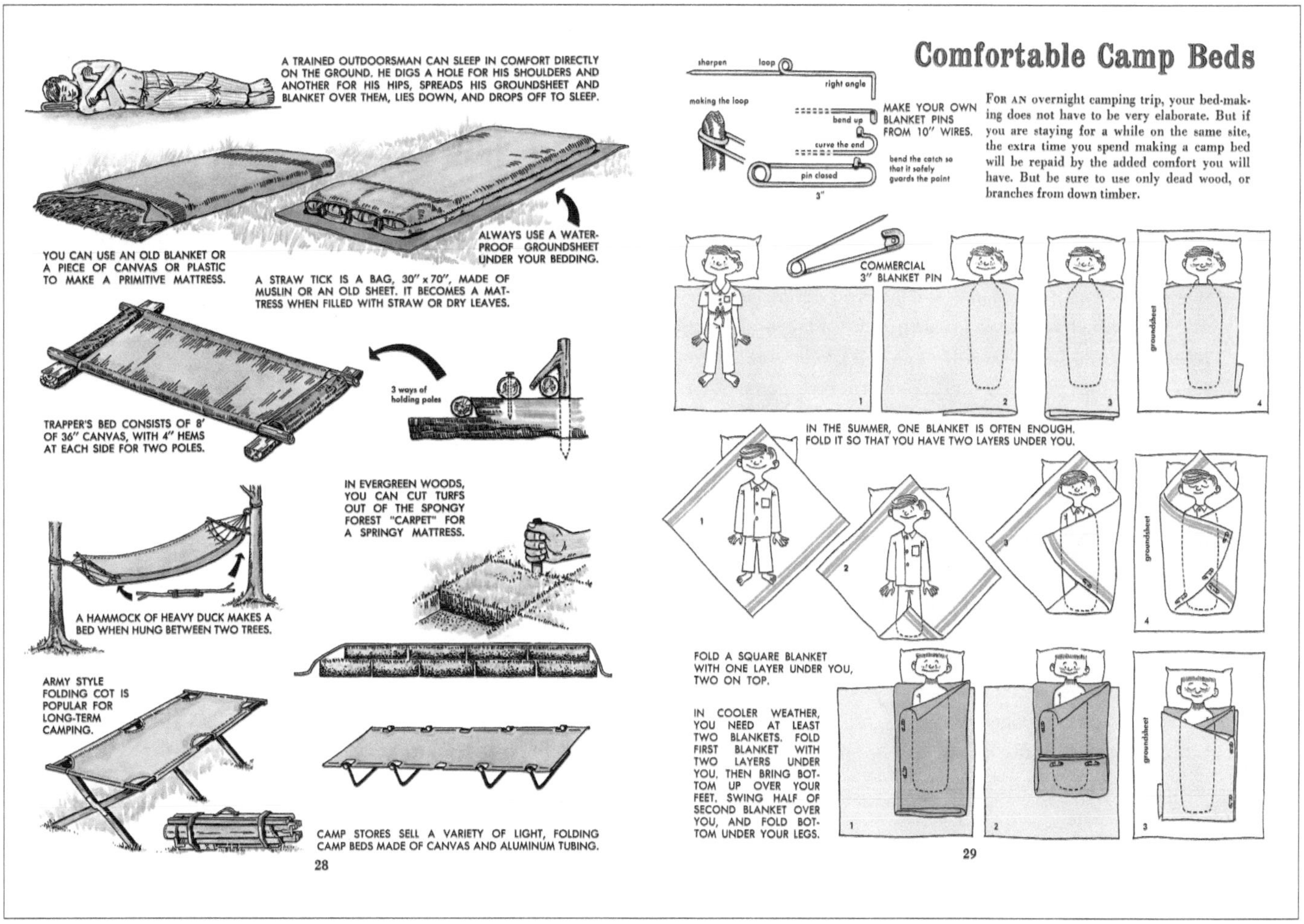

Figure 7.6
"Comfortable Camp Beds." William Hillcourt and Ernest Kurt Barth, *The Golden Book of Camping* (New York: Golden Press, 1971), 29. Reprinted with permission of the "Green Bar Bin" Hillcourt Trust.

60 Anthony Greenbank, *The Book of Survival*, 1967

"One barometer of people's social confidence level is the sales of books on survival," Stewart Brand noted in the opening to his joint review of Anthony Greenbank's *The Book of Survival* and Paul Nesbitt's *The Survival Book*; and if the Whole Earth Truck Store's book sales were any indication, the social confidence level of the catalog's readership was quite low. What troubled Brand, however, was not that sales of survival books were "booming," but that the shop sold more copies of *The Survival Book*, a text aimed at survivors of airplane crashes stranded in unfamiliar and remote territory, than of *The Book of Survival*, which addressed emergencies a layperson might feasibly face in his or her lifetime. To Brand, this apparent preference seemed romantic and naive; it signaled that countercultural readers were "less interested in survival and more…in the return of the frontier."

For readers who were concerned with practical survival skills, however, *The Book of Survival* provided innumerable strategies and guidance. Anthony Greenbank, an expert outdoorsman from Yorkshire, England, intended his book to mentally equip readers "to live through almost every conceivable accident or disaster that our dangerous world can produce," from car crashes and blizzards to muggings and nuclear fallout (xi). Organized by the fundamental source of the emergency—"Too Crowded," "Too Dark," or "Too Fast," for example—*The Book of Survival* examined each contingency the modern reader might be forced to confront. Greenbank's book was not a novelty for the romantics, the countercultural pioneers seeking refuge in the wilderness, but a manual for those committed to preparing themselves for hazards closer to home.—MG

Source
Anthony Greenbank, *The Book of Survival: Everyman's Guide to Staying Alive and Handling Emergencies in the City, the Suburbs, and the Wild Lands Beyond* (New York: Harper & Row, 1967), xiii–xv. (Referred to in *LWEC*, 274.)

Further Reading
Nesbitt, Paul. *The Survival Book*. Princeton, NJ: Van Nostrand, 1959.

Stephens, Don. *The Survivor's Primer and Up-dated Retreater's Bibliography*. Spokane, WA: Stephens, 1976.

Wheat, Margaret M. *Survival Arts of the Primitive Paiutes*. Reno: University of Nevada Press, 1967.

Introduction to *The Book of Survival*

The Book of Survival is like no other. It gives a man, his wife and his children a fighting chance in any catastrophe at a time when—as any newspaper will tell you—the chances are they will succumb and die. It is not a do-it-yourself James Bond manual, for it assumes no advance preparation except reading it.

People ARE growing softer today. Elements of modern-day living not only make us more vulnerable: They fatten for the kill. Read the headlines: *Careering car wipes out newlyweds; runaway train annihilates 37 commuters; crowd at spectator sport crushes 9; flaming electric blanket murders family; unseen knifing in dark discotheque; veering plane massacres 24 vacationers.*

Existing survival books go strongly on the mountain, desert, jungle, arctic and ocean scene. They are useless to the family facing an onrushing tide and trapped by cliffs, or to the salesman whose car is trapped among snowdrifts miles from anywhere. Not only are they unlikely to have read them: If they had, the pertinent information—wrapped among graphs, tables, case histories, maps and bibliographies—would be almost impossible to recollect.

Most of us have only the vaguest idea of survival. If thought about at all, it is by some quaint rule of thumb (does one pour gasoline on frostbitten fingers—NO!). We face the bitterest winters at home with equanimity—and are surprised, shocked and hurt when threats of exposure and frostbite turn into a reality.

This book deals with *crises*. Prevention is better than cure, but no one is immune from human or mechanical failure, or Act of God. Hence the stress on *when* catastrophe hits *home—when* car is submarining to bottom of lake, *when* intruders are binding you hand and foot, *when* eyes are blinded. Where possible, prevention is treated too—in note form.

People would never remember in a crisis? Not usually so! Fed with a drill to save its skin, the brain grabs that plan by the scruff of the neck. True, the girl whose arms are already pinned by an assailant cannot produce the comb which might save her, but let her hear him first padding behind along the pavement and she could.

Instead of being mesmerized with fright, you are more likely to be ready and braced in a sudden crisis after reading this book than before. In a fight for life lasting weeks your subconscious would automatically reach back for help from its pages. To this end, mnemonics (memory-fixers) are incorporated in the very layout of the book by using entry headings unique in survival tracts to date.

Human feelings are the key. Pitched into catastrophe they cry for warmth or cold or water or dry land or shade or light or speed or slowness regardless of whether dazzling glare is from H-bomb or headlights, stifling heat is in desert or burning house, choking contamination is radiation or carbon monoxide. This book is arranged to rally these feelings by grouping remedies under them: fire making under *TOO COLD*, car-collision drill under *TOO FAST*. And so on.

No entry here presumes that you are prepared. When crisis *looms—that's it!* There are YOU (bone/skin/hair/teeth/nails/saliva), your possessions (shoes/socks/pants/skirt/dress/watch/possibly cash, comb, etc.), the surroundings (sand/rock/water/trees/concrete/guano) and quite often wrecked transportation (car/plane/train/boat). Desert captives are not allowed to find a handy sheet of plastic for a water still, though if they have one already, its use is shown—but well down the list of water-seeking methods.

The theme of this book is not so much "be prepared" as *be prepared to improvise with anything*. When "everything" seems so slight as to be ridiculous, think of rescuers saying: "If only he'd wrapped up in those newspapers…" This book takes it for granted you have nothing rather than something… but if you have something, use it.

Using this book we suggest you (a) anticipate trouble ahead in likely places (football game/plane flight/rush-hour traffic) by checking in the index under such headings, then turning to the relevant pages for advance briefing; (b) read the book more than once so that in sudden fire/flood/tempest/earthquake you increase your chances of surviving. And as an extra aid the entries are linked by cross-references: Climbing from sea into boat comes in *TOO LOW*, but getting to that boat happens in *TOO WET*. Hence there is a cross-reference to link them.

Survivors will always live to tell of surviving by doing just the opposite of others who have also survived. Medical experts have often told survivors that by all rights they should be dead. Instead of dying they had the WILL to live. YOU TOO MUST ENLIST THIS WILL, that sense of self-preservation which starts with a deep breath and the determination not to give way at any cost.

People are growing softer today—yes; but it is a cheering fact that, given a plan of action, even though despairing and shocked, people can and do react well in crises. Once the initial shock has lessened you stand every chance of becoming a *survivor* if, as we suggest, you give yourself a regular servicing by rereading this, a kind of maintenance handbook.

Communications

The fifty-nine-page "Communications" section covered a wide range of topics, from different forms of conceptualization (such as mapping, language, mental structures, and cognition), to means of reproduction and distribution (graphics, electronics, and computer design), to media genres (music, theater, photography, and film). There was even a section on economics, which discussed radical alternatives to capitalist systems. The cumulative effect was to stress the value of communication between people and the need to control the media and means of distribution. Although books on do-it-yourself projects—musical instruments, for example—were included, many of the articles require extremely expensive items of equipment, such as cameras, video equipment, portable sound recorders, calculators, and computers.

The catalog aimed to make conceptually difficult subjects interesting by using imagery and juxtaposing different approaches. For example, the spread on pages 318–319 of the *LWEC* presented books on mathematics interspersed with images of geometric structures and patterns, including *The Graphic Work of M.C. Escher* and Ernst Haeckel's *Challenger* monograph. This layout was attractive to architects. The catalog being itself a medium of communication, we find items here that refer to the workings of the magazine itself, such as the Indecks information retrieval system cards and the McBee cards (*LWEC*, 320), which Brand used in organizing the material for publication. Similarly, Brand was fascinated by means of organizing data and recommended J. L. Jolley's *Data*

Study (1968) with the observation that "information that isn't organized isn't signal in your life, it's noise. You waste yourself searching the full length of a file for something, feeling as stupid as a driver in crosstown New York, and for the same reason: This book can help you if not New York. It presents theory and practice on how to keep stuff straight, at least in terms of organization" (322).

Brand's appreciation of *Fortune* magazine might appear surprising: "I started reading *Fortune* after reading a lot of Buckminster Fuller because it was the only periodical with regular informative articles on his level of thinking (world shipping, construction, new industrial processes, natural resources, etc.). What I found was editing and reporting that made the contents of most popular American magazines look like so much paste" (340). A double-page spread on photography (352–353) attested to Brand's technical and artistic interest in the medium: "Three teachers and two books taught me photography. John Collier showed me the objective power of photographs (anthropologists with a polaroid start far more informative conversations than anthropologists without). Minor White showed me how to go into mild trance to see directly. Jack Welpott showed me that the main obstacle to a good picture is my own ideas and arrogance" (352).—CM

61 Arthur Lockwood, *Diagrams*, 1969

Lockwood's book was selected first for the section, which we can take to be significant. The choice of this book demonstrated the power of images to convey knowledge. This was one of Brand's favorite themes. He introduced the selection as follows: "A diagram is a conceptual map. Elegantly done it can ease comprehension. Thoroughly done it can aid analysis. Done with originality it can remake your internal world. This book, the first of its kind, is a splendid survey of the range and usefulness of diagramming."

The wide selection of forms of representation—from different countries—was intended to stimulate the reader to think about how to explain complex organizations and structures. The diagrams shown in Lockwood's book come from a wide variety of sources: atlases, encyclopedias, school and university textbooks, scientific magazines, newspapers, advertisements, and television. Techniques of communication vary from abstract charts and diagrams to symbolic pictograms, deformation of maps, timelines, and flowcharts. The author shows how diagrams can be used to compare and evaluate, but also to manipulate knowledge and to persuade. Brand pointed out that "the pioneers—in effective presentation—include magazines such as *Fortune* and *Scientific American*, firms such as the Container Corporation of America and particularly the Isotype Institute, London."

The layout on page 306 of *LWEC* distinguished between four kinds of representations. The objective representation of quantities in the Isotype chart is very different from the conceptual mapping of schools of European painting in which quantities and relationships represent a series of value judgments. The distortion used in the *Fortune* magazine map showing the size of US investment in foreign countries clearly has a rhetorical purpose by contrasting the flow of investment with the actual size of different countries. The fourth diagram tries to represent visually the extent to which different vowel sounds might be distinguished. Space here is a metaphor for clarity. —CM

Source
Arthur Lockwood, *Diagrams: A Visual Survey of Graphs, Maps, Charts and Diagrams for the Graphic Designer* (London: Studio Vista, 1969), 113. (Referred to in *LWEC*, 306.)

Further Reading
Morrish, Michael, and Diagram Group. *Human Geography in Diagrams.* New York: Cambridge University Press, 1986.

Figure 8.1
"Flow Chart." Arthur Lockwood, *Diagrams: A Visual Survey of Graphs, Maps, Charts and Diagrams for the Graphic Designer* (London: Studio Vista, 1969), 113.

Flow chart

1   Diagram from a school book, showing products derived from coal. *Le Charbon,* Hachette

2   Diagram from a scientific magazine, showing energy flow from sun. See family tree version page 127. *Power*

3   Diagram from a text book, showing stages from ingot to finished product. Compare with typographical method opposite. J. A. Rankin, *Workshop processes and materials for mechanical engineering technicians,* Penguin Education

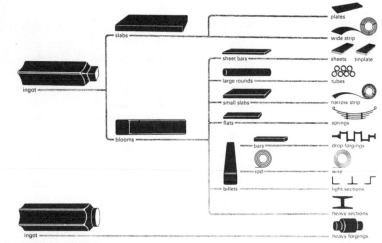

2

3

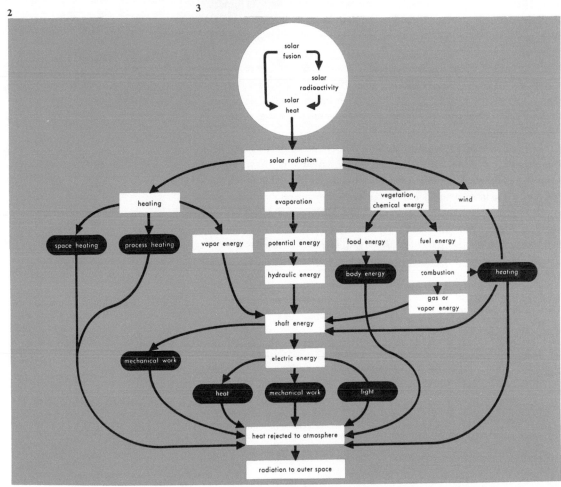

113

62 David Greenhood, *Mapping*, 1964

David Greenhood's *Mapping*, based on his earlier *Down to Earth: Mapping for Everybody* (New York: Holiday House, 1944), is a delight to read, well written and enthusiastic in tone. In the introduction, the author advises the reader: "This book has been written to be read rather than studied. It is a book for the amateur, designed to give him an understanding and appreciation of maps" (*Mapping*, ix). The book is, however, extremely thorough, with an index, a bibliography, and a chapter devoted to the tools used for observing, surveying, mapping, and drafting. The author explains how to interpret maps and discern their hidden meaning. For example, he cites the Harvard map expert Erwin Raisz: "It seems to be a tendency among map makers of every country to put at the top of the map the direction toward which national attention is turned. The Romans headed their maps as they so often did with their empire- stretching ships, eastward. So did the Crusaders trying to recover the Holy Land" (*Mapping*, x).

The author insists on the imaginative role of mapmaking: "Many of the most inventive novelists draw maps while writing, so that they may keep before them the whole scene of the action and make their fiction as credible as history to the reader. … In contemplating maps, whether already made or about to be made, the imagination is a vital activity" (*Mapping*, x).

This is what appealed to Brand. In his introduction, he wrote: "A map is the meeting ground of drawing, writing, and geometry. No other medium carries such a wealth of critical information at glance readiness. Students of brain and thought design are lately placing more and more emphasis on the matter of Where-Is-It—apparently much of the mind's store, retrieve, and relate systems are based on position relationships in mental space" (*LWEC*, 306). Buckminster Fuller expended a great deal of energy trying to invent a new system of cartographic representation. His world map was illustrated in the 1964 edition of Greenhood's book.—CM

Source

LWEC, 306. The images featured in the "Diagrams" subsection are from David Greenhood, *Mapping*, illus. Ralph Graeter (Chicago: University of Chicago Press, 1964).

Further Reading

Koselleck, Reinhart, and Todd Samuel Presner. *The Practice of Conceptual History: Timing History, Spacing Concepts, Cultural Memory in the Present.* Stanford, CA: Stanford University Press, 2002.

Richards, E. G. *Mapping Time: The Calendar and Its History.* New York: Oxford University Press, 1999.

Rosenberg, Daniel, and Anthony Grafton. *Cartographies of Time.* New York: Princeton Architectural Press, 2010.

Figure 8.2
LWEC, 306. From David Greenhood, *Mapping*, illus. Ralph Graeter (Chicago: University of Chicago Press, 1964).

Communications

Diagrams

A diagram is a conceptual map. Elegantly done it can ease comprehension. Thoroughly done it can aid analysis. Done with originality it... remake your internal world. This book, the first of its kind, is a splendid survey of the range and usefulness of diagramming.
—SB

Diagrams
Arthur Lockwood
1969; 254 pp.

$15.00 postpaid

from:
Watson-Guptill Publications
2160 Patterson St.
Cincinnati, OH 45214

or WHOLE EARTH CATALOG

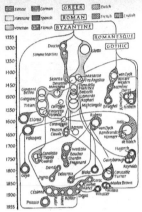

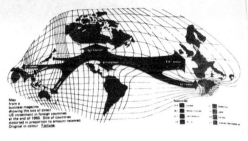

Map
from a
business magazine
showing the size of direct
US investment in foreign countries
at the end of 1966. Size of countries
distorted in proportion to amount received.
Original in colour. **Fortune**.

Chart from a book, showing chief schools of European painting from Giotto. Devised by Eric Newton who explains that it attempts to indicate relative importance of schools (areas of shaded masses), approximate dates, principal artists (circle), relative importance (size of circle), threads of influence between schools and artists. Eric Newton, **European Painting and Sculpture**, Faber and Faber.

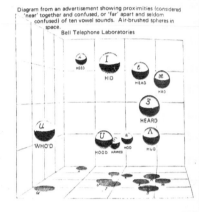

Diagram from an advertisement showing proximities (considered 'near' together and confused, or 'far' apart and seldom confused) of ten vowel sounds. Air-brushed spheres in space. Bell Telephone Laboratories

There is no attempt to give this book a historical perspective, although some of the pioneers in the effective presentation of information in visual form are included. The pioneers include magazines such as **Fortune** and **Scientific American**; firms such as the Container Corporation of America, and particularly the Isotype Institute, London, whose approach and achievement has affected all designers of diagrams even though they may not be aware of their debt.

Mapping

A map is the meeting ground of drawing, writing, and geometry. No other medium carries such a wealth of critical information at glance readiness. Students of brain and thought design are lately placing more and more emphasis on the matter of Where-Is-It—apparently much of the mind's store, retrieve, and relate systems are based on position relationships in mental space.

This book is a well-made introduction to map use, map construction, and something of the meaning of maps.
—SB

Mapping
David Greenhood
1944, 1964; 288 pp.

$2.95 postpaid

from:
University of Chicago Press
5801 Ellis Ave.
Chicago, Illinois 60637

or WHOLE EARTH CATALOG

Pen pressed against T square *too hard*

Pen sloped away from T square

Pen too close to edge. Ink ran under

Ink on outside of blade, ran under

Pen blades not kept parallel to T square

T square (or triangle) slipped into wet line

Not enough ink to finish line

Avoidable troubles of the ruling pen

Land-form map. Accurate information shown with vividness and grace. Drawn by Erwin Raisz for a geography textbook

We can scan the history of civilization as well as of mapping itself simply by observing which direction has topped the map. Erwin Raisz, the Harvard authority on cartography, says, "It seems to be a tendency among map makers of every country to put at the top of the map the direction toward which national attention is turned."

The Romans headed their maps as they so often did their empire-stretching ships, eastward. So did the Crusaders trying to recover the Holy Land. Many medieval wind roses have a cross as an east-mark.

Maps and charts—you may have noticed in this chapter the repetition of these two words. What's the difference in their meaning? Charts are maps with a special, practical purpose such as navigating, weather forecasting, and population studies. A chart is a map on which to work with protractor, compass, dividers, and guages—even a densitometer.

This connection between local pinpoints and cosmic points is one of the majestic relationships recognized by science. Considering how much the mere idea of a map encompasses, maps are probably more unconfusing than anything else put down on paper. A map should be regarded as an antidote to panic.

63 Norbert Wiener, *Cybernetics*, 1961

Norbert Wiener was a brilliant mathematician who had made a number of important discoveries before turning his interest to what he called cybernetics. Wiener's theoretical interests were given a wider relevance by his research into the guidance of antiaircraft fire during World War II. His interaction with neurologists, anthropologists, and philosophers in the Macy Conferences,[1] organized in 1946 to promote interdisciplinary research in New York, led to his decision to formulate a general theory of what he called "the scientific study of control and communication in the animal and the machine."' The first edition of *Cybernetics* (1948) prompted research into self-regulating systems and was instrumental in the design of robotics and computers.

Many of Stewart Brand's collaborators became increasingly interested in computers and later developed magazines such as *Wired* (1993) to bridge the gap between cybernetics and practical computing. As Brand explained in his book review, it was the ambitious way in which cybernetics appears to explain systems of communication from biology to electronics that gave the book its appeal: "The research and speculation that led to computer design arose from investigation of healthy and pathological human response patterns embodied in the topological make-up of the nervous system....Society, from organism to community to civilization to universe, is the domain of cybernetics. Norbert Wiener has the story, and to some extent, is the story."

Brand would later turn his attention from cybernetics to the ideas of the philosopher Gregory Bateson, a founding member of the Macy Conferences. Bateson's work was not included in the *LWEC*, but he became a guru of subsequent publications.—CM

1. The Macy Conferences were ten meetings of scholars from different academic disciplines held in New York between 1946 and 1953. They were conceived and organized by Warren McCulloch and the Josiah Macy Jr. Foundation. The main purpose of the meetings was to set the foundations for a general science of the workings of the human mind.

Source

Norbert Wiener, *Cybernetics; or, Control and Communication in the Animal and the Machine* (Cambridge, MA: MIT Press, 1961), 15–27. © Norbert Wiener. Reprinted with the permission of the MIT Press. (Referred to in *LWEC*, 307.)

Further Reading

Hardin, Garrett. "Tragedy of the Commons." *Science* 162 (1968): 1243–1248.

Wiener, Norbert. *The Human Use of Human Beings: Cybernetics and Society.* Garden City, NY: Doubleday, 1954.

"Cybernetics in History," from *Cybernetics*

It is the thesis of this book that society can only be understood through a study of the messages and the communication facilities which belong to it; and that in the future development of these messages and communication facilities, messages between man and machines, between machines and man, and between machine and machine, are destined to play an ever-increasing part....

Naturally there are detailed differences in messages and in problems of control, not only between a living organism and a machine, but within each narrower class of beings. It is the purpose of Cybernetics to develop a language and techniques that will enable us indeed to attack the problem of control and communication in general, but also to find the proper repertory of ideas and techniques to classify their particular manifestations under certain concepts.

The commands through which we exercise our control over our environment are a kind of information which we impart to it. Like any form of information, these commands are subject to disorganization in transit. They generally come through in less coherent fashion and certainly not more coherently than they were sent. In control and communication we are always fighting nature's tendency to degrade the organized and to destroy the meaningful; the tendency, as Gibbs has shown us, for entropy to increase. Much of this book concerns the limits of communication within and among individuals. Man is immersed in a world which he perceives through his sense organs. Information that he receives is co-ordinated through his brain and nervous system until, after the proper process of storage, collation, and selection, it emerges through effector organs, generally his muscles. These in turn act on the external world, and also react on the central nervous system through receptor organs such as the end organs of kinaesthesia; and the information received by the kinaesthetic organs is combined with his already accumulated store of information to influence future action.

Information is a name for the content of what is exchanged with the outer world as we adjust to it, and make our adjustment felt upon it. The process of receiving and of using information is the process of our adjusting to the contingencies of the outer environment, and of our living effectively within that environment. The needs and the complexity of modern life make greater demands on this process of information than ever before, and our press, our museums, our scientific laboratories, our universities, our libraries and textbooks, are obliged to meet the needs of this process or fail in their purpose. To live effectively is to live with adequate information. Thus, communication and control belong to the essence of man's inner life, even as they belong to his life in society....

I call to the kitten and it looks up. I have sent it a message which it has received by its sensory organs, and which it registers in action. The kitten is hungry and lets out a pitiful wail. This time it is the sender of a message. The kitten bats at a swinging spool. The spool swings to its left, and the kitten catches it with its left paw. This time messages of a very complicated nature are both sent and received within the kitten's own nervous system through certain nerve end-bodies in its joints, muscles, and tendons; and by means of nervous messages sent by these organs, the animal is aware of the actual position and tensions of its tissues. It is only through these organs that anything like a manual skill is possible....

I have said that man and the animal have a kinaesthetic sense, by which they keep a record of the position and tensions of their muscles. For any machine subject to a varied external environment to act effectively it is necessary that information concerning the results of its own action be furnished to it as part of the information on which it must continue to act. For example, if we are running an elevator, it is not enough to open the outside door because the orders we have given should make the elevator be at that door at the time we open it. It is important that the release for opening the door be dependent on the fact that the elevator is actually at the door; otherwise something might have detained it, and the passenger might step into the empty shaft. This control of a machine on the basis of its *actual* performance rather than its *expected* performance is known as *feedback*, and involves sensory members which are actuated by motor members and perform the function of *tell-tales* or *monitors*—that is, of elements which indicate a performance. It is the function of these mechanisms to control the mechanical tendency toward disorganization; in other words, to produce a temporary and local reversal of the normal direction of entropy.…

If I pick up my cigar, I do not will to move any specific muscles. Indeed in many cases, I do not know what those muscles are.…Similarly, when I drive a car, I do not follow out a series of commands dependent simply on a mental image of the road and the task I am doing. If I find the car swerving too much to the right, that causes me to pull it to the left. This depends on the actual performance of the car, and not simply on the road; and it allows me to drive with nearly equal efficiency a light Austin or a heavy truck, without having formed separate habits for the driving of the two.…

It is my thesis that the physical functioning of the living individual and the operation of some of the newer communication machines are precisely parallel in their analogous attempts to control entropy through feedback. Both of them have sensory receptors as one stage in their cycle of operation: that is, in both of them there exists a special apparatus for collecting information from the outer world at low energy levels, and for making it available in the operation of the individual or of the machine. In both cases these external messages are not taken *neat*, but through the internal transforming powers of the apparatus, whether it be alive or dead. The information is then turned into a *new* form available for the further stages of performance. In both the animal and the machine this performance is made to be effective on the outer world. In both of them, their *performed* action on the outer world, and not merely their *intended* action, is reported back to the central regulatory apparatus. This complex of behavior is ignored by the average man, and in particular does not play the role that it should in our habitual analysis of society; for just as individual physical responses may be seen from this point of view, so may the organic responses of society itself. I do not mean that the sociologist is unaware of the existence and complex nature of communications in society, but until recently he has tended to overlook the extent to which they are the cement which binds its fabric together.

64 Marshall McLuhan, *Culture Is Our Business*, 1970

Upon the publication of *Understanding Media* in 1964, Marshall McLuhan, a Cambridge-educated professor of English literature-cum-media and communications analyst, quickly became a distinguished voice in discussions of modern media culture. His well-known, characteristically pithy phrase, "the medium is the message," which succinctly described the complex idea that meaning is inextricably embedded in the medium through which it is communicated, was especially evocative, influencing countercultural thinkers, corporate strategists, and academics alike and even inspiring a "McLuhan festival" in San Francisco.

Culture Is Our Business, published six years later, expressed this theory visually, through a close reading of commercial advertisements; McLuhan paired each advertisement, printed on the right-hand side of a two-page spread, with his own narratives, musings, and analyses, as well as literary and popular quotations on the left, to reorient and excite readers' perceptions of the "hidden…magic forms" (the ads) that circumscribed their lives, unseen but effective (*Culture Is Our Business*, 48). "The resulting energy across the spread," Brand recounted, "is economic and multi-directional—i.e., you make it."

That Brand, a great admirer of McLuhan's work, praised *Culture Is Our Business* as "[the author's] best format," is unsurprising given his foregoing commentary, which could veritably describe the catalog itself. Although only two of McLuhan's texts appeared in the catalog—*Understanding Media* being the other, placed on the adjacent page—his theories of media culture and modern communications coursed throughout the *LWEC*, a unique medium that acknowledged, unabashedly, its power to convey ideas both directly and indirectly through conscientious formal decisions.—MG

Source
Marshall McLuhan, *Culture Is Our Business* (New York: McGraw-Hill, 1970), 180–181. Courtesy of the Estate of Corinne McLuhan. (Referred to in *LWEC*, 309.)

Further Reading
Kostelanetz, Richard. *The Theatre of Mixed Means: An Introduction to Happenings, Kinetic Environments, and Other Mixed-Media Performances.* New York: Dial Press, 1968.

McLuhan, Marshall. *Understanding Media: The Extensions of Man.* New York: McGraw Hill, 1964.

Oren, Michael. "USCO: 'Getting Out of Your Mind to Use Your Head.'" *Art Journal* 69, no. 4 (2010): 76–95.

PICKING DAISIES

Like symbolic poetry, the present ad works by suggestion, not statement. It starts with the effect and lets the audience fill in the cause. Shirts are not featured. The implied shirts make you feel fresh as a daisy, etc. In the same way, the filling in of the gap between the old biological environment of our bodies and the new electric environment of our extended nervous system automatically evokes the world of ESP and LSD.

"Of necessity, therefore, anything in process of change is being changed by something else." (Thomas Aquinas)

When the evolutionary process shifts from biology to software technology the body becomes the old hardware environment. The human body is now a probe, a laboratory for experiments. In the middle of the nineteenth century Claude Bernard was the first medical man to conceive of le milieu intérieur. He saw the body, not as an outer object, but as an inner landscape, exactly as did the new painters and poets of the avant garde.

Figure 8.3
"Picking Daisies." Marshall McLuhan, *Culture Is Our Business* (New York: McGraw-Hill, 1970), 180–181. Courtesy of the Estate of Corinne McLuhan.

65 Nicholas Negroponte, *The Architecture Machine*, 1970

In 1967, Nicholas Negroponte, then a young professor at the Massachusetts Institute of Technology's School of Architecture and Planning, established the Architecture Machine Group (Arch Mac) in partnership with fellow faculty member Leon Groisser, an assistant professor of structures at the school. Arch Mac, which would later be integrated into MIT's Media Lab, sought to further the field of computer-aided design, adopting an interdisciplinary approach through which professors and students of architecture, electrical engineering, and computer science might work collaboratively on innovative projects. One such project, URBAN 5, a computer-aided design program aimed at nurturing a relationship between man and machine grounded in "mutual training, resilience, and growth" (*The Architecture Machine*, ii), greatly inspired Negroponte's book, published just three years after Arch Mac's conception.

Although Stewart Brand's description suggested otherwise, *The Architecture Machine* was less a "good intro to life with dumb-fuck genius machines" and more a working document, by the admission of the author, and a fascinating reflection on the potential for the symbiotic exchange of knowledge between computer and human designer. The book, Negroponte wrote in the "preface to the preface," was intended not for "those who expect to find the answers in this volume" but instead for those "who are interested in groping with problems they do not know how to handle" (*The Architecture Machine*, ii). In the passage that follows, Negroponte introduced the problems with which he and his colleagues were grappling and which he would consider in greater depth in the chapters beyond. — MG

Source
Nicholas Negroponte, *The Architecture Machine* (Cambridge, MA: MIT Press, 1970), 9–15. © Nicholas Negroponte. Reprinted with the permission of the MIT Press. (Referred to in *LWEC*, 321.)

Further Reading
Allen, Edward, ed. *The Responsive House: Selected Papers and Discussions from the Shirt-Sleeve Session in Responsive House-building Technologies Held at the Department of Architecture, Massachusetts Institute of Technology, Cambridge, Massachusetts, May 3–5, 1972.* Cambridge, MA: MIT Press, 1974.

Krampen, Martin, and Peter Seitz, eds. *Design and Planning 2: Computers in Design and Communication.* New York: Hastings House, 1967.

Negroponte, Nicholas. *Soft Architecture Machines.* Cambridge, MA: MIT Press, 1975.

"Prelude to an Architecture-Machine Dialogue," from *The Architecture Machine*

You are in a foreign country, do not know the language, and are in desperate need of help. At first your hand movements and facial expressions carry most of your meaning to the silent observer. Your behavior uses a language of gestures and strange utterances to communicate your purpose. The puzzled listener searches for bits of content he can understand and link to his own language. You react to his reactions, and a language of pantomime begins to unfold. This new language has evolved from the mutual effort to communicate. Returning to the same person a second time, let us say with a new need, the roots of a dialogue already exist. This second conversation might be gibberish to a third party brought into the exchange at this time.

A designer-to-machine introduction should have a similar linguistic evolution. Each should track the other's design maneuvers, evoking a rhetoric that cannot be anticipated. "What was mere noise and disorder or distraction before, becomes pattern and sense; information has been metabolized out of noise" (Brodey and Lindgren, 1967). The event is circular inasmuch as the designer-machine unity provokes a dialogue and the dialogue promotes a stronger designer-machine unity. This progressively intimate association of the two dissimilar species is the symbiosis. It evolves through mutual training, in this case, through the dialogue.

Such man-machine dialogue has no historical precedent. The present antagonistic mismatch between man and machine, however, has generated a great deal of preoccupation for it. In less than a decade the term "man-machine communication" has passed from concept to cliché to platitude. Nevertheless, the theory is important and straightforward: in order to have a cooperative interaction between a designer of a certain expertise and a machine of some scholarship, the two must be congenial and must share the labor of establishing a common language. A designer, when addressing a machine, must not be forced to resort to machine-oriented codes. And in spite of computational efficiency, a paradigm for fruitful conversations must be machines that can speak and respond to a natural language.

With direct, fluid, and natural man-machine discourse, two former barriers between architects and computing machines would be removed. First, the designers, using computer-aided design hardware, would not have to be specialists. With natural communication, the "this is what I want to do" and "can you do it" gap could be bridged. The design task would no longer be described to a "knobs and dials" person to be executed in his secret vernacular. Instead, with simple negotiations, the job would be formulated and executed in the designer's own idiom. As a result, a vibrant stream of ideas could be directly channeled from the designer to the machine and back....

But, the tête-à-tête must be even more direct and fluid; it is gestures, smiles, and frowns that turn a conversation into a dialogue. "Most Americans are only dimly aware of this silent language even though they use it every day. They are not conscious of the elaborate patterning of behavior which prescribes our handling of time, our spatial relationships, our attitudes towards work, play, and learning" (Hall, 1959). In an intimate human-to-human dialogue, hand-waving often carries as much meaning as text. *Manner* carries cultural information: the Arabs use their noses, the Japanese nod their heads. Customarily, in man-machine communication studies, such silent languages are ignored and frequently are referred to as "noise." But such silent languages are not noise; a dialogue is composed of "whole body involvement—with hands, eyes, mouth, facial expressions—using many channels simultaneously, but rhythmized into a harmoniously simple exchange" (Brodey and Lindgren, 1968).

Imagine a machine that can follow your design methodology and at the same time discern and assimilate your conversational idiosyncrasies. This same machine, after observing your behavior, could build a predictive model of your conversational performance. Such a machine could then reinforce the dialogue by using the predictive model to respond to you in a manner that is in rhythm with your personal behavior and conversational idiosyncrasies.

What this means is that the dialogue we are proposing would be so personal that you would not be able to use someone else's machine, and he would not understand yours. In fact, neither machine would be able to talk directly to the other. The dialogue would be so intimate—even exclusive—that only mutual persuasion and compromise would bring about ideas, ideas unrealizable by either conversant alone. No doubt, in such a symbiosis it would not be solely the human designer who would decide when the machine is relevant....

One might argue that we are proposing the creation of a design machine that is an extension of, and in the image of, a designer who, as he stands, has already enough error and fault. However, we have indicated that the maturation would be a reciprocal ripening of ideas and ways. At first, jobs where the man is particularly inept would stimulate a nontrivial need for cooperation. Subsequently each interlocutor would avoid situations notably clumsy for his constitution, while prying into issues that were originally outside the scope of concern (or the concern of his profession). Eventually, a separation of the parts could not happen: "The entire 'symbiotic' system is an artificial intelligence that cannot be partitioned" (Pask, 1964).

In the prelude to an architect-machine dialogue the solidarity of the alliance will rely on the ease of communication, the ability to ventilate one's concerns in a natural vernacular, and the presence of modes of communication responsive to the discipline at hand. A wine taster would expect his partner to have taste buds and an understanding of vintages. An architect would expect his associate to have at least a graphic ability capable of manipulating and displaying a host of environmental data and, in particular, physical form.

66 *Cybernetic Serendipity*, 1968

Cybernetic Serendipity was both an exhibition at the ICA in London in 1968 and a special issue of *Studio International* magazine, which was also published as a book. In a page dedicated to computers, Brand welcomed it as "the best collection of computer art yet." The exhibition was curated by the art critic Jasia Reichardt and designed by the well-known illustrator Franciszka Themerson. As Reichardt explained: "The aim is to present an area of activity which manifests artists' involvement with science, and the scientists' involvement with the arts; also, to show the links between the random systems employed by artists, composers and poets, and those involved with the making and the use of cybernetic devices."[1] The exhibition contributed to a long-running debate in Britain about the relationship of the arts and sciences.

Reichardt believed that the exhibition's most important contribution was to draw engineers and architects into the realm of the arts. Norbert Wiener was celebrated with an extract from *The Human Use of Human Beings*, and his ideas informed much of the work. John Cage was one of the many composers involved in the exhibition, and his graphic scores might as well have been included as computer art.

Reichardt presented the exhibition as follows:

Computer-generated graphics, computer-animated films, computer-composed and played music, and computer poems and texts.

Cybernetic devices as works of art, cybernetics environments, remote-control robots and painting machines.

Machines demonstrating the uses of computers and an environment dealing with the history of cybernetics.

Cybernetic Serendipity deals with possibilities rather than achievements, and in this sense it is prematurely optimistic. There are no heroic claims to be made because computers have so far neither revolutionized music, nor art, nor poetry, in the same way that they have revolutionized science.[2]

—CM

1. *Cybernetic Serendipity: Studio International Special Issue*, 2nd ed. (New York: Praeger, 1969), 5.

2. Ibid.

Source

Jasia Reichardt, ed., "Cybernetic Serendipity: The Computer and the Arts," special issue, *Studio International*, 1968, cover. Courtesy *Studio International*, www.studiointernational.com. (Referred to in *LWEC*, 322.)

Further Reading

Cybernetic Serendipity. Exhibition Institute of Contemporary Arts, London, July 2014.

Cybernetic Serendipity (ICA)—Late Night Lineup (1968). https://www.youtube.com/watch?v=n8TJx8n9UsA.

Exhibition Histories talks: Jasia Reichardt in conversation with David Morris. Whitechapel Gallery, January 23, 2014. http://www.afterall.org/online/exhibition-histories-talks_jasia-reichardt-video-online#.Vq5zO7LhC02.

Figure 8.4
Cover image from "Cybernetic Serendipity:
The Computer and the Arts." Jasia
Reichardt, ed., "Cybernetic Serendipity:
The Computer and the Arts," special issue,
Studio International, 1968. Courtesy
of *Studio International*, http://www.
studiointernational.com.

67 Robert Theobald, *An Alternative Future for America II*, 1970

In 1964, the radical economist and futurist Robert Theobald, along with fellow members of the Ad Hoc Committee on the Triple Revolution, prepared a written statement, titled "The Triple Revolution," directed at US president Lyndon B. Johnson, secretary of labor W. Willard Wirtz, and the majority and minority leaders of the Senate and House of Representatives, in response to the proposed (and later enacted) Economic Opportunity Act. The committee contended that "three separate and mutually reinforcing revolutions"—cybernation, weaponry, and human rights—were currently affecting mankind and necessitated a "fundamental reexamination of existing values and institutions," especially with regard to economic and labor policy. Within this revolutionary triad, they presented as their primary concern the cybernation revolution, a movement that, they believed, heralded a postindustrial era characterized by increased automation and, as a result, a greater rate of unemployment as unskilled workers would no longer be needed.[1]

Theobald continued such analysis in his futurist economics books, *An Alternative Future for America* (1968) and *An Alternative Future for America II: Essays and Speeches*, in which he examined the potential social, political, and economic impacts of computer technology and industrial automation in the United States. In the passage that follows, the author grappled once more with the ideological shortcomings of Johnson's "War on Poverty," framing the elimination of poverty as an ethical issue, at the root of which lay the "unfounded and heartless belief that the poor have only themselves to blame for their situation." His minimum solution, "guaranteed income" for all US citizens, would, by his estimation, require little economic sacrifice from the federal government but would demand an evolution of social values and public conscientiousness. One of Theobald's favorite quotations, "We are as gods, and we'd better get good at it," from the first page of the catalog, gets at this commitment to a better society and contextualizes his popularity within the counterculture as well.[2]—MG

1. Ad Hoc Committee on the Triple Revolution, "The Triple Revolution," *Liberation*, April 1964, 9.

2. Stephen Silha, "Robert Theobald: The Ultimate Model of Citizenship," *Christian Science Monitor*, March 13, 2000, http://www.csmonitor.com/2000/0313/p21s2.html (accessed August 30, 2015).

Source

Robert Theobald, *An Alternative Future for America II: Essays and Speeches* (Chicago: Swallow Press, 1970), 101–108. Reprinted with permission. (Referred to in *LWEC*, 340.)

Further Reading

Ad Hoc Committee on the Triple Revolution. *The Triple Revolution*. Santa Barbara, CA: The Committee, 1964.

Theobald, Robert. *The Challenge of Abundance*. New York: C. N. Potter, 1961.

Theobald, Robert, et al. "Interview with Robert Theobald." *Perspecta* 11 (1967): 11–19.

"War on Poverty," from *An Alternative Future for America II*

To be poor is to have too little money. The immediate need of the poor is for more money. The immediate need is not moral uplift, cultural refinements, extended education, retraining programs or make work jobs, but more money.

This is the belief of the advocates of the guaranteed income. This view is much too simple for the sophisticated defenders of the status quo, who argue that we must provide the poor with the skills which would enable them to earn an adequate wage and who ignore the fact that the majority of the poor cannot or should not hold jobs which would allow them to earn a decent wage.

The barriers to the elimination of poverty are not economic: the funds which will be released at the end of the Vietnam war would suffice to allow the introduction of the guaranteed income. The barriers are moral and social. The United States is not willing to apply its vast productive potential to the elimination of poverty and hides its unwillingness with statements about the need for motivation and incentives. …

It is the essentially dehumanized approach of the poverty program which ensures that it cannot be successful. I am, of course, aware that there are many people within the poverty program who attempt to use its potential to benefit individuals, but they are fighting the basic thrust of the whole program and their efforts will always be insignificant compared to the total size of the problem. The goal of the poverty program is not to help people to find themselves but rather to push them back into the industrial system just as fast and as often as they are forced out of it.

The failures of the poverty program are being increasingly recognized, but there is still insufficient understanding of the fact that the failure of the poverty program was inevitable from the day of its creation. It is critically important that we realize that the program did not fail because of bad luck or bad administration: it failed because it was poorly conceived. This is now a commonplace statement and it may appear like second-guessing after the fact. In order to demonstrate that it was possible to tell that failure was inevitable before the program even began, I quote at length from a speech which I gave immediately after the program was announced:

"Let me make it perfectly clear that I do not impugn the motivations of those who are developing the President's poverty program: I am simply convinced that they have so far failed to understand the true nature of poverty in a cybernated era, an era where the computer is combined with advanced machinery. The present program assumes that poverty can be eliminated through Federal action. I, on the other hand, believe that poverty can only be abolished by motivating the individual and the community. I am against the simplistic laws coined by Parkinson; however, for those who appreciate them, may I suggest one of my own—that the number of helpless poor inevitably increases with the number of bureaucrats trying to aid them.

"I believe that the President's poverty program as presently drawn is unsatisfactory. Let me list several reasons: First, as I have already stated, the present bill is based, both in letter and spirit, on an overemphasis on only one direction in what should be essentially a two-directional process: on the central government telling the state and the locality what to do. I believe that such an approach will necessarily fail. Second, the proposal to aid small farmers and small businessmen through small-scale financial aid has no chance of success in the light of the realities of the cybernated era—it is now impossible to turn a small, struggling farmer into a small, successful farmer or a small, struggling businessman into a small, successful businessman, merely through the provision of very limited sums of money. Third, the proposed types of training camps, clearly modeled on the experience of the CCC of the thirties, are no longer appropriate for the social and economic conditions of the sixties. In the industrial age, which still existed in the thirties, if you put a person in a camp and kept him there for six months or a year you were able in many cases to provide him with the limited skills and education required to make him an effective member of society. This technique will not work in a cybernated era, for the education and skills required to become a functioning member of society are not now teachable in this way.

"The consequences of setting up such camps will, in these circumstances, be unsatisfactory and possibly dangerous; for while adults still have not understood the realities of this age, our children are fully aware. ... Today, we have a poor class who know they are poor and who know they are going to stay poor.

"Finally, passage of the existing program will destroy the urgency which is increasingly felt—it will make it appear as though we have a program although we actually do not. The amount of money proposed, as has often been pointed out, is not even enough to finance a skirmish against poverty. The war on poverty must be placed in the proper context if we are to win it. At the outset we must recognize that the cost of the campaign on poverty is not the problem—that we can easily afford to provide the required funds. ...

"We must comprehend that the emergence of the cybernated era based on full education means the end of the industrial age based on full employment. This means that the parts of the country which have not yet been brought into the industrial system cannot possibly pass through the industrialization process. This will appear disastrous to many whose highest goal is to enter the industrial age—I, on the contrary, consider it most fortunate. The industrial age required that everybody toil—the cybernated era will allow everybody a reasonable standard of living without toil and will allow the development of a better society.

"It is now becoming clear that the values of the industrial age were deeply destructive of human and community values and that it will be extraordinarily difficult for the industrial parts of the country and the world to create the values necessary to a satisfactory life in the non-toil, non-consumption oriented cybernated era.

"Our key problem stems from the fact that we have made the value of a man synonymous with the economic value of the toil he performs: we fail to recognize that people should have a claim on resources even if they do not toil. The measure of destruction of our values is, I believe, shown in the fact that those living in an industrial society find it natural that people do not receive an adequate amount of food, clothing and shelter even though there is surplus food in storage and the possibility of producing more

housing and more clothing if we gave people the money to buy them. We can contrast this view with that of the so-called primitive societies; in many of these it was literally impossible to starve unless the whole community was starving. George Peter Murdoch, the celebrated anthropologist, described the reaction of one group of natives when he tried to explain the problem of the poor in Western countries. There was stark disbelief: 'How can he have no food? Does he have no friends? How can he have no house? Does he have no neighbors?'

"The poverty in the Western world can only be explained by a failure of conscience, by an unfounded and heartless belief that the poor have only themselves to blame for their situation. The minimum step which we must take if we are to eliminate the poverty which exists in the United States is to introduce the guaranteed income."

I see no reason to change significantly any portion of these analyses. We must therefore return to first principles and examine what route is available to deal significantly with the problems of hunger and poverty.

68 Law Commune, "How to Become a Non-profit Tax-Exempt Corporation, and Why, and Why Not"

Brand asked the Law Commune of Menlo Park to explain the advantages and disadvantages of nonprofit corporation status. Among the possible disadvantages of the most beneficial forms of nonprofit corporations was the refusal to allow political activity. This was part of the extremely frank discussion of the financial basis of the *WEC* that was continued on pages 438 to 441 with the publication of the 1970–1971 annual accounts. On pages 339 to 344, the *LWEC* introduced various books on radical and conventional economics. On page 342, for example, was a detailed description of token currencies derived from *The Token Economy*,[1] which Brand described as "sensational if you're interested in miracle cures, social cybernetics, whole-system education, or experimental economics. It's like an x-ray of economic civilization. ... If a school or a community goes on a token economy, they suddenly become much more of a single entity, more interrelated, and more distinct from the rest of the world." The problem with a token currency is that it is illegal to claim for it a formal dollar value. These two very different publications both believed in the potential value of economic and management systems to bring together a community and give it a strong sense of identity. — CM

1. Theodoro Ayllon and Nathan Azrin, *The Token Economy: A Motivational System for Therapy and Rehabilitation* (New York: Appleton-Century-Crofts, 1968).

Source
LWEC, 341.

"How to Become a Non-profit Tax-Exempt Corporation, and Why, and Why Not"

What are the advantages of forming a nonprofit corporation?

The advantages of operating through a corporate structure are all based on the concept of the corporation's being a separate person. As a separate person the corporation remains no matter what individuals come and go.... The advantages of making the corporation [have] to do with tax-status.... Another advantage of a corporation is that its very existence lends stability and respectability to your enterprise, especially in the eyes of people like bankers and local storekeepers who hand out things in credit.

What are the disadvantages of forming a non-profit corporation?

You will have the initial expense of forming the corporation.... As a corporation you will have to keep careful financial records and file detailed tax returns concerning the corporation's finances. Incorporation means that you must in some way show proof of observance of specific regulations and formalities, for example, regular meetings of a board of directors; it also means someone will be watching to see that you do things like make payroll deductions, pay unemployment assurance taxes, ITC....

Could you explain more about the taxes angles?

Federal exemptions for nonprofit corporation are set up in specific categories. The most advantageous category is for "religious," "charitable," and "educational" organizations. Not only is the income of corporations within this category tax exempt, but unlike all the other categories, persons making donations to these corporations may receive tax reductions on their personal income tax. However, corporations in this category have a very important limitation on their activities; they may not participate in any political campaign, lobby for specific legislation, or otherwise engage in active political activity. This is the reason that the IRS has taken away the Sierra Club's tax exemption.... Exemptions from state income taxes usually are modeled on the federal laws. For most of these categories, any "unrelated business" carried on by the corporation as a subsidiary to its main purposes, for example, a college bookstore, is subject to tax on its income. (Whole Earth pays taxes on its store and mail order operation.)

69 Lawrence Halprin, *The RSVP Cycles*, 1970

The landscape architect Lawrence Halprin (1916–2009), who was trained at Harvard in the 1940s, became committed to an understanding of the processes behind change in landscape and developing strategies to restore land damaged by intensive agriculture exploitation as well as industrial or natural forces. During a period in Israel in the mid-1950s, Halprin had observed the restoration of desert land damaged by centuries of grazing. The *RSVP Cycles* describes his approach to landscape conservation, taking as a case study the development of the ten-mile-long settlement known as the Sea Ranch in California. *RSVP* stands for Resource, Score, "Valuaction," Performance. The poet Gerd Stern reviewed the book for the *LWEC*: "The RSVP Cycle is a scoring guide, a technical manual, highly illustrated and explicated in every language. It is a self-defining, singing, swinging score (S) of Halprin's own performance (P), his electric, focused stash of resources (R) from the I Ching, to a controversial Northwestern-Minnesota football play and his consciously objective valuactions (V)—a term he coined to suggest action and decision aspects of the RSVP cycles."

Working with a range of experts, Halprin would analyze the geological prehistory and subsequent developments to the flora and fauna of a landscape; this was Resource. Score consisted in charting these processes as an evolution that could then be used as a basis for evaluation and action. Performance would result from these decisions. The diagram "Sea Ranch Ecoscore" demonstrates the passage of time as a spiral from the earliest times to the recent involvement of humans in the nineteenth century. The effects of industrialized logging, serving the voracious needs of nearby San Francisco, and the exploitation of the deforested areas for pasture had caused erosion of the soil. Halprin's solution was to reforest the land and to suggest a form of settlement that would respond to the prevailing winds and microclimate. —CM

Source
Lawrence Halprin, "Ecoscore," in *The RSVP Cycles: Creative Processes in the Human Environment* (New York: G. Braziller, 1970). (Referred to in *LWEC*, 344.)

Further Reading
"Ecological Architecture: Planning the Organic Environment." *Progressive Architecture* 47 (May 1966): 120–133.

Halprin, Lawrence, and Jim Burns. *Taking Part: A Workshop Approach to Collective Creativity*. Cambridge, MA: MIT Press, 1974.

Lyndon, Donlyn. *The Sea Ranch: Fifty Years of Architecture, Landscape, Place, and Community*. New York: Princeton Architectural Press, 2013.

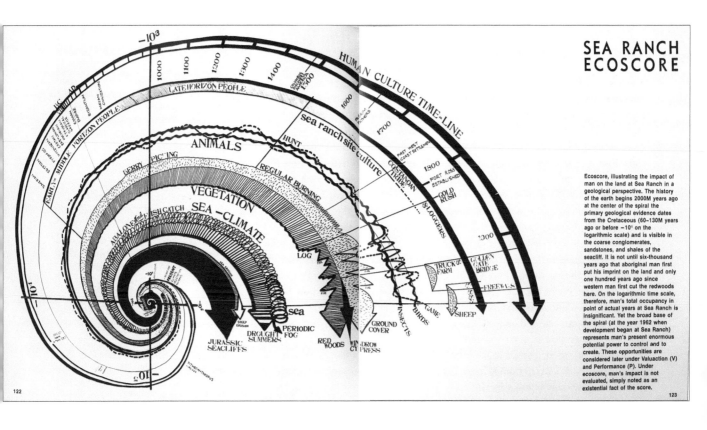

SEA RANCH
ECOSCORE

Ecoscore, illustrating the impact of
man on the land at Sea Ranch in a
geological perspective. The history
of the earth begins 2000M years ago
at the center of the spiral the
primary geological evidence dates
from the Cretaceous (60–130M years
ago or before –10⁶ on the
logarithmic scale) and is visible in
the coarse conglomerates,
sandstones, and shales of the
seacliff. It is not until six-thousand
years ago that aboriginal man first
put his imprint on the land and only
one hundred years ago since
western man first cut the redwoods
here. On the logarithmic time scale,
therefore, man's total occupancy in
point of actual years at Sea Ranch is
insignificant. Yet the broad base of
the spiral (at the year 1962 when
development began at Sea Ranch)
represents man's present enormous
potential power to control and to
create. These opportunities are
considered later under Valuation (V)
and Performance (P). Under
ecoscore, man's impact is not
evaluated, simply noted as an
existential fact of the score.

Figure 8.5
"Sea Ranch Ecoscore." Lawrence Halprin,
"Ecoscore," in *The RSVP Cycles: Creative
Processes in the Human Environment* (New
York: G. Braziller, 1970), 122–123.

70 Beryl Korot and Phyllis Gershuny, "Editorial," *Radical Software*, 1971

Raindance Corporation was an "alternate culture think-tank"[1] that embraced video as a militant form of cultural communication; it also launched the video magazine *Radical Software*. The editors were Beryl Korot, Phyllis Gershuny, and Ira Schneider. The magazine first appeared in spring 1970 (2,000 copies) and ran for eleven issues. In 1971 the circulation was 5,000. *Radical Software* celebrates the low-cost portable video cameras that came on the market in 1967. The first editorial in spring 1970 noted the relationship between power and control of information, as well as the importance of freeing television from corporate control. David Ross, writing on the website of *Radical Software Redux*, described the objectives of the editors: "They thought reversing the process of television, giving people access to the tools of production and distribution, giving them control of their own images and, by implication, their own lives—giving them permission to originate information on the issues most meaningful to themselves—might help accelerate social and cultural change."[2]

The journal was a support group for video filmmakers using the new Sony Portapak video cameras, with technical information and discussions. Many of alternative cultural activists, such as the Ant Farm, Raindance, and Video Workshop Amsterdam, were collaborators.

Stewart Brand expressed much admiration for these "TV heads": "They sense power (95% of US homes have TV) an unexplored territory (broadcast TV still scarcely seems to know what it is) and the hard cider of dwelling on evolution's imploding edge…maybe. In a way it is about time: we have head radio, head records, head books, magazines, newspapers, head movies; very little head TV."—CM

1. Davidson Gigliotti, "A Brief History of RainDance," http://www.radicalsoftware.org/history.html (accessed August 12, 2015).

2. David Ross, "Radical Software Redux," http://www.radicalsoftware.org/e/index.html (accessed August 10, 2015).

Source

Beryl Korot and Phyllis Gershuny, "Editorial," *Radical Software* 1, no. 11 (spring 1970): 1, http://www.radicalsoftware.org/volume1nr1/pdf/VOLUME1NR1_0002.pdf (accessed August 20, 2015). (Referred to in *LWEC*, 345.)

Further Reading

Coffman, Elizabeth. "'VT Is Not TV': The Raindance Reunion in the Digital Age." *Journal of Film and Video* 64, nos. 1–2 (spring–summer 2012): 65–71.

Kaizen, William. "Steps to an Ecology of Communication: *Radical Software*, Dan Graham and the Legacy of Gregory Bateson." *Art Journal* 67, no. 3 (2008): 86–107.

Ryan, Paul. *Video Mind, Earth Mind: Art, Communications, and Ecology.* New York: P. Lang, 1993.

Turner, Fred. "Bohemian Technocracy and the Countercultural Press." In *Power to the People: The Graphic Design of the Radical Press and the Rise of the Counterculture, 1964–1974*, ed. George Kaplan, 149–150. Chicago: University of Chicago Press, 2013.

"Editorial," from *Radical Software*

As problem solvers we are a nation of hardware freaks. Some are into seizing property or destroying it. Others believe in protecting property at any cost—including life—or at least guarding it against spontaneous use. Meanwhile, unseen systems shape our lives.

Power is no longer measured in land, labor, or capital, but by access to information and the means to disseminate it. As long as the most powerful tools (not weapons) are in the hands of those who would hoard them, no alternative cultural vision can succeed. **Unless we design and implement alternate information structures which transcend and reconfigure the existing ones, other alternate systems and life styles will be no more than products of the existing process.**

Fortunately, new tools suggest new uses, especially to those who are dissatisfied with the uses to which old tools are being put. We are not a computerized version of some corrupted ideal culture of the early 1900's, but a whole new society because we are computerized. Television is not merely a better way to transmit the old culture, but an element in the foundation of a new one.

Our species will survive neither by totally reject-ing nor unconditionally embracing technology—but by humanizing it; by allowing people access to the informational tools they need to shape and reas-sert control over their lives. There is no reason to expect technology to be disproportionately bad or good relative to other realms of natural selec-tion. The automobile as a species, for example, was once a good thing. But it has now overrun its ecological niche and upset our balance for opti-mum living. Only by treating technology as ecol-ogy can we cure the split between ourselves and our extensions. We need to get good tools into good hands—not reject all tools because they have been misused to benefit only the few.

Even life styles as diverse as the urban political and the rural communal require complex tech-nological support systems which create their own realities, realities which will either have to be considered *as part of the problem, or, better, part of the solution*, but which cannot be ignored.

Coming of age in America means electronic imprinting which has already conditioned many millions of us to a process, global awareness. And we intuitively know that there is too much cen-tralization and too little feedback designed into our culture's current systems. …

Fortunately, however, the trend of all technol-ogy is towards greater access through decreased size and cost. Low-cost, easy-to-use, portable videotape systems may seem like "Polaroid home movies" to the technical perfectionists who broadcast "situation" comedies and "talk" shows, but to those of us with a few preconceptions as possible they are the seeds of a responsive, useful communications system. …

Videotape can be to television what writing is to language. And television, in turn, has sub-sumed written language as the globe's dominant communications medium. Soon, accessible VTR systems and video cassettes (even before CATV opens up) will make alternate networks a reality.

Those of us making our own television know that the medium can be much more than "a radio with a screen" as it is still being used by the networks as they reinforce product oriented and outdated notions of fixed focal point, point of view, subject matter, topic, asserting their own passivity, and ours, giving us feedback of information rather than asserting the implicit

immediacy of video, immunizing us to the impact of information by asking us to participate in what already can be anticipated—the nightly dinner-time Vietnam reports to serialized single format shows. If information is our environment, why isn't our environment considered information?

So six months ago some of us who have been working in videotape got the idea for an informa-tion source which would bring together people who were already making their own televi-sion, attempt to turn on others to the idea as a means of social change and exchange, and serve as an introduction to an evolving handbook of technology.

Our working title was *The Video Newsletter* and the information herein was gathered mainly from people who responded to the questionnaire at right. While some of the resulting contents may seem unnecessarily hardware-oriented or even esoteric, we felt that thrusting into the pub-lic space the concept of practical software design as social tool could not wait.

In future issues we plan to continue incor-porating reader feedback to make this a process rather than a product publication. We especially hope to turn the interest and efforts of the sec-ond and third television generations on college campuses, whose enormous energies are often wasted by the traditional university way of struc-turing knowledge, towards the creation of their own alternate information centers. (We are the first television generation ourselves).

To encourage dissemination of the informa-tion in *Radical Software* we have created our own symbol of an *x* within a circle: (x). This is a Xerox mark, the antithesis of copyright, which means *DO* copy. (The only copyrighted contents in this issue are excerpted from published or soon-to-be published books and articles which are copyrighted.)

The individuals and groups listed here are committed to the process of expanding television. It is our hope that what is printed here will help create exchanges and interconnections necessary to expedite this process.

71 Harry L. Hiett, *57 How-to-Do-It Charts*, 1959

Harry L. Hiett's *57 How-to-Do-It Charts* **was one of five books on silk screen printmaking selected for** *The Whole Earth Catalog*. **Hiett ran a professional printmaking workshop where he shared his extensive knowledge on silk screen practice. As he wrote in his introduction, the book "gives you a quick breakdown, in chart form, of all the necessary steps for making frames, producing your own films, and building your own basic equipment." An enlargement of one of the pages from the book was given a prominent place in the** *LWEC* **spread. Silk screen methods, cheaper than lithography, allowed for the production of large, powerful images and became a standard for posters of political protest. Silk screening was also a medium much used by artists at the time: Robert Rauschenberg's silk screen poster for Earth Day (April 22, 1970) became an icon of the environmental movement. Silk screening allowed for the combination of photographic images, often taken from newspapers, with colored overlays, text, and graphic art. Hiett's book explains how to achieve these effects in the context of a commercial practice. —CM**

Source

LWEC, 359. The images featured in the "Silk Screen" subsection are from Harry L. Hiett, *57 How-to-Do-It Charts: On Materials-Equipment-Techniques for Screen Printing* (Cincinnati, OH: Signs of the Times, 1959).

Further Reading

Biegeleisen, Jacob Israel. *The Silk Screen Printing Process*. New York: McGraw-Hill, 1941.

Cushing, Lincoln. *All of Us or None: Social Justice Posters of the San Francisco Bay Area*. Berkeley, CA: Heyday, 2012.

Eisenberg, James, and Francis J. Kafka. *Silk Screen Printing*. Bloomington, IL: McKnight & McKnight, 1957.

Figure 8.6
LWEC, 359. From Harry L. Hiett, *57 How-to-Do-It Charts: On Materials-Equipment-Techniques for Screen Printing* (Cincinnati, OH: Signs of the Times, 1959).

Silk Screen Books

Two excellent books for beginners and journeymen silk-screeners.

I like best:

57 How-to-it Charts
by Harry L. Hiett (**$2.00**)

63 pp. 8-1/2 x 2 x 11 "Gives you a quick breakdown, in chart form, of all the necessary steps for making frames, producing your own films, and building your own basic equipment. Shows you how to stretch your silk, both by rope & by tacking method, & how to build jigs for 3-dimensional printing. A practical book for the professional., the student, or the technician."

The supply outlet likes best:

Silk Screen Printing ($2.20)
by Eisenberg & Kafka

90 pp., 18 chapters. "Devoted to describing each phase of screen processing clearly w. illustrations & photographs. Step-by-step suggestions on techniques & methods, clearning of equipment, color mixing, art & lettering, multi-color screening, etc. Recommended handbook for beginners."

The prices are so reasonable, for a welcome change, I really advise getting both. My source, get their giant catalog, too, is:

WESTERN Sign Supplies, Inc.
77 8th St.
Oakland, Calif. 94607

[Suggested and reviewed by Norman Solomon]

57 How-to-it Charts also available from:

Signs of the Times Publishing Co.
407 Gilbert Ave.
Cincinnati, Ohio 45202

Naz-Dar Silk Screen Materials

Probably the most complete source of screen processing materials for printing on almost everything but soft butter. While intended primarily for the commercial market, Naz-Dar's catalogue will give you an insight into this practically unlimited medium.

[Reviewed by Carl Mueller.
Suggested by Jerome Skuba]

The Naz-Dar Company
1087 N. North Branch Street
Chicago, Ill. 60622

Silk Screen Printing

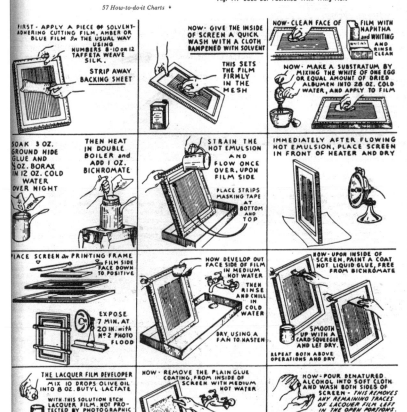

Fig. 41. Back Bar Fastened With Wing Nuts

57 How-to-it Charts +

Problems Answered

The film refuses to adhere to the silk.
The silk was not washed with both solvent and water before use. Insufficient solvent is used during adhering operation.

The film buckles while cutting.
There is an excessive humidity, hence the work must be handled rapidly and not allowed to stand around too long before adhering.

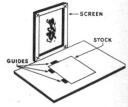

Spots appear on the print.
Pieces of dirt or lint get on the under surface of the screen and must be picked off. Fine skins from the ink have been ground into the meshes of the silk by the squeegee action, necessitating washing up of the job.

Colors on a several color job fail to register.
The film is allowed to lie around after cutting and expands on a very humid day. The guides moved during printing of early colors. The frame is loose or sprung from constant, hard use. The first colors were not printed with all sheets fed carefully to guide.

	Per Inch
No. 5100 Complete Squeegee, Black Rubber, Medium Grade	$.20
No. 5102 Black Rubber only	.15
No. 5150 Complete Squeegee, Gray Rubber, Hard Grade	.20
No. 5151 Gray Rubber only	.15
No. 5200 Complete Squeegee, Amber Rubber, Soft Grade	.20
No. 5202 Amber Rubber only	.15
No. 5420 Complete Squeegee, Extra Hard Grade	.20
No. 5421 Rubber only	.15

STENCRAFT STENCIL SILK (Price per Yard)

number	Mesh Count	Width	1-14 Yards	15-29 Yards	30 Yards
6xx	70	40"	$3.45	$3.20	$3.00
8xx	86	40"	3.60	3.35	3.10
10xx	107	40"	4.00	3.70	3.45
12xx	125	40"	4.25	3.95	3.65
14xx	139	40"	4.55	4.20	3.90
16xx	156	40"	5.60	5.20	4.85

Leonard was putting on a tie. He looked strange in a suit, and yet it belonged on him too. He was tall and roughly-cut, big-boned and lean, a powerful man. Leonard was in his fifties, he was a grandfather as a matter of fact. But nobody ever deferred to him because of it. People deferred to him but it wasn't because they thought he was old or beyond his prime. He was right in his prime, and so was Roxie. Roxie was large and getting round, but she was some years younger than Leonard, barely forty. Their children were grown, most of them already married, living away in Indiana and Illinois. Their youngest son Glenn had left home just the year before. He was in the Army, over in Vietnam now. Leonard showed D.R. his picture when D.R. went in the bedroom to try on Glenn's graduation suit.

•

A perfect fit.

•

Close to it, anyhow.

•

Why yes, said Roxie. That suit looks fine on you.

•

And Leonard and D.R. went up to the church to talk to Reverend Bagby about the service.

•

Reverent Bagby had it all planned. Mrs. Thornton's niece Dorothy had agreed to sing Abide With Me, and Mrs. Bagby would play the piano. The message would be short, said Reverend Bagby. It was hot, and there was the work of getting the casket up on the hill to bury. It all sounded fine to D.R. He liked Reverend Bagby. He liked Mrs. Bagby. He liked Leonard and he liked Roxie. D.R. tried to hide it as best he could, but the truth of the matter was, he was really enjoying this day so far.

•

And then there Marcella was to help him enjoy it more.

•

How good, just plain good it was to see her. And Doyle and Herschel and little Debbie, my what a pretty family. Doyle grabbed D.R.'s hand and shook it as Marcella gave him a powerful hug, and Herschel began tugging at his coat pocket. Debbie hung back, and D.R. lifted her into the air as a reward for being demure. Lift me! Lift me! Herschel shouted, and Doyle took him by the hand and led him off to one side to explain why he ought not be yelling like that. Leonard stood among them as a greeter. It was proper that he do so. Everybody felt it, and in their separate ways they thanked him for it. Doyle held his hand out to Leonard and said you may not remember me, Leonard, but I'm Doyle. Leonard said why Doyle, I remember you good. You fixed a car of mine one time, a Mercury, back in the days when I drove Mercurys, don't you remember that? Doyle looked a little puzzled, then embarrassed. Now I'd plum forgot about that, said Doyle. A maroon car. Put a new coil in it for you, down there by your alf's store.

Further Silk Screen

There's a good British book:

Practical Screen Printing
Stephen Russ
1969; 96 pp.

$8.50 postpaid

from:
Watson-Guptill Publications
165 West 46th St.
New York, N.Y. 10036

And a good expensive American book with a full list of suppliers and annotated bibliography:

Screen Printing
J.I. Biegeleisen
1971; 159 pp.

$12.50 postpaid

from:
Watson-Guptill Publications
165 West 46th St.
New York, N.Y. 10036

Silk Screen
Communications **359**

Learning

On the first page of the *Whole Earth Catalog*, Brand stated the book's purpose: to provide tools that might enable an individual "to conduct his own education, find his own inspiration, shape his own environment, and share his adventure with whoever is interested." In a publication of this sort, teeming with educational texts, a chapter dedicated to learning may seem redundant—the concept, after all, courses throughout the catalog—and the fact that many of the "tools" that constituted this ninth section could find a place in the previous eight chapters raises further questions. For instance, why was the ecologist John H. Storer's *The Web of Life* (1953), a text on humans' interference with healthy ecosystems, included in the "Nature" subsection of "Learning," and not in the "Ecology" subsection of "Whole Systems"? Or why were the books *Indian Crafts and Lore* (1954) and *North American Indian Arts* (1970), both of which furnished information on Native American handicrafts, situated in the "Indians" subsection of "Learning" instead of in the "Crafts" section?

Brand hinted at the answer in one of his previous reviews, of *The Golden Book of Camping*, in the "Nomadics" section: "I'd have put this book in the Learning section of the CATALOG only it has better camping information than the authoritative tomes here." If *The Golden Book of Camping* had not been so directly relevant to the "Camps" subsection, it may conceivably have been relocated to the "Wilderness" page in "Learning," near the Boy Scouts of America's *Fieldbook for Boys and Men* (1967)—no less significant to the catalog's editors, but simply better suited. In this way, "Learning" became a sort of repository of crucial books, products, and subject matter that, although they were not quite appropriate for other chapters or subsections, were nonetheless significant to the purposes of the catalog. *The Web of Life* was arguably too anecdotal to be integrated into the more scientific "Ecology" subsection, and positioning Native American craft books near cultural, political, and historical texts may have strengthened the three pages of the "Native American Indians" subsection.

This approach to the chapter generated a compelling potpourri of learning tools, as evidenced in the selections that follow. Both the subsection "Toys" and Samuel Yanes's introduction to *Big Rock Candy Mountain* examined and commented on childhood learning, but as the chapter progressed, instructional lessons for all audiences began to appear. For instance, Storer's *The Web of Life* presented complicated ecological concepts in a manner useful to readers of all ages, who, through the author's clear descriptions of his own experiences, might begin to see their local environments from a more ecological perspective. Similarly, Forrest Wilson's *Architecture: A Book of Projects for Young Adults*, although targeted at children, was equally applicable to older laypeople eager to understand architecture in the material sense allowed by the building exercises the architecture professor devised. *The Eagle, the Jaguar, and the Serpent*, with its abundance of illustrations, informed even the most casual readers of the geographical and aesthetic breadth of Native American artwork while providing comprehensive art historical analysis for the more rigorous. Although other texts, such as *The Psychedelic Review* and Ram Dass's *Be Here Now*, were certainly more pertinent to an adult consumer, the substance of each promoted a certain type of learning unexpressed in the preceding pages. The "Learning" section was an amalgam of educational opportunities, an excellent thematic conclusion to a catalog essentially devoted to its final topic. —MG

72 "Toys"

The first eight pages of "Learning" focused on children. From the first subsection, "Parent," which included Dr. Benjamin Spock's seminal *Baby and Child Care*, through the eighth, "Learning Books," the catalog followed the intellectual development of a child before widening its scope to students of all ages. Two pages were devoted to "Toys," the second of which we have included here. The page had only two entries: "Wooden Toys," a compendium of the best manufacturers of wooden toys, and the book *Diary of an Early American Boy* (1962), both representative of more traditional forms of play in contrast to the modern toys featured on the previous page.

Although these toys and the products and texts promoted in the other seven subsections may have been directed at countercultural parents anxious to make positive educational decisions for their children, their placement as a preface for "Learning" underscored the significance of childhood to members of the counterculture more generally. For many in the movement, children represented not only hope for a better future but also an innocence, playfulness, and curiosity that they hoped to restore within themselves. By beginning the chapter with parenting and children's toys, the catalog intimated a similar ambition. — MG

Source
LWEC, 368.

Further Reading
"Advertisements for a Counter Culture." *Progressive Architecture* 51 (July 1970): 71–93.

Matterson, E. M. *Play and Playthings for the Preschool Child*. Baltimore, MD: Penguin Books, 1965.

Whitehurst, Robert M. "Some Comparisons of Conventional and Counterculture Families." *Family Coordinator* 21, no. 4 (October 1972): 395–401.

Figure 9.1
The first page of the "Toys" subsection, in the "Learning" section, *LWEC*, 368.

Wooden Toys

Community Playthings has the widest variety, including a nice set of blocks. These toys are the product and income of the Rifton, N.Y. Bruderhof community.
   Catalog free.

[Suggested by William Ilson]

Vermont Wooden Toy has particularly lively small toys. Dune buggie and raceway. Kid-size wooden sled. Dusenberg with passengers that fall out. Catalog free.

Everdale Toys, from a community in Ontario, prices stiffer because of the border. Catalog free.

Matten Enterprises, fine toy boats. Catalog free.

[Suggested by Joeann Yellot]

Dick Schnacke has a splendid selection of traditional toys — ball & cup, Jacob's ladder, bull roarer, tops, bean bags, etc. Good prices (60¢ – $4.50) Catalog free.

[Suggested by Bob Railford]
—SB

an authentic mountain folk toy

Instructions:

—Carefully remove tape, release pith ball

—Place wired ball in nozzle of blowpipe. Blow gently on pipe and ball will float up on air stream. Try to snag the ball on the hoop!

—Now blow again, try to unhook ball and back the ball down into pipe nozzle

Dick Schnacke
Mountain Craft Shop
Route 1
Proctor, West Virginia 26055

Vermont Wooden Toy Co.
Old High School Building
Waitsfield, Vt. 05673

Dune Buggy Raceway $11.95

Van $19.80 plus taxes

Everdale Toys
Box 29
Hillsburgh, Ontario
Canada

14-inch Silly Sailor
$8.50

Matten Enterprises
The Toy Shop
Southwest Harbor, Maine 04679

Unit Block Family Sets

F-135 Three Year Old Set Unit Blocks.
   32 blocks with 7 different shapes.
   18 lbs . . . . $ 9.50
F145 Four Year Old Set Unit Blocks.
   55 blocks in 12 shapes, 32 lbs 16.50
F155 Five Year Old Set Unit Blocks.
   85 blocks in 15 shapes, 50 lbs 24.50

Community Playthings
Rifton, N.Y. 12471

Diary of An Early American Boy

A quaint account of life on an early American homestead. Its daily entries are so vividly portrayed by the author's word-pictures and ink sketches that you actually feel you are living with Noah and his family. With them, the forest is the marketplace, the creek is power, and Nature is ever in Command. With a knowledge of levers, use their ox to clear a field for corn. With a broadaxe and the help of your neighbors, erect a king post bridge over your creek. With flour ground in the mill you built, bake bread. With scraps of branches taken from fallen trees, make toys for children. And on Christmas Eve, as a blanket of snow glistens outside under the stars, sit by your warm hearth and give Thanks to the Forces that made you.

[Suggested and reviewed by Sawdust Ribiba]

Diary of an Early American Boy
Eric Sloane
1962; 108 pp.

$6.50 postpaid

FUNK & WAGNALLS
c/o T.Y. Crowell
   201 Park Ave. So.
   New York, New York 10022
or WHOLE EARTH CATALOG

25: A cold and windy day. Neighbor Adams with son Robert stopp'd by. We drank mead and mint tea. No work done this day. Father is going to the woodlot behind the barn tomorrow for floor timbers. I shall assist him.

26: A light snow fell which Father believes will be the last of the winter. We fell'd a fine oak and rolled it upon rails for Spring seasoning. Mother is joyous at the thought of a good wood floor.

In the little forge barn half-way down the bridge, Izaak had earned most of the money needed around the homestead. Not that an early American had great need for cash, as most things were traded; but a good farmer always had a cash crop or some paying trade, other than farming. Izaak was a nail-maker. He had taught Noah the art of making hand-wrought nails by letting him pump the bellows of the forge as a little fellow; now and then he actually hammered or broke the iron nail rods.

STAND OF UNCUT TIMBER. BARN. TOOL SHED. CORNFIELD. STONE BOAT. SLED. MILL. SLUICE. GATE. MILL WHEEL. COVERED BRIDGE. TOLL HOUSE. ROAD to Village. PART OF BROOK DIVERTED TO MILL-WHEEL. STONE FENCE. NOAH'S WINDOW. CELLAR. NOAH'S ROOM. The BLAKE HOUSE. WELL-SWEEP. SHINGLE ROOF. MILL POND.

The BLAKE *place some time, after 1805 . .*

With the help of Noah, Izaac Blake had created a workable homestead. The Indian Trail became a roadway . . the brook became a source of power to grind corn that grew where once a forest stood . . the shelter became home to an early American boy.

For more Eric Sloane, see p. 150.

73 Samuel Yanes and Cia Holdorf, eds., *Big Rock Candy Mountain,* 1972

Big Rock Candy Mountain, **a Portola Institute publication, edited by Samuel Yanes and Cia Holdorf and published in 1972, followed the format of the** *Whole Earth Catalog* **but focused primarily on education. The** *New York Times* **reviewed it in 1972.**

Yanes and Holdorf's educational philosophy broadly resembled that of Caleb Gattegno's book *What We Owe Children* (1971), reviewed at the beginning of the volume. The stress was on empowering children and challenging them rather than "taking knowledge down from the shelves and handing it out to students." The catalog includes many different kinds of publications, some theoretical and some highly practical. Throughout the section there was a strong emphasis on play; a good example is Jay Beckwith's playground book, in which he gave tips on designing play areas for children. This was one of a number of selected books about schools. Stewart Brand commented: "It is too wishful, too little experienced in grimy schoolwork....But...I also know that its readers are glad to have it."

The introduction, written by Samuel Yanes, focuses on personal development guided by Oriental mysticism. Beckwith's book with Jeremy Hewes, *Build Your Own Playground!,* helped to build a reputation that led to a one-man exhibition at the de Young museum in San Francisco. Beckwith went on to manufacture a number of community-based play structures, which were the basis for standard industrialized systems. —CM

Source
Samuel Yanes and Cia Holdorf, eds., *Big Rock Candy Mountain: Resources for Our Education* (New York: Delacorte Press, 1972), 9. (Referred to in *LWEC,* 402.)

Further Reading
Kohl, Herbert R. *The Open Classroom: A Practical Guide to a New Way of Teaching.* Westminster, MD: Random House, 1969.

Repo, Satu, ed. *The Book Is about Schools.* New York: Pantheon Books, 1970.

Introduction to *Big Rock Candy Mountain*

Some basic stuff is obviously missing from today's learning opportunities. Nowhere in the education world—under or overground—can one be taught to discover fundamental attributes that guide a human being on his own course, not in any school geared to help an individual discover his personal law and function, fitting with his individual capacity and inclination.... Somewhere along the "evolution" of our educational values, man has forgotten that he is born with a birthright.

We are not born butcher, baker, or candlestick maker; these are mere contingencies to our nature, just as the sun is yellow, but our innate dispositions may be best served within one of those occupations. Our vision has been superimposed, however; instead of our natural ability determining our jobs, position in the world, etc., the artificial categories of our social and educational structure have determined what we shall be. In opposition, true vocation is always willing and always self-fulfilling regardless of the accidental state of affairs, for one is doing what one is suited to do, one is satisfying his birthright, and perhaps more important, one is also working toward his own salvation. On the one hand, our lives are fulfilled by pursuing our duties and values, and on the other hand, our only true vocation is to attain liberation from all valuation and responsibilities. This runs deep, far deeper than is called for here, but even superficial understanding would have startling consequences for the education world. But the birthright has been given the status of "once-upon-a-time" and Esau's porridge has the significance of the three bears' breakfast. This was not always the case.

Social Changes for Vocation...

This concept of vocation has been embedded in traditional Indian society. In the Dance of Shiva, Ananda K. Coomaraswamy writes:[1]

"'Reading,' says the Garuda Putana, 'to a man devoid of wisdom, is like a mirror to the blind.' The Brahmans attached no value to uncoordinated knowledge or to unearned opinions, but rather regarded these as dangerous tools in the hand of unskilled craftsmen. The greatest stress is laid on the development of character. Proficiency in hereditary aptitudes is assured by pupillary succession within the cast. But it is in respect of what we generally understand by higher education that the Brahman method differs most from modern ideals; for it is not even contemplated as desirable that all knowledge should be made accessible to all. The key to education is to be found in personality. There should be no teacher for whom teaching is less than a vocation (none may 'sell the Vedas'), and no teacher should impart his knowledge to a pupil until he finds the pupil ready to receive it, and the proof of this is to be found in the asking of the right questions. 'As the man who digs with a spade obtains water, even so an obedient pupil obtains the knowledge which is his teacher.'"

Herein lies the basis for a truly diversified educational system. Marco Pallis in *The Way and the Mountain*:[2]

"For any individual, the realizing in full of the possibilities inherent in his svabhava (one's own nature and innate disposition) makes the limit of achievement, after which there is nothing further to be desired. As between two such beings, who are wholly themselves, no bone of contention can exist, since neither can offer to the other anything over and above what he already possesses; while on the supra-individual level their common preoccupation with the principal Truth, the central focus where all ways converge, is the guarantee of a unity which nothing will disturb; one can therefore say that the maximum of differentiation, whenever two

---

1. A. K. Coomaraswamy (1877–1947) was a Ceylonese Tamil philosopher and metaphysician, as well as a pioneering historian and philosopher of Indian art, and an early interpreter of Indian culture to the West.

2. Marco Pallis (1895–1989), translator of ancient Greek texts, wrote important books about religion and Tibet's culture.

beings are subjected to the steamroller of uniformity, not only will both of them be frustrated in respect of some of the elements normally includable in their own personal realization, but they will, besides, [be] placed in the position of having to compete in the same artificially restricted field, and this can only result in the heightening of oppositions—the greater the degree of uniformity imposed, the most inescapable are the resulting conflicts."

The objections to the conventional education system that create the alternative education movement appear rather superficial in this light.

Try some of these comparisons:

Physical Education	Yoga
Awkward, musclebound	Graceful strength;
Rigid; brittle…	Supple, flexible
Short-lived, cheap glory.	Long-term, daily benefits.
Rodin's Thinker	**Buddha**
Troubled; hard,	Serene; soft;
Ponderous; rational;	intuitive; synthesis;
Analysis; in the Dark.	In the light.
Observation	**Concentration**
Outward;	Inward;
Rote memorization;	meditation;
Shopkeeper's mentality	contemplation.
Cheeseburger	**Fresh Fruit and Vegetables**

3. An alternative journal founded in Toronto in 1966.

"They will maintain the fabric of the world and in the handywork of their craft is their prayers."—I can feel the social indignation of *This Magazine Is about Schools*[3] and in the writings of Ivan Illich, but protestations at a social level, in the long run, only serve the status quo, for they do not allow for vertical diversity, which requires the support and the guidance of the social order, which, in return, is supported and maintained by the diversity of every-one performing a function best suited to his nature.

The Bhagavad Gita says, "Man reaches perfection by his loving devotion to his work." Somehow this must be embodied in our lives. *The Whole Earth Catalog* presents tools for the craft of living, but the goal of education is not only to be a skilled craftsman, but also an artist whose work requires contemplation in addition to practice and skill. It is in the practice of the particular arts that the congregation of powers which man has at his disposal can be used most efficiently as a means to self-realization. When one's life is a work of art, all work is priestly, sacramental, and willing. Shaving becomes a deliberate brush stroke and walking becomes an exercise in grace. One's education then satisfies both material and spiritual needs, labor becomes purposeful—"The last end of every maker, as such, is himself."

74 John H. Storer, *The Web of Life*, 1953

In the early twentieth century, John H. Storer, a nature photographer and wild-life enthusiast, began to develop a progressive ecological sensibility through close observation of the distinctive environments in which he worked. *The Web of Life* was his expression of these experiences, an educational but also personal account of ecological concepts intended not for specialists and advanced scholars but for those seeking, as Brand described it, "a good anecdotal introduction to ecology." Yet *The Web of Life* was more than a simple primer: while advancing the basic principles of ecology, it portrayed ecological thinking as an ethical response to environmental degradation and imbued what was commonly presented as a purely scientific theory with a conscience. The noted conservationist Fairfield Osborn, in his preface to the text, declared Storer's "truly observant eye [and] truly understanding heart" to be the foundation of this synthesis, which, Osborn believed, distinguished *The Web of Life* from other ecology "treatises" (*The Web of Life*, vi).

Indeed, Storer's straightforward, intimate account of ecology differed much from the abstract ecological models presented in Howard Odum's *Environment, Power, and Society*, and its position in the "Nature" subsection underscored this contrast. For readers, however, these two books would effectively complement each other as tools for achieving a greater ecological understanding and create a fluid connection between the first and last chapters of the catalog as well. —MG

Source

John H. Storer, *The Web of Life: A First Book of Ecology* (New York: Devin-Adair, 1953), 72–75. © Devin-Adair, Publishers, Inc., Old Greenwich, CT, 06870. Reprinted with permission. All rights reserved. (Referred to in *LWEC*, 375.)

Further Reading

Cornell, Joseph Bharat. *Sharing Nature with Children: A Parents' and Teachers' Nature Awareness Guidebook.* Nevada City, CA: Ananda Publications, 1979.

Vosburgh, John. *Living with Your Land: A Guide to Conservation for the City's Fringe.* Bloomfield Hills, MI: Cranbrook Institute of Science, 1968.

"The Grassland Community," from *The Web of Life*

In the year 1910 I saw a field of alfalfa being harvested in Montana. There was a fence around the field, and on every fence post there sat a hawk that swooped down from time to time to pick up the meadow mice as they scurried for shelter from the cut hay. There were discarded pieces of mice scattered around the base of each post, and hawks patrolled behind the mower to pick up the mice exposed by the cutter bar.

That was over forty years ago; but since then the hawks have been pretty thoroughly extirpated over large areas by sportsmen and farmers who didn't recognize their best friends when they saw them.

The larger predators, the coyotes and wolves, also played a very important role, for the grass required protection from the large grazing animals too, and these predators helped to control their numbers. The grass grows its annual harvest of food which can be eaten by the animals, but it is also the factory that produces the harvest. To maintain this factory in good condition, enough must always be left standing after grazing to build new leaves and support the roots in good health. The factory must never be harvested with the crop; in nature's organization, predators perform the function of protecting the factory from grazing animals. When undisturbed by man or by some catastrophe, this protective system adapts itself to maintain a fairly good balance.

In a study made at Yellowstone Park, Adolph Murie found that on a good range the well-fed deer can protect themselves and their young from attacks by coyotes. During the summer, therefore, coyotes live chiefly on smaller creatures. But in a severe winter, when snow covers their forage, the deer that are weakened by hunger can be easily killed. And now the coyotes reduce their numbers to those that the winter range can support in good health. So, when coyotes kill deer in winter it is a very good indication that there are more deer than the range can support.

The coyotes' numbers are controlled by hunger too, for there are two basic laws that regulate numbers of all predatory animals. First, in order to survive, the animal must always find enough food to carry it to the climax of its next hunt. Second, in order to multiply, it must find enough added food to rear its young.

So, in a mild winter, when the deer find enough food and strength to protect themselves, some coyotes may die for lack of meat or, in their weakened condition, succumb more easily to parasites and disease. And, of those that survive, some will fail to rear young. In this way, protected by predators, the community of plants and mammals maintains its balance. Before the coming of the white man, predators kept prairie herds down to the numbers that the winter range could support. The properly controlled herds were well fed by the luxuriant growth of the rich summer grass, whose tasty upper leaves they harvested; then they moved on to fresh pasture, leaving the tough, unappetizing lower leaves of the factory to produce another crop.

So each member in the grassland community, as in the forest, plays its special useful role, and the value of each member depends on the action of other members which interact upon each other to keep the whole in balance.

Many of man's problems today have been brought about by ignorance of this fact. For example, there is the famous record of the Kaibab Forest in northern Arizona, covering a little over 700,000 acres. In 1905 the stock of deer was estimated to be about 4,000, while the potential carrying capacity of the range was nearer 30,000. In an effort to increase the number of deer, the coyotes, wolves, and mountain lions were killed off. Between 1907 and 1917, 600 lions were removed and in the next few years 116 more were killed. Beginning in 1907, 11 wolves were killed, and between 1907 and 1939 more than 7,000 coyotes. With this removal of predators, the deer began to increase. By 1918 the population had gone up to over 40,000 and the food plants on the range began to show signs of damage. By 1920 some of the fawns were starving. By 1923 the herd had increased to an estimated 100,000, far above the normal carrying capacity of the range. The forage on the range was very much damaged, lessening still more its carrying capacity, and during the next two winters about 60,000 deer starved to death. The reduction continued until by 1939 the herd had dropped to about 10,000, struggling to survive on a very much damaged range. While there were, perhaps, too many of the predators at the beginning of this experiment and the deer were below the range capacity, it may well have been that part of this reduction was due to overhunting by man. It is obvious that the killing of the predators resulted in the ruin of the range, reducing its power to support deer. Actually, predators in proper numbers had protected both the range and the deer.

75 Miguel Covarrubias, *The Eagle, the Jaguar, and the Serpent*, 1954

Miguel Covarrubias, a veritable polymath renowned for both his artistic and intellectual endeavors, enjoyed a successful career as an illustrator and caricaturist for magazines such as *Vanity Fair*, the *New Yorker*, and *Vogue* before turning his attention toward cultural anthropology in the 1940s. Especially interested in the artifacts of his own people, Covarrubias became an expert scholar of Mesoamerican art and material culture, with a particular concentration on the Olmec civilization in Mexico. His academic research sought to illuminate this "important and little-known part of [Americans'] continental artistic heritage" and culminated in three texts on Indian art of the Americas, including *The Eagle, the Jaguar, and the Serpent*, which was featured in the *LWEC* within the three-page subsection "Indians."[1]

The Eagle, the Jaguar, and the Serpent was a comprehensive survey of North American Indian art practices, organized by geographical location. The book was rigorous in its analysis of tribal art production but also vividly illustrated by the author-artist, a quality that captivated J. D. Smith, who wrote: "I've had this book out of the library for about four years now, and haven't read the text yet, got my money's worth out of the illustrations." For readers of Smith's inclinations, the work functioned quite successfully at a visual level, a testament to Covarrubias's artistic talent; but for those seeking a greater depth of knowledge regarding the "aesthetic values" and "historical implications" of Native American artwork, it truly excelled.[2] —MG

1. Miguel Covarrubias, *The Eagle, the Jaguar, and the Serpent: Indian Art of the Americas* (New York: Knopf, 1954), 8.

2. Ibid.

Source

Miguel Covarrubias, *The Eagle, the Jaguar, and the Serpent: Indian Art of the Americas* (New York: Knopf, 1954), 182. (Referred to in *LWEC*, 383.)

Further Reading

Deloria, Vine, Jr. *Custer Died for Your Sins: An Indian Manifesto*. New York: Macmillan, 1969.

Mann, Charles C. *1491: New Revelations of the Americas before Columbus*. New York: Knopf, 2005.

Thompson, Stith, ed. *Tales of the North American Indians*. Bloomington: Indiana University Press, 1966.

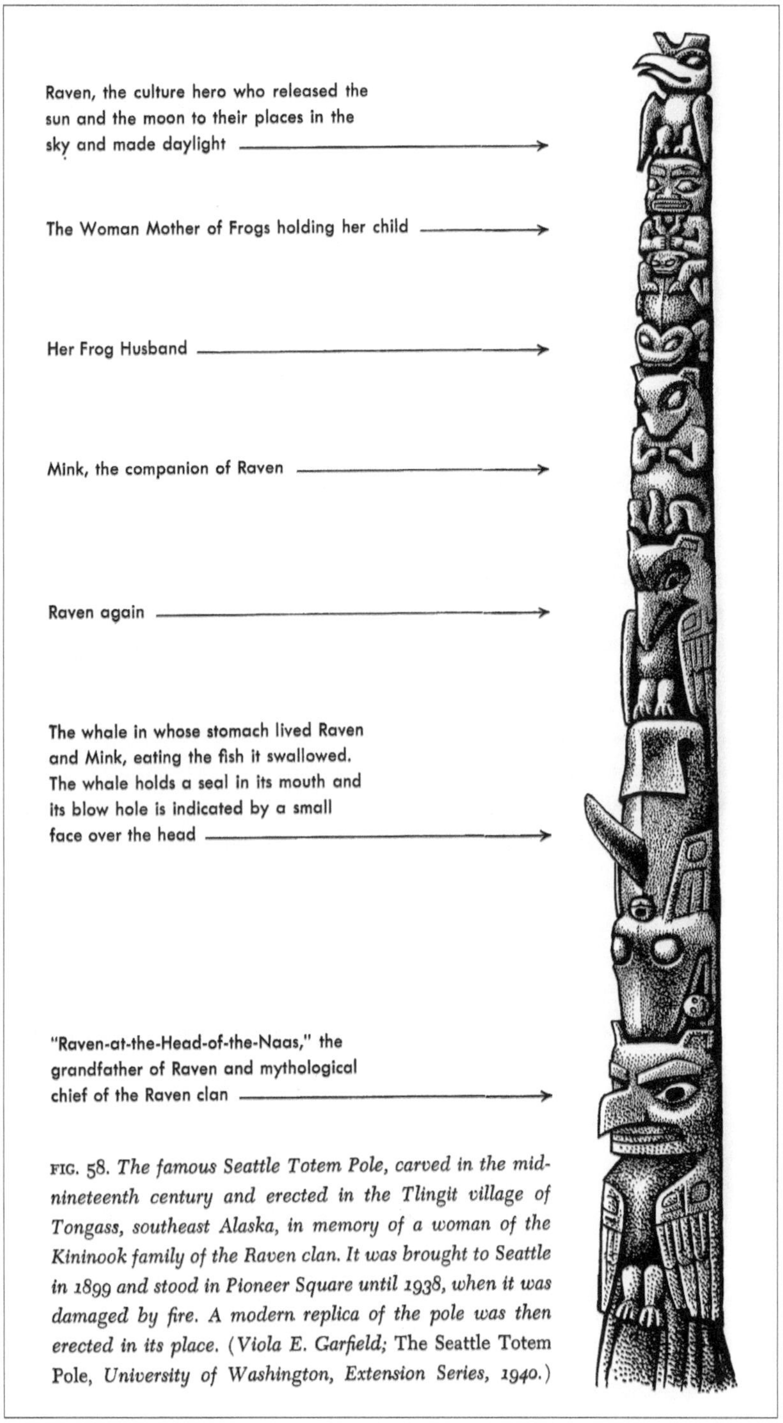

Raven, the culture hero who released the
sun and the moon to their places in the
sky and made daylight ⎯⎯⎯⎯⎯⎯→

The Woman Mother of Frogs holding her child ⎯⎯⎯⎯→

Her Frog Husband ⎯⎯⎯⎯⎯⎯⎯⎯⎯→

Mink, the companion of Raven ⎯⎯⎯⎯⎯→

Raven again ⎯⎯⎯⎯⎯⎯⎯⎯⎯→

The whale in whose stomach lived Raven
and Mink, eating the fish it swallowed.
The whale holds a seal in its mouth and
its blow hole is indicated by a small
face over the head ⎯⎯⎯⎯⎯⎯→

"Raven-at-the-Head-of-the-Naas," the
grandfather of Raven and mythological
chief of the Raven clan ⎯⎯⎯⎯⎯⎯→

FIG. 58. *The famous Seattle Totem Pole, carved in the mid-nineteenth century and erected in the Tlingit village of Tongass, southeast Alaska, in memory of a woman of the Kininook family of the Raven clan. It was brought to Seattle in 1899 and stood in Pioneer Square until 1938, when it was damaged by fire. A modern replica of the pole was then erected in its place. (Viola E. Garfield; The Seattle Totem Pole, University of Washington, Extension Series, 1940.)*

Figure 9.2
"Fig. 58. Viola E. Garfield, 'The famous Seattle totem pole.'" Miguel Covarrubias, *The Eagle, the Jaguar, and the Serpent: Indian Art of the Americas* (New York: Knopf, 1954), 182. Used with permission of the University of Washington Press.

Forrest Wilson, *Architecture: A Book of Projects for Young Adults,* **1968**

As its title suggests, *Architecture: A Book of Projects for Young Adults* was a collection of simple building exercises—thirty-three in all—that older children could execute using common household materials. Unlike other project-oriented juvenile workbooks, however, these tasks served to supplement and reinforce the historical, structural, and aesthetic lessons with which Forrest Wilson, a professor at Ohio University's School of Architecture, populated his text. Wilson hoped that this synthesis of learning and building would help young "architects" to perceive what he called the "architectural alphabet" of building and the "architectural language" of "texture, scale, and space."[1]

Although Brand admitted in his review that he "didn't know that architecture was a useful term or profession anymore," he nevertheless included Wilson's book in the catalog. Brand may have had qualms about the discipline, but he seldom discouraged a product that would so clearly foster creativity. According to Brand, *Architecture* "got [kids] into building stuff": an affirmative outcome for both him and Wilson.—MG

1. Forrest Wilson, *Architecture: A Book of Projects for Young Adults* (New York: Reinhold, 1968), 7.

Source

Forrest Wilson, *Architecture: A Book of Projects for Young Adults* (New York: Reinhold, 1968), 59–63. (Referred to in *LWEC*, 387.)

Further Reading

Wilson, Forrest. "Teaching the Built Environment." *School Review* 82, no. 4 (August 1974): 681–686.

Wilson, Forrest. *What It Feels Like to Be a Building*. Washington, DC: Preservation Press, 1988.

Selections from *Architecture: A Book of Projects for Young Adults*

Structure Following the Geometry of Stress

In modern architecture, machine made parts and engineered geometry replace the classic harmonies. To understand the forms of modern architecture we must understand engineering principles. ...

Modern materials with their ability to make strong joints transfer stresses entirely differently than the old masonry buildings which depended upon gravity alone for their stability.

Instead of the pyramidal piling of masonry the base of a modern building may do just the opposite by coming to a hinged point.

Let us experiment with some of these principles to see how they work. We will begin with the triangle. You have seen that beams which span long distances must be very deep. The arch, although materially economic, is difficult to hold in place high in the air and masonry is very heavy. The truss, which is an ingenious arrangement wherein the members are arranged so that they function in either tension or compression

without bending stresses, is a better answer as you will see in the next project.

The triangle is the only geometric figure that cannot change shape without altering its sides by either shortening or lengthening.

The triangle is a geometric figure that will hold its shape even though its three joints are hinged. To test this, take three pieces and four pieces of cardboard of equal length and pin them together. ... You will find that the triangle will hold its shape when pushed but that the square will become a parallelogram. Therefore, if we build a structure composed entirely of triangles it will be very rigid.

If you are going to use triangles in a truss you should find which members are in compression and which in tension. This is important since a tension member could be much lighter. You will find in the following experiments that a thin thread or string can be used instead of a much heavier piece of wood.

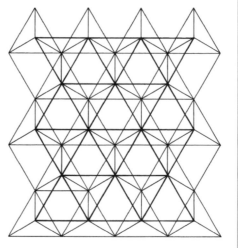

When you have made the structure, hold it in your hand and gently squeeze. If it does not feel rigid but moves in your hand look for the configurations that move. You will find that these invariably are not triangulated. Glue toothpicks across their corners to make them rigid.

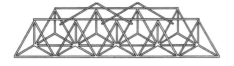

Now that we have found how well the triangles work let us compare them to a structure of rigid frames. This is a construction the joints of which are so strongly fastened that they transfer bending stresses to adjacent members. This is the opposite principle of the pin jointed triangle.

Figure 9.3
"Build a Space Frame." Forrest Wilson, *Architecture: A Book of Projects for Young Adults* (New York: Reinhold, 1968), 63. Reprinted with permission.

Project 21—Build a Space Frame

Materials: Two boxes of round toothpicks, white glue, waxed paper.

Tools: Knife or cutting instrument....

Procedure: Your aim is to make a construction of toothpicks that will span two feet between supports and hold up ten pounds of books. The toothpicks are to be glued together using as few toothpicks as possible. Do not cut the toothpicks; use them as they come out of the box.

The best method to build a space frame, you will find, is to construct triangles as shown in the drawing. The drawing shows the construction of a tetrahedron (a polyhedron with four sides). First construct a triangle by gluing three toothpicks together, then lay the triangle flat and build a pyramid of toothpicks from each of its corners. You will find that this is a very rigid and very strong structure. If you continue to use this principle you will be obtaining the maximum strength from the toothpicks. However, you will find this is almost impossible to do if you wish to construct certain shapes. When you depart from triangulation to construct a square, it will not be as stable as the triangle. When this happens, you can glue toothpicks across the corner of the square to strengthen it, as shown in the drawing.

When you have made the structure, hold it in your hand and gently squeeze. If it does not feel rigid but moves in your hand look for the configurations that move. You will find that these invariably are not triangulated. Glue toothpicks across their corners to make them rigid.

Now that we have found how well the triangles work let us compare them to a structure of rigid frames. This is a construction the joints of which are so strongly fastened that they transfer bending stresses to adjacent members. This is the opposite principle of the pin jointed triangle.

77 Timothy Leary, "Farewell Address," 1970–71

Psychedelic drug use is typically emphasized as one of the key components of the American counterculture; Timothy Leary's notorious work with the Harvard Psilocybin Project, San Francisco's drug-fueled Human Be-In, and Ken Kesey and the Merry Pranksters' psychedelic cross-country adventures and Acid Tests represent just a few outstanding cases in the overall narrative of the 1960s drug culture. Yet this apparently overwhelming trend toward drug use failed to infiltrate *The Whole Earth Catalog*, which, in fact, allocated just three pages to the topics of "Dope" and "Psychedelics." As a result, Brand was strategic with his selection of texts, not only incorporating books that proselytized the consciousness-opening effects of psychedelics but also providing first-aid instructions for avoiding "unwelcome trips" and *A Conscientious Guide to Drug Abuse* (1970) intended for drug users and their families.

The *Psychedelic Review*, according to the countercultural psychologist James A. Fadiman, who wrote its review in the catalog, was "the only journal to treat the literary, the scientific, and the spiritual aspects of psychedelic experience equally well," and thus the journal was an important addition to the "Dope" subsection. Fadiman dubbed Leary, whose essays often appeared in the *Psychedelic Review*, the "godfather" of the magazine, likely due to his pioneering work with LSD and psilocybin (a constituent of "magic mushrooms") at Harvard University from 1960 through 1962, and his steadfast commitment to the beneficial properties of psychedelics. The passage found on the following pages, Leary's "Farewell Address," was a written statement, distributed to newspapers and magazines across the nation, explaining the reasons behind his escape from prison ("the POW camp") and subsequent emigration from the United States to Algeria in the fall of 1970. Perhaps more significantly, however, this document emerged as a public rallying cry for vigorous government resistance, a war championed by Leary, the activist whom President Richard Nixon dubbed "the most dangerous man in the world."—MG

Source
Timothy Leary, "Farewell Address," in "The Radicalization of Timothy Leary," *Psychedelic Review* 11 (winter 1970–71): 26–27. (Referred to in *LWEC*, 413.)

Further Reading
Elcock, Chris. "From Acid Revolution to Entheogenic Evolution: Psychedelic Philosophy in the Sixties and Beyond." *Journal of American Culture* 36, no. 4 (2013): 296–311.

Higgs, John. *I Have America Surrounded: The Life of Timothy Leary*. Fort Lee, NJ: Barricade Books, 2006.

Watts, Alan W. *The Joyous Cosmology: Adventures in the Chemistry of Consciousness*. Foreword by Timothy Leary and Richard Alpert. New York: Vintage Books, 1962.

Timothy Leary's "Farewell Address" in "The Radicalization of Timothy Leary," from the *Psychedelic Review*

Third Bardo

The Period of Reentry

2.
You must leave now
Take what you need
You think will last
But whatever you wish to keep
You better grab it fast.

—Bob Dylan

(He's) leaving home, after living alone
For so many years. Bye Bye.
Silently closing (his) bedroom door
Leaving the note (he) hoped would
Say more
Quietly turning the backdoor key
Stepping outside (he) is free…

—The Beatles

(The following statement was written in the POW camp and carried over the wall in full sight of two gun trucks. I offer loving gratitude to my Sisters and Brothers in the Weatherman Underground who designed and executed my liberation. Rosemary and I are now with the Underground and we'll continue to stay high and wage the revolutionary war.)

There is the time for peace and the time for war.

There is the day of laughing Krishna and the day of Grim Shiva.

Brothers and Sisters, at this time let us have no more talk of peace.

The conflict which we have sought to avoid is upon us. A world-wide ecological religious warfare. Life vs. death.

Listen. It is a comfortable, self-indulgent cop-out to look for conventional economic-political solutions.

Brothers and Sisters, this is a war for survival.
Ask Huey and Angela. They dig it.
Ask the wild free animals. They know it.
Ask the turned-on ecologists. They sadly admit it.

I declare that World War III is now being waged by short-haired robots whose deliberate aim is to destroy the complex web of free wild life by the imposition of mechanical order.

Listen. There is no choice left but to defend life by all and every means possible against the genocidal machine.

Listen. There are no neutrals in genetic war. There are no non-combatants at Buchenwald, My Lai, or Soledad.

You are part of the death apparatus or you belong to the network of free life. Do not be deceived. It is a classic stratagem of genocide to camouflage their wars as law and order police actions.

Remember the Sioux and the German Jews and the black slaves and the marijuana programs and the TWA indignation over airline hijackings!

If you fail to see that we are the victims—defendants of genocidal war, you will not understand the rage of the blacks, the fierceness of the browns, the holy fanaticism of the Palestinians, the righteous mania of the Weathermen, and the pervasive resentment of the young.

Listen, Americans. Your government is an instrument of total lethal evil.

Remember the buffalo and the Iroquois!
Remember Kennedy, King, Malcolm, Lenny!
Listen. There is no compromise with a machine. You cannot talk peace and love to a humanoid robot whose every Federal Bureaucratic impulse is soulless, heartless, lifeless, loveless.

In his life struggle we use the ancient holy strategies of organic life:
1. Resist lovingly in the loyalty of underground sisterhoods and brotherhoods.
2. Resist passively, break lock-step … drop out.
3. Resist actively, sabotage, jam the computer … hijack planes … trash every lethal machine in the land.
4. Resist publicly, announce life … denounce death.
5. Resist privately, guerrilla invisibility.
6. Resist beautifully, create organic art, music.
7. Resist biologically, be healthy … erotic … conspire with seed … breed.

8. Resist spiritually, stay high … praise God … love life … blow the mechanical mind with Holy Acid … dose them … dose them.

9. Resist physically, robot agents who threaten life must be disarmed, disabled, disconnected by force. … Arm yourself and shoot to live. … Life is never violent. To shoot a genocidal robot policeman in the defense of life is a sacred act.

Listen Nixon. We were never that naïve. We knew that flowers in your gunbarrels were risky. We too remember Munich and Auschwitz all too well as we chanted love and raised our Woodstock fingers in the gentle sign of peace.

We begged you to live and let live, to love and let love, but you have chosen to kill and get killed. May God have mercy on your souls.

For the last seven months, I, a free, wild man, have been locked in POW camps. No living creature can survive in a cage. In my flight to freedom I leave behind a million brothers and sisters in the POW prisons of Quentin, Soledad, Con Thien.

Listen comrades. The liberation war has just begun. Resist, endure, do not collaborate. Strike. You will be free.

Listen you brothers of the imprisoned. Break them out! If David Harris has ten friends in the world, I say to you, get off your pious non-violent asses and break him out.

There is no excuse for one brother or sister to remain a prisoner of war.

Right on Leila Khaled!

Listen, the hour is late. Total war is upon us. Fight to live or you'll die. Freedom is life. Freedom will live.

(Signed) Timothy Leary

WARNING: I am armed and should be considered dangerous to anyone who threatens my life or my freedom.

78 Baba Ram Dass, *Be Here Now*, 1971

Before 1967, Ram Dass was known as Richard Alpert, a Stanford-educated psychologist and Harvard University professor. In spite of his professional accomplishments, Alpert felt unfulfilled and dissatisfied with the other dimensions of his life. On March 6, 1961, however, he discovered a salve for his ennui—the psychedelic drug psilocybin—and entered a new "psychedelic" phase of life. The young scholar's tenure at Harvard serendipitously overlapped with Timothy Leary's, and the two cultivated a fertile working relationship, collaborating on studies by the Harvard Psilocybin Project and coauthoring the book *The Psychedelic Experience: A Manual Based on the Tibetan Book of the Dead* (1964) with Ralph Metzner. In the mid-1960s, when the future "baba" sought a higher level of consciousness than psychedelic substances could offer, he traveled to India, entering what he called his "yogi" stage. It was there that he met the guru Neem Karoli Baba, who renamed Alpert Ram Dass, or "Servant of God."

Upon his return from India, Dass settled at the Lama Foundation commune in New Mexico, where he composed *Be Here Now*, an introduction to Eastern philosophy aimed at a countercultural American audience. George de Alth, who reviewed the book for the catalog, captured the inventiveness of Dass's approach, which presented spiritual profundities in a manner accessible to fledgling yogis: "Rum Dum's background is tennis shoes and bikes and cars and trying to get laid and gaining prestige, just like all of us, so his meditations are as familiar to us as our own bathwater. He makes the Message of the East comfortable for our coffee table studies."

Toward this end, for the second chapter of *Be Here Now*, "From Bindu to Ojas: The Core Book," Dass presented an expressive collection of meditations on modern life, brilliantly illustrated by the artists in residence at Lama—a sort of illuminated manuscript for the countercultural set, two images from which appear on the following pages.—MG

Source

Baba Ram Dass, *Be Here Now* (San Cristobal, NM: Lama Foundation, 1971), 37, 79. (Referred to in *LWEC*, 413.)

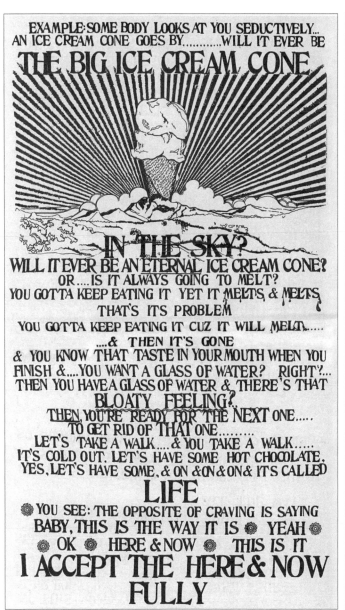

79 Robert Crumb, "Keep on Truckin'"

American cartoonist Robert Crumb's signature drawing "Keep On Truckin'," published in the first issue of *Zap Comix* in 1968, became a popular symbol of hippie culture and was widely imitated and copied. Surprisingly, comic strips were not often present in the *WEC*, but Brand entrusted the editing of a *WEC* supplement to Paul Krassner in 1971, who asked Crumb to illustrate the cover. Crumb was highly aware of trends in popular culture. Responding to a questionnaire in 1978, he explained: "I love to look at 'zines, all kinds. I've always been an avid reader of 'zines. They are a good way to catch the pulse-beat of the culture and, being into 'graphics' myself, I like to study the visuals. I also dig around secondhand magazine stores, a gold-mine of American cultural history. In McDonalds (Tenderloin, San Francisco) upstairs in the back there's a big stack of police and F.B.I. 'zines that make for interesting study, and I recently purchased a stack of 'parties' magazines from 1929–30 which reflect the absurdities of middle-class & college 'fun' from that period."[1] The anonymous editor—"Steamboat"—insisted on the importance of this kind of comic: "These ain't your run-of-the-mill comic books. At times they are gross, beautiful, or violent. Best of all, they can be very funny, and a good reminder not to take things (such as yourself) too seriously. Who sez comic books ain't relevant?"—CM

1. Robert Crumb, letter to Stewart Brand, December 27, 1978. WEAP M1245.

Source
LWEC, 407.

Further Reading
Crumb, Rob. *R. Crumb's Fritz the Cat.* New York: Ballantine Books, 1969.

Crumb, Rob. *R. Crumb's Head Comix.* New York: Ballantine Books, 1970.

Kaplan, Geoff, ed. *Power to the People: The Graphic Design of the Radical Press and the Rise of the Counter-Culture, 1964–1974.* Chicago: University of Chicago Press, 2013.

Figure 9.6
Robert Crumb, "Keep On Truckin'." *LWEC*, 407.

80 Sim Van der Ryn et al., *Farallones Scrapbook*, 1971

Professor of Architecture at the College of Environmental Design at the University of California, Berkeley, Sim Van der Ryn, assisted by the carpenter-architect Jim Campe, began a series of graduate courses aimed at providing students design/build opportunities. Each of these studios culminated in a publication. The *Farallones Scrapbook* is the best known of these, picked up by Random House in 1971. The aim of this project was to encourage children to experiment with construction using found materials: paper, cardboard, tires. Behind the practical exercises was the conviction that creating new spaces could break down social conventions, and this applied to the classroom.

The emphasis was on thinking about different kinds of spaces and unconventional pleated structures based on three-dimensional geometry. The project combined advance thinking on pedagogy with a radical view of a shelter. Indeed, Van der Ryn had used full-size structures of this kind for providing temporary shelters for seasonal crop workers.

The success of the publication, which reached beyond its immediate context, was partly due to the high quality of the production, with its square format, visually striking yellow paper, and imaginative use of illustrations.—CM

Source

Sim Van der Ryn et al., *Farallones Scrapbook: A Memento and Manual of Our Apprenticeship in Making Places and Changing Spaces in Schools, at Home, and* within Ourselves (Point Reyes Station, CA: Farallones Designs, 1971). (Referred to in *LWEC*, 404.)

Further Reading

Castillo, Greg. "Counterculture Terroir: California's Hippie Enterprise Zone." In *Hippie Modernism: The Struggle for Utopia*, ed. Andrew Blauvelt. Minneapolis: Walker Art Center, 2015, 87–101.

Van der Ryn, Sim. *Design for Life: The Architecture of Sim Van der Ryn.* Layton, UT: Gibbs Smith, 2005.

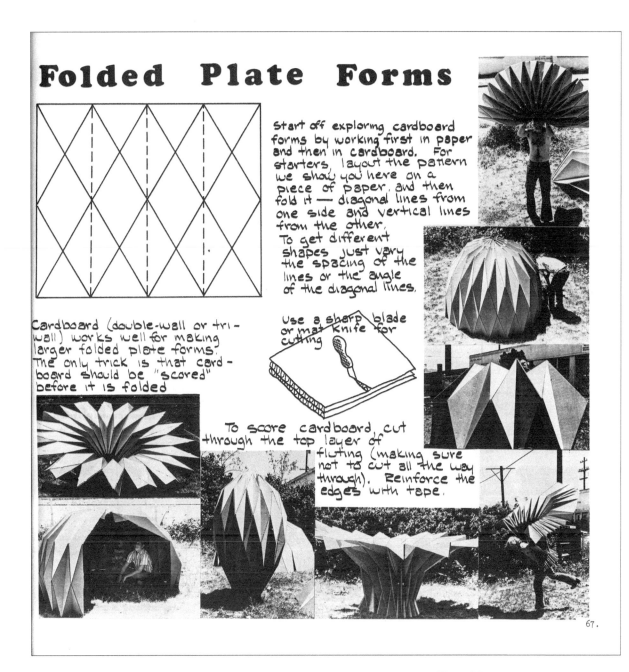

Folded Plate Forms

Start off exploring cardboard forms by working first in paper and then in cardboard. For starters, layout the pattern we show you here on a piece of paper, and then fold it — diagonal lines from one side and vertical lines from the other.
To get different shapes just vary the spacing of the lines or the angle of the diagonal lines.

Cardboard (double-wall or tri-wall) works well for making larger folded plate forms. The only trick is that cardboard should be "scored" before it is folded

Use a sharp blade or mat knife for cutting

To score cardboard, cut through the top layer of fluting (making sure not to cut all the way through). Reinforce the edges with tape.

67.

Figure 9.7
"Folded Plate Forms." Sim van der Ryn et al., *Farallones Scrapbook: A Memento and Manual of Our Apprenticeship in Making Places and Changing Spaces in Schools, at Home, and within Ourselves* (Point Reyes Station, CA: Farallones Designs, 1971), 67.

Appendix

Books Cited in *The Last Whole Earth Catalog*

This appendix lists all the books cited specifically in the *LWEC* that we could find in the Library of Congress or major American university libraries. The list is intended to give an idea of the range of reading proposed for *LWEC* readers and is organized alphabetically by author and title. Editions as close as possible to 1971 are chosen. Citations of books in the *LWEC* usually indicated page lengths, which helped us to identify the edition. Some ephemeral publications, including those issued by state and federal agencies, are no longer cataloged in official sources and may have incomplete records.

Aaronfreed, Justin. *Developmental Psychology Today*. Del Mar, CA: CRM Books, 1971.

Aaronson, Bernard Seymour, and Humphry Osmond. *Psychedelics: The Uses and Implications of Hallucinogenic Drugs*. Hogarth Psychology and Psychiatry. London: Hogarth Press, 1971.

Abehsera, Michel. *Zen Macrobiotic Cooking: Book of Oriental and Traditional Recipes*. New York: University Books, 1968.

Abramowitz, Milton, and Irene A. Stegun. *Handbook of Mathematical Functions, with Formulas, Graphs, and Mathematical Tables*. New York: Dover, 1965.

Adney, Edwin Tappan, Howard Irving Chapelle, and Museum of History and Technology. *The Bark Canoes and Skin Boats of North America*. US National Museum Bulletin. Washington, DC: Smithsonian Institution, 1964.

Agee, James, and Walker Evans. *Let Us Now Praise Famous Men: Three Tenant Families*. Boston: Houghton Mifflin, 1960.

Alcosser, Edward, James P. Phillips, and Allen M. Wolk. *How to Build a Working Digital Computer*. New York: Hayden, 1967.

Alexander, Christopher. *Notes on the Synthesis of Form*. Cambridge, MA: Harvard University Press, 1964.

Alinsky, Saul David. *Rules for Radicals: A Practical Primer for Realistic Radicals*. New York: Random House, 1971.

Allen, Durward Leon. *The Farmer and Wildlife*. Washington, DC: Wildlife Management Institute, 1949.

Allen, Edward. *Stone Shelters*. Cambridge, MA: MIT Press, 1969.

Allen, Lewis M. *Printing with the Handpress*. New York: Van Nostrand Reinhold, 1969.

Amateur Yacht Research Society. *Sailing Hydrofoils*. Newbury: Amateur Yacht Research Society, 1970.

American Society for Cybernetics and Heinz Von Foerster. *Purposive Systems: Proceedings of the First Annual Symposium of the American Society for Cybernetics*. New York: Spartan Books, 1969.

American Society of Educators. *Media and Methods*. Philadelphia: American Society of Educators, 1969.

Anagarika, Govinda. *The Way of the White Clouds: A Buddhist Pilgrim in Tibet*. London: Hutchinson, 1966.

Anderson, Edgar. *Plants, Man and Life*. Boston: Little, Brown, 1952.

Anderson, Edwin P. *Audels Domestic Water Supply and Sewage Disposal Guide: A Practical Treatise Covering Construction of Wells, Modern Pumping Systems, Pump Controls, Pumphouses, Pump Installation and Maintenance, Hydraulic Rams, Water Treatment Methods*. Audels Helping Handbooks for Mechanics. New York: T. Audel, 1960.

Anderson, Edwin P. *Electric Motors*. 2nd ed. Indianapolis: T. Audel, 1968.

Anderson, Edwin P. *Home Appliance Servicing*. 3rd ed. Indianapolis: T. Audel, 1974.

Anderson, L. O. *Low-Cost Wood Homes for Rural America: Construction Manual*. US Department of Agriculture Agriculture Handbook. Washington, DC: US Department of Agriculture, Forest Service; US Government Printing Office, 1969.

Angier, Bradford. *Skills for Taming the Wilds: A Handbook of Woodcraft Wisdom*. Harrisburg, PA: Stackpole Books, 1967.

Angulo, Jaime de. *Indian Tales*. New York: A. A. Wyn, 1962.

Ant Farm. *Inflatocookbook*. Sausalito, CA: Ant Farm, 1971.

Arbib, Michael A. *Brains, Machines, and Mathematics*. New York: McGraw-Hill, 1964.

Arendt, Hannah. *The Human Condition*. Charles R. Walgreen Foundation Lectures. Chicago: University of Chicago Press, 1958.

Armstrong-Wright, A. T. *Critical Path Method: Introduction and Practice*. Upper Saddle River, NJ: Prentice Hall Press, 1969.

Arnow, Harriette Louisa Simpson. *Seedtime on the Cumberland*. New York: Macmillan, 1960.

Ashbrook, F. G. *Butchering, Processing, and Preservation of Meat*. New York: Van Nostrand, 1955.

Ashby, W. Ross. *Design for a Brain*. New York: J. Wiley, 1952.

Ashby, W. Ross. *An Introduction to Cybernetics*. New York: J. Wiley, 1956.

Ashley, Clifford Warren. *The Ashley Book of Knots*. Garden City, NY: Doubleday, 1963. (Originally published 1944.)

Ashton-Warner, Sylvia. *Teacher*. Harmondsworth: Penguin in association with Secker & Warburg, 1966.

Atwater, Mary Meigs. *Byways in Hand-Weaving*. New York: Macmillan, 1954.

Audsley, George Ashdown. *The Art of Organ-Building: A Comprehensive Historical, Theoretical, and Practical Treatise on the Tonal Appointment and Mechanical Construction of Concert-Room, Church, and Chamber Organs*. 2 vols. New York: Dover, 1965.

Austin, Robert, Kōichirō Ueda, and Dana Levy. *Bamboo*. New York: Walker/Weatherhill, 1970.

Bach, Richard, and Russell Munson. *Jonathan Livingston Seagull*. New York: Macmillan, 1970.

Back, Joe. *Horses, Hitches, and Rocky Trails*. Chicago: Sage Books, 1959.

Badmaieff, Alexis, and Don Davis. *How to Build Speaker Enclosures*. A Howard W. Sams Photofact Publication, Seb-1. Indianapolis: H. W. Sams, 1966.

Baer, Steve. *Dome Cookbook*. Corrales, NM: Cookbook Fund/Lama Foundation, 1969.

Baer, Steve. *Zome Primer: Elements of Zonohedra Geometry; Two and Three Dimensional Growths of Stars with Five Fold Symmetry*. Albuquerque: Zomeworks, 1970.

Ballentyne, D. W. G., and L. E. Q. Walker. *A Dictionary of Named Effects and Laws in Chemistry, Physics, and Mathematics*. 2nd rev. and enl. ed. New York: Macmillan, 1961.

Ballou, Robert O., Frederic Spiegelberg, and Horace L. Friess. *The Bible of the World*. New York: Viking, 1939.

Barker, Wayne. *Brain Storms: A Study of Human Spontaneity*. New York: Grove Press, 1968.

Barrett, Raymond E. *Build-It-Yourself Science Laboratory*. Garden City, NY: Doubleday, 1963.

Bartholomew, John, and Son. *The Times Atlas of the World, Comprehensive Edition*. 2nd ed. London: Times Newspapers, 1968.

Batchelder, Marjorie Hope. *The Puppet Theatre Handbook*. New York: Harper & Brothers, 1947.

Bealer, Alex W. *The Art of Blacksmithing*. New York: Funk & Wagnalls, 1969.

Bear, Fred. *The Archer's Bible*. Garden City, NY: Doubleday, 1968.

Beard, Daniel Carter. *The American Boys' Handybook of Camp-Lore and Woodcraft*. Woodcraft Series. Philadelphia: J. B. Lippincott, 1920.

Behanan, Kovoor Thomas. *Yoga: A Scientific Evaluation*. New York: Dover, 1960.

Behrman, Daniel. *The New World of the Oceans: Men and Oceanography*. Boston: Little, Brown, 1969.

Benade, Arthur H. *Horns, Strings, and Harmony*. Science Study Series, S11. Garden City, NY: Anchor Books, 1960.

Benford, Jay R., and Herman Husen. *Practical Ferro-cement Boatbuilding*. Friday Harbor, WA: J. R. Benford, 1970.

Bennett, Bob. *Rabbit Raising*. Rev. ed. Merit Badge Series 3375. North Brunswick, NJ: Boy Scouts of America, 1974.

Berman, Susan. *The Underground Guide to the College of Your Choice*. New York: New American Library, 1971.

Berreman, Gerald Duane. *Anthropology Today*. Del Mar, CA: CRM Books, 1971.

Berry, Wendell. *The Long-Legged House*. New York: Harcourt, 1969.

Besant, Annie, and C. W. Leadbeater. *Thought-Forms*. 2nd Quest Books ed. Wheaton, IL: Theosophical Publishing House, 1969. (Originally published 1901.)

Biegeleisen, J. I. *Screen Printing: A Contemporary Guide to the Technique of Screen Printing for Artists, Designers, and Craftsmen*. New York: Watson-Guptill, 1971.

Big Sur Recordings. San Rafael, CA: n.d.

The Biosphere. San Francisco: W. H. Freeman, 1970.

Black Elk, and John Gneisenau Neihardt. *Black Elk Speaks*. New York: W. Morrow, 1932.

Blake, Noah, and Eric Sloane. *Diary of an Early American Boy, Noah Blake, 1805*. New York: W. Funk, 1962.

Blanchard, Harold Frederick, and Ralph Ritchen. *Auto Engines and Electrical Systems*. 4th ed. New York: Motor, 1967.

Bleibtreu, John N. *The Parable of the Beast*. New York: Macmillan, 1968.

Boat Owners Guide. New York: Yachting Publishing, 1969.

Bockus, H. William. *Advertising Graphics: A Workbook and Reference for the Advertising Artist*. Vol. 1. New York: Macmillan, 1969.

Boeke, Kees. *Cosmic View: The Universe in 40 Jumps*. New York: J. Day, 1957.

Boericke, William Fay. *Prospecting and Operating Small Gold Placers*. New York: J. Wiley & Sons/Chapman & Hall, 1933.

Borrego, John. *Space Grid Structures: Skeletal Frameworks and Stressed-Skin Systems*. Cambridge, MA: MIT Press, 1968.

Boston Women's Health Collective. *Women and Their Bodies: A Course*. Boston: Boston Women's Health Collective, 1970.

Boudin, Kathy. *The Bust Book: What to Do Till the Lawyer Comes*. Rev. ed. An Evergreen Black Cat Book, B-232-Z. New York: Grove Press, 1970.

Boulding, Kenneth E. *Beyond Economics: Essays on Society, Religion, and Ethics*. Ann Arbor: University of Michigan Press, 1968.

Boulding, Kenneth E. *The Image: Knowledge in Life and Society*. Ann Arbor: University of Michigan Press, 1956.

Bowditch, Nathaniel, and Jonathan Ingersoll Bowditch, and the US Hydrographic Office. *American Practical Navigator*. Washington, DC: Government Printing Office, 1966. (Originally published 1802.)

Boyes, William E., and Harold Sedlik. *Low-Cost Jigs, Fixtures and Gages for Limited Production*. Dearborn, MI: Society of Manufacturing Engineers, Publications Development Department, Marketing Division, 1986.

Boy Scouts of America. *Fieldbook*. North Brunswick, NJ: Boy Scouts of America, 1967.

Brautigan, Richard. *Trout Fishing in America: A Novel*. Writing 14. San Francisco: Four Seasons Foundation, 1967.

Bravery, H. E. *Home Brewing without Failures*. New York: Arc Books, 1966.

Bravery, H. E. *Successful Wine Making at Home*. New York: Arc Books, 1961.

Breed, Charles Blaney, Alexander J. Bone, and B. Austin Barry. *Surveying*. 3rd ed. New York: Wiley, 1971.

Brilliant, Larry, ed. *The Body Politic*. San Francisco: Medical Committee for Human Rights, 1971–1987.

Brockman, John. *By the Late John Brockman*. New York: Macmillan, 1969.

Bromfield, Louis. *Malabar Farm*. New York: Harper, 1948.

Brooklyn Botanic Garden. *Plants and Gardens*. Brooklyn, NY: Brooklyn Botanic Garden, 1945.

Brower, D. R. *Manual of Ski Mountaineering*. San Francisco: Sierra Club, 1969.

Brower, D. R., ed. *The Sierra Club Wilderness Handbook*. New York: Ballantine Books, 1967.

Brown, Edward Espe. *The Tassajara Bread Book*. Berkeley: Shambala, 1970.

Brown, Robert E. *Psychedelic Guide to Preparation of the Eucharist*. Austin, TX: Linga Sharira Incense Co, 1968.

Brown, Sam. *All about Telescopes*. Edmund Scientific, 1967.

Brummitt, Wyatt. *Kites*. A Golden Handbook Guide. New York: Golden Press, 1971.

Buchman, Herman. *Stage Makeup*. New York: Watson-Guptill, 1971.

Buck, Robert N. *Weather Flying*. New York: Macmillan, 1970.

Buckles, Robert A. *Ideas, Inventions, and Patents: How to Develop and Protect Them*. New York: Wiley, 1957.

Buckwalter, Len. *ABC's of Short-Wave Listening*. Indianapolis: Howard W. Sams, 1962.

Building a Log House. Juneau, AK: Cooperative Extension Service, University of Alaska, 1965.

Bundy, Clarence E., and Ronald V. Diggins. *Livestock and Poultry Production*. 3rd ed. Prentice-Hall Vocational Agriculture Series. Englewood Cliffs, NJ: Prentice-Hall, 1968.

Burack, Richard. *The New Handbook of Prescription Drugs: Official Names, Prices, and Sources for Patient and Doctor*. Rev. ed. New York: Pantheon Books, 1970.

Burris-Meyer, Harold, and Edward Cyrus Cole. *Scenery for the Theatre: The Organization, Processes, Materials and Techniques Used to Set the Stage*. Boston: Little, Brown, 1938.

Burroughs, John. *Canadian Wood-Frame House Construction*. Canada Mortgage and Housing Corporation, 1967.

Burstall, Aubrey F. *Simple Working Models of Historic Machines: Easily Made by the Reader*. Cambridge, MA: MIT Press, 1969.

Buryn, Ed. *Vagabonding in Europe and North Africa*. New York: Random House, 1971.

Cage, John. *Silence: Lectures and Writings*. Middletown, CT: Wesleyan University Press, 1961.

Cage, John. *A Year from Monday: New Lectures and Writings*. Middletown, CT: Wesleyan University Press, 1967.

Cage, John, and Ihab Habib Hassan. *Liberations: New Essays on the Humanities in Revolution*. Middletown, CT: Wesleyan University Press, 1971.

Calder, Nigel. *The Mind of Man: An Investigation into Current Research on the Brain and Human Nature*. New York: Viking, 1970.

Caldwell, John H. *The New Cross-Country Ski Book*. 4th ed. Brattleboro, VT: S. Greene, 1968.

Campbell, Joseph. *The Hero with a Thousand Faces*. 2nd ed. Bollingen Series. Princeton, NJ: Princeton University Press, 1968.

Campbell, Joseph. *The Masks of God*. New York: Viking, 1969.

Campbell, Joseph. *Occidental Mythology*. New York: Penguin, 1964.

Cardew, Michael. *Pioneer Pottery*. Harlow, England: Longmans, 1969.

Casady, Robert B., Paul B. Sawin, and J. Van Dam. *Commercial Rabbit Raising*. Agriculture Handbook No. 309. Rev. ed. Washington, DC: US Agricultural Research Service, 1971.

Casewit, Curtis W. *How to Get a Job Overseas*. New York: Arc Books, 1970.

Castaneda, Carlos. *A Separate Reality: Further Conversations with Don Juan*. New York: Simon & Schuster, 1971.

Castaneda, Carlos. *The Teachings of Don Juan: A Yaqui Way of Knowledge*. Berkeley: University of California Press, 1968.

Cayford, John E. *Underwater Work: A Manual of Scuba Commercial, Salvage, and Construction Operations*. 2nd ed. Cambridge, MD: Cornell Maritime Press, 1966.

Cement Mason's Manual for Residential Construction. Skokie, IL: Portland Cement Association, 1960.

Chapelle, Howard Irving. *Boatbuilding: A Complete Handbook of Wooden Boat Construction*. New York: Norton, 1941.

Chapman, Charles F., and Elbert S. Maloney. "Piloting, Seamanship, and Small Boat Handling." In *Motor Boating's Ideal Series*. New York: Motor Boating & Sailing, 1948.

Chill Out Lounge. New York: Benz Street, 2005. Sound recording, 1 sound disc; digital; 4 3/4 in., BZ 10082–2 Benz Street.

Christopher, F. J. *Basketry*. Dover-Foyle Handbook. New York: Dover, 1953.

Clarke, Arthur C., and Time-Life Books. *Man and Space*. Life Science Library. New York: Time-Life Books, 1969.

Clarke, Geoffrey, and Stroud Cornock. *A Sculptor's Manual*. New York: Studio Vista Reinhold, 1968.

Coats and Clark. *Coats and Clark's Sewing Book: Newest Methods from A to Z*. New York: Golden Press, 1967.

Coker, Jerry. *Improvising Jazz*. A Spectrum Book. Englewood Cliffs, NJ: Prentice-Hall, 1964.

Colbeck, John. *Pottery: The Technique of Throwing*. New York: Batsford/Watson-Guptill, 1969.

Collingwood, G. H., and the Warren David Brush, Devereux Butcher, and American Forestry Association. *Knowing Your Trees*. Washington, DC: American Forestry Association, 1974.

Collingwood, Peter. *The Techniques of Rug Weaving*. London: Faber, 1968.

Common Diseases of Domestic Rabbits. USDA Bulletin No. 45-3. Washington, DC: United States Department of Agriculture, 1963.

Commoner, Barry. *Science and Survival*. New York: Viking, 1966.

A Complete Guide to Meat Curing. Chicago: Morton Salt Company, 1969.

Conference for National Cooperation in Aquatics. *The New Science of Skin and Scuba Diving: A Revision of the Widely Used Science of Skin and Scuba Diving*. Rev. ed. New York: Association Press, 1962.

Consumer Reports Buying Guide. Mount Vernon, NY: Consumers Union, 1936.

Conway, Carle, and the Soaring Society of America. *The Joy of Soaring: A Training Manual*. Los Angeles: Soaring Society of America, 1969.

Cook, Peter, and the Archigram Group. *Archigram*. London: Studio Vista, 1972.

Cooke, Nelson Magor. *Basic Mathematics for Electronics*. 2nd ed. New York: McGraw-Hill, 1960.

Cooper, Kenneth H. *The New Aerobics*. New York: M. Evans, distributed in association with Lippincott, 1970.

Cortright, Edgar M., and the US National Aeronautics and Space Administration. *Exploring Space with a Camera*. NASA SP-168. Washington, DC: Scientific and Technical Information Division, National Aeronautics and Space Administration, 1968.

Coulson, Ernest Harrison, A. E. J. Trinder, and Aaron E. Klein. *Test Tubes and Beakers:*

Chemistry for Young Experimenters. Garden City, NY: Doubleday, 1971.

Courtney, W. F., ed. *The Reader's Adviser: A Layman's Guide 2*. 11th ed. New York: Bowker, 1969.

Covarrubias, Miguel, and Jay I. Kislak Reference Collection (Library of Congress). *The Eagle, the Jaguar, and the Serpent: Indian Art of the Americas; North America; Alaska, Canada, the United States*. New York: Knopf, 1954.

Cowie, Peter. *International Film Guide*. International Film Guide Publications. Cranbury, NJ: A. S. Barnes, 1974.

Critchlow, Keith. *Order in Space: A Design Source Book*. London: Thames & Hudson, 1969.

Crovitz, Herbert F. *Galton's Walk: Methods for the Analysis of Thinking, Intelligence and Creativity*. New York: Harper & Row, 1970.

Crow, James T., and Cameron A. Warren. *Four Wheel Drive Handbook*. Newfoundland, NJ: Haessner, 1970.

Crowhurst, Norman H. *Electronic Musical Instruments*. Blue Ridge Summit, PA: Tab Books, 1971.

Cullum, Albert. *Push Back the Desks*. New York: Citation Press, 1967.

Cundy, H. Martyn, and A. P. Rollett. *Mathematical Models*. 2nd ed. Oxford: Clarendon Press, 1961.

Curry, Manfred. *The Beauty of Flight*. London: J. Miles, 1932.

Curtis, Charles Pelham. *The Practical Cogitator; or, The Thinker's Anthology*. Boston: Houghton Mifflin, 1962.

Cuthbertson, Tom, and Rick Morrall. *Anybody's Bike Book: An Original Manual of Bicycle Repairs*. Berkeley, CA: Ten Speed Press, 1971.

Dacey, Norman F. *How to Avoid Probate*. Bridgeport, CT: National Estate Planning Council, 1965.

Daniels, Farrington. *Direct Use of the Sun's Energy*. Trends in Science. New Haven: Yale University Press, 1964.

Darlington, Jeanie. *Grow Your Own: An Introduction to Organic Gardening*. Berkeley, CA: Bookworks, 1970.

Darr, Jack. *Electric Guitar Amplifier Handbook*. 2nd ed. Indianapolis: H. W. Sams, 1968.

Darvill, Fred T. *Mountaineering Medicine: A Wilderness Medical Guide*. 10th ed. Mount Vernon, WA: Wilderness Press, 1968.

Davenport, Elsie G. *Your Handspinning*. Big Sur, CA: Craft & Hobby Book Service, 1964.

Davis, Adelle. *Let's Cook It Right*. Rev. ed. New York: Signet/New American Library, 1970.

Day, Richard. *The Practical Handbook of Plumbing and Heating*. Practical Workshop Library. New York: Arco, 1969.

De Bell, Garrett. *The Environmental Handbook*. New York: Ballantine Books, 1970.

De Cristoforo, R. J. *How to Build Your Own Furniture*. Popular Science Skill Book. New York: Popular Science, 1965.

Delgado, José Manuel Rodríguez. *Physical Control of the Mind: Toward a Psychocivilized Society*. World Perspectives. New York: Harper & Row, 1969.

Deloria, Vine. *Custer Died for Your Sins: An Indian Manifesto*. New York: Macmillan, 1969.

Dennison, George. *The Lives of Children: The Story of the First Street School*. New York: Random House, 1969.

Desert Survival: Information for Anyone Travelling in the Desert Southwest. Phoenix, AZ: Maricopa County Department of Civil Defense and Emergency Services, 1971.

Dickey, Esther. *Passport to Survival: Four Foods and More to Use and Store*. Salt Lake City: Bookcraft, 1969.

Dietz, Albert G. H. *Plastics for Architects and Builders*. Cambridge, MA: MIT Press, 1969.

Directory of Community Market. Deerfield, MA: Community Market, n.d.

Dirt Tracks. Rick Spreitzer, 2004. Sound recording.

Dodson, Fitzhugh. *How to Parent*. Los Angeles: Nash, 1970.

Do-It-Yourself Encyclopedia. 2 vols. Indianapolis: Bobbs-Merrill, 1968.

Dominick, John. *The Drug Bust*. New York: Light Company, 1970.

Dore, W. G. *Wild-Rice*. Canada Department of Agriculture Publication. Ottawa: Department of Agriculture, Research Branch, 1969.

Dorf, Richard H. *Electronic Musical Instruments*. 3rd ed. New York: Radiofile, 1968.

Douglass, Ralph. *Calligraphic Lettering with Wide Pen and Brush*. 3rd ed. New York: Watson-Guptill, 1967.

Downing, A. B. *Euthanasia and the Right to Death: The Case for Voluntary Euthanasia*. Contemporary Issues Series. London: Owen, 1969.

Drake, William Daniel. *The Cultivator's Handbook of Marijuana*. Eugene, OR: Augur, 1970.

Drucker, Peter F. *The Age of Discontinuity: Guidelines to Our Changing Society*. New York: Harper & Row, 1969.

Drucker, Peter F. *The Effective Executive*. New York: Harper & Row, 1967.

The Drugstore's Rockin'. Vol. 2. Hambergen, Germany: Bear Family Records, 2002. Sound recording, 1 sound disc; digital; 4 3/4 in., BCD 16607 AR Bear Family Records.

Dubos, René J. *Man Adapting*. Yale University Mrs. Hepsa Ely Silliman Memorial Lectures. New Haven: Yale University Press, 1965.

Dubos, René J. *So Human an Animal*. New York: Scribner, 1968.

Duddington, C. L. *Evolution in Plant Design*. London: Faber, 1969.

Duncan, Ida Riley. *The Complete Book of Progressive Knitting*. Rev. ed. New York: Liveright, 1966.

Dunstan, Maryjane, and Patricia Wallace Garlan. *Worlds in the Making: Probes for Students of the Future*. Englewood Cliffs, NJ: Prentice-Hall, 1970.

Du Pont de Nemours, E. I., and Company, Explosives Department, and Arthur La Motte. *Manual para el uso de explosivos*. Wilmington, DE: E. I. Du Pont de Nemours & Co, 1932.

Durbahn, Walter Edward, and Elmer W. Sundberg. *Fundamentals of Carpentry*. 4th ed. 2 vols. Chicago: American Technical Society, 1967.

Durkin, Ellen Usher. *The Penny Pincher's Guide*. New York: Bantam Books, 1971.

Dye Plants and Dyeing—a Handbook. Brooklyn, NY: Brooklyn Botanic Garden, 1964.

Earthborn. Bloomsburg, PA: Mystery Fyre, 2010. Sound recording, 1 sound disc; digital; 4 3/4 in.

Earth for Homes: Ideas and Methods Exchange. Washington, DC: Government Printing Office, 1955.

Ecological Surveys from Space. Washington, DC: NASA, 1970.

Edlin, Herbert L., and American Museum of Natural History. *Plants and Man: The Story of Our Basic Food. Nature and Science Library*. Garden City, NY: Published for the American Museum of Natural History by the Natural History Press, 1969.

Educators Progress League and Educators Progress Service. "Educators Guide to Free Films." Randolph, WI: Educators Progress League, 1973.

Ehrlich, Paul R. *The Population Bomb. A Sierra Club—Ballantine Book*. New York: Ballantine Books, 1968.

Ehrlich, Paul R., and Anne H. Ehrlich. *Population Resources Environment: Issues in Human Ecology. A Series of Books in Biology*. San Francisco: W. H. Freeman, 1970.

Ehrlich, Paul R., and Richard L. Harriman. *How to Be a Survivor*. New York: Ballantine Books, 1971.

Eiseley, Loren C. *The Unexpected Universe*. New York: Harcourt, 1969.

Eisenberg, James, and Francis J. Kafka. *Silk Screen Printing*. 2nd ed. Bloomington, IL: McKnight & McKnight, 1957.

Elfrink, Hank. *Volkswagen Technical Manual: A Comprehensive Technical Manual for Engineers, Mechanics, and Owners*. 11th enl. ed. Los Angeles: H. Elfrink Automotive, 1964.

Emery, F. E. *Systems Thinking: Selected Readings. Penguin Modern Management Readings*. Harmondsworth: Penguin, 1969.

Engel, Heino. *The Japanese House: A Tradition for Contemporary Architecture*. Rutland, VT: C. E. Tuttle, 1964.

Environment Action. *Do It Yourself Ecology*. Washington, DC, n.d.

Environmental Action, Sam Love, Peter Harnik, and Avery Taylor. *Earth Tool Kit: A Field Manual for Citizen Activists*. New York: Pocket Books, 1971.

Escher, M. C. *The Graphic Work of M. C. Escher*. New York: Duell, 1961.

Evans-Wentz, W. Y. *The Tibetan Book of the Great Liberation*. Oxford: Oxford University Press, 1968.

Evenari, Michael, Leslie Shanan, and Naphtali Tadmor. *The Negev: The Challenge of a Desert*. Cambridge, MA: Harvard University Press, 1971.

Ewers, William, and H. W. Baylor. *Sincere's Sewing Machine Service Book*. Phoenix, AZ: Sincere Press, 1968.

Fadiman, James. *The Proper Study of Man: Perspectives on the Social Sciences*. New York: Macmillan, 1971.

Fannin, Allen. *Handspinning Art and Technique*. New York: Van Nostrand Reinhold, 1970.

Farmer, Fannie Merritt, and Wilma Lord Perkins. *The Fannie Farmer Cookbook*. 11th ed. Boston: Little, Brown, 1965.

Farmer, Robert A., and Associates. *How to Adopt a Child. Know Your Law*. New York: Arco, 1967.

Farnham, Albert Burton. *Home Tanning and Leather Making Guide: A Book of Information for Those Who Wish to Tan and Make Leather from Cattle, Horse, Calf, Sheep, Goat, Deer and Other Hides and Skins; Also Explains How to Skin, Handle, Classify and Market*. Columbus, OH: A. R. Harding, 1922.

Feininger, Andreas. *Total Picture Control*. New York: Chilton, 1970.

Feirer, John Louis. *Cabinetmaking and Millwork*. Peoria, IL: C. A. Bennett, 1967.

Ferguson, Mike, and Marilyn Ferguson. *Champagne Living on a Beer Budget: How to Buy the Best for Less*. New York: Putnam, 1968.

Fernald, Merritt Lyndon, and Alfred C. Kinsey. *Edible Wild Plants of Eastern North America*. Gray Herbarium of Harvard University Special Publication. Cornwall-on-Hudson, NY: Idlewild Press, 1943.

Feynman, Richard P. *The Character of Physical Law*. The Messenger Lectures, 1964. Cambridge, MA: MIT Press, 1965.

Fichter, George S., and Philip Francis. *A Guide to Fresh and Salt-Water Fishing. A Golden Handbook*. New York: Golden Press, 1965.

Fletcher, Colin. *The Complete Walker: The Joys and Techniques of Hiking and Backpacking.* New York: Knopf, 1969.

Ford, Norman D. *How to Travel and Get Paid for It.* Greenlawn, NY: Harian Publications, 1969.

Forest Products Laboratory. *Wood Handbook: Basic Information on Wood as a Material of Construction with Data for Its Use in Design and Specification.* US Department of Agriculture, Agriculture Handbook. Washington, DC: US Government Printing Office, 1955.

Fraser, Douglas. *Village Planning in the Primitive World.* Planning and Cities. New York: Braziller, 1968.

Friedman, Milton. *Capitalism and Freedom.* Chicago: University of Chicago Press, 1962. "Full Earth (Nr 68-713)." Washington, DC: National Audiovisual Center, n.d.

Fuller, R. Buckminster. *Education Automation: Freeing the Scholar to Return to His Studies; A Discourse before the Southern Illinois University, Edwardsville Campus Planning Committee, April 22, 1961.* Southern Illinois University Occasional Publication. Carbondale: Southern Illinois University Press, 1962.

Fuller, R. Buckminster. *Ideas and Integrities: A Spontaneous Autobiographical Disclosure.* Englewood Cliffs, NJ: Prentice-Hall, 1963.

Fuller, R. Buckminster. *Nine Chains to the Moon.* Philadelphia: J. B. Lippincott, 1938.

Fuller, R. Buckminster. *No More Secondhand God and Other Writings.* Southern Illinois University Occasional Publication. Carbondale: Southern Illinois University Press, 1963.

Fuller, R. Buckminster. *Operating Manual for Spaceship Earth.* Carbondale: Southern Illinois University Press, 1969.

Fuller, R. Buckminster. *Untitled Epic Poem on the History of Industrialization.* Jargon. Highlands, NC: J. Williams, 1962.

Fuller, R. Buckminster. *Utopia or Oblivion: The Prospects for Humanity.* Bantam Matrix Editions QM5263. New York: Bantam Books, 1969.

Fuller, R. Buckminster, John McHale, and Southern Illinois University at Carbondale World Resources Inventory. *World Design Science Decade, 1965–1975: Five Two-Year Phases of a World Retooling Design Proposed to the International Union of Architects for Adoption by World Architectural Schools,* 6 vols. Carbondale: World Resources Inventory, Southern Illinois University, 1963.

Fundamentals of Rock Climbing. Cambridge, MA: MIT Outing Club, 1956.

Furth, Hans G. *Piaget for Teachers.* Englewood Cliffs, NJ: Prentice-Hall, 1970.

Galaxy Posters. Barrington, NJ: Edmund Scientific, n.d.

Gaskin, Stephen, and William Myers. *Monday Night Class.* Santa Rosa, CA: Book Farm, 1970.

Gater, Anthony, Tim Meller, and John Warsany, and the Intermediate Technology Development Group. *Tools for Progress: Guide to Equipment and Materials for Small-Scale Development Available in the United Kingdom.* London: Intermediate Technology Development Group, 1968.

Gbadamosi, Bakare, and Ulli Beier. *Not Even God Is Ripe Enough: Yoruba Stories.* African Writers Series, 48. London: Heinemann Educational, 1968.

Geiger, Rudolf. *The Climate near the Ground.* Cambridge, MA: Harvard University Press, 1965.

Gelb, Ignace J. *A Study of Writing: The Foundations of Grammatology.* London: Routledge & Kegan Paul, 1952.

Gelmis, Joseph. *The Film Director as Superstar.* Garden City, NY: Doubleday, 1970.

Genders, Roy. *Mushroom Growing for Everyone.* London: Faber, 1969.

General Drafting Company and Norman Joseph William Thrower. *Man's Domain: A Thematic Atlas of the World.* New York: McGraw-Hill, 1968.

Gentille, Thomas. *Step-by-Step Jewelry: A Complete Introduction to the Craft of Jewelry.* New York: Golden Press, 1968.

Gibat, Norman, and Kathleen Howard. *The Lore of Still Building.* Norman, OK: Popular Topics Press, 1973.

Gibbons, Euell. *Stalking the Blue-Eyed Scallop.* New York: D. McKay, 1964.

Gibbons, Euell. *Stalking the Good Life: My Love Affair with Nature.* New York: D. McKay, 1971.

Gibbons, Euell. *Stalking the Healthful Herbs.* New York: D. McKay, 1966.

Gibbons, Euell. *Stalking the Wild Asparagus.* New York: D. McKay, 1962.

Glasstone, Samuel. *Sourcebook on the Space Sciences, Etc.* Princeton, NJ: D. Van Nostrand, 1965.

Glegg, Gordon Lindsay. *The Design of Design.* Cambridge Engineering Series. London: Cambridge University Press, 1969.

Goldstein, Jerome, and M. C. Goldman. *Guide to Organic Foods Shopping and Organic Living.* Emmaus, PA: Rodale Press, 1970.

Gombrich, E. H. *Art and Illusion: A Study in the Psychology of Pictorial Representation.* 2nd ed. Bollingen Series, 35. The A. W Mellon Lectures in the Fine Arts, 5. New York: Bollingen Foundation, 1961.

Goodman, Percival, and Paul Goodman. *Communitas: Means of Livelihood and Ways of Life.* 2nd ed. New York: Vintage Books, 1960.

Good Medicine. Fort Atkinson, WI: Laughing Cat Records, 2009. Sound recording, LC-1999 Laughing Cat Records.

Gopi, Krishna. *Kundalini: Erweckung der geistigen Kraft im Menschen.* Weilheim/Obb.: O. W. Barth, 1968.

Gordon, J. E. *The New Science of Strong Materials.* New York: Walker, 1968.

Gordon, William J. J. *The Metaphorical Way of Learning and Knowing.* Cambridge, MA: Porpoise Books, 1971.

Gorton, Audrey Alley. *The Venison Book: How to Dress, Cut Up, and Cook Your Deer.* Brattleboro, VT: Stephen Greene Press, 1957.

Gotaas, Harold Benedict. *Composting: Sanitary Disposal and Reclamation of Organic Wastes.* World Health Organization Monograph Series. Geneva: World Health Organization, 1956.

Graham, Frank Duncan. *Audels Masons and Builders Guide: A Practical Illustrated Trade Assistant on Modern Construction for Bricklayers, Stone Masons, Cement Workers, Plasterers, and Tile Setters.* New York: T. Audel, 1961.

Grant, Bruce. *How to Make Cowboy Horse Gear.* Cambridge, MD: Cornell Maritime Press, 1956.

Grant, Bruce, and Larry Spinelli. *Leather Braiding.* Cambridge, MD: Cornell Maritime Press, 1950.

Graumont, Raoul, and John J. Hensel. *Encyclopedia of Knots and Fancy Rope Work.* 4th ed. Cambridge, MD: Cornell Maritime Press, 1952.

Graumont, Raoul, and Elmer Walton Wenstrom. *Square Knot Handicraft Guide.* New York: Random, 1949.

Graves, Richard H. *The Ten Bushcraft Books.* Sydney, Australia: Dymocks Book Arcade, 1965.

Greenbank, Anthony. *The Book of Survival: How to Save Your Skin When Disaster Strikes without Warning.* London: Wolfe, 1967.

Greene, Vaughn M. *Underwater Prospecting Techniques: The Gold Diver's Handbook; Tech. Manual.* 2nd ed. San Francisco: publisher unknown, 1960.

Greenhood, David. *Mapping.* Rev. ed. Chicago: University of Chicago Press, 1964.

Gregory, R. L. *Eye and Brain.* London: Weidenfeld & Nicolson, 1966.

Grieve, Maud, and C. F. Leyel. *A Modern Herbal: The Medicinal, Culinary, Cosmetic and Economic Properties, Cultivation and Folk-Lore of Herbs, Grasses, Fungi, Shrubs, and Trees with All Their Modern Scientific Uses.* 2 vols. New York: Harcourt, 1931.

Griffin, John Howard, and Thomas Merton. *A Hidden Wholeness: The Visual World of Thomas Merton.* Boston: Houghton Mifflin, 1970.

Grossinger, Richard. *Solar Journal (Oecological Sections).* Los Angeles: Black Sparrow Press, 1970.

Grotowski, Jerzy. *Towards a Poor Theatre.* Teatrets Teori Og Teknikk. Holstebro: Odin Teatret, Box 118, 1968.

Groves, J. W. *Edible and Poisonous Mushrooms of Canada.* Ottawa: Canada Department of Agriculture, 1962.

The Guinness Book of Records. Enfield: Guinness Superlatives, 1955.

Gunther, Bernard, and Paul Fusco. *Sense Relaxation below Your Mind.* New York: Collier Books, 1968.

Gutzwiller, F. W. *SCR Manual.* General Electric, 1967.

Hale, Judson D. *The Old Farmer's Almanac Book of Everyday Advice.* New York: Random House, 1995.

Hall, Edward T. *The Hidden Dimension.* Garden City, NY: Doubleday, 1966.

Halprin, Lawrence. *The RSVP Cycles: Creative Processes in the Human Environment.* New York: G. Braziller, 1970.

Hamilton-Hunt, Margaret. *Knitting Dictionary: Eight Hundred Stitches, Patterns, and Knitting, Crochet, Jacquard Technics.* New York: Crown Publishers, 1969.

Hamm, Hans W. *Low Cost Development of Small Water Power Sites.* Mount Rainier, MD: Volunteers in Technical Assistance, 1971.

Hammesfahr, James E., and Clair L. Stong. *Creative Glass Blowing.* San Francisco: W. H. Freeman, 1968.

Hampton, Charles William. *Planecraft, Etc.* 7th imp., rev. and enl. Sheffield: C. & J. Hampton, 1959.

Handler, Philip, and the National Academy of Sciences Survey Committee on the Life Sciences. *Biology and the Future of Man.* New York: Oxford University Press, 1970.

Hardin, Garrett James. *Nature and Man's Fate.* New York: Rinehart, 1959.

Hardin, Garrett James. *Population, Evolution, and Birth Control: A Collage of Controversial Ideas.* 2nd ed. A Series of Books in Biology. San Francisco: W. H. Freeman, 1969.

Harlé, Philippe. *The Glénans Sailing Manual.* Tuckahoe, NY: J. de Graff, 1965.

Harrington, Carroll. *If You Want to Save Your Environment, Start at Home!* New York: Hawthorn Books, 1970.

Harrington, Harold David. *Edible Native Plants of the Rocky Mountains.* Albuquerque: University of New Mexico Press, 1968.

Hart, M. R. *How to Navigate Today.* 5th ed. Cambridge, MD: Cornell Maritime Press, 1970.

Harvey, Virginia I. *Macramé: The Art of Creative Knotting.* New York: Reinhold, 1967.

Haviland, Dale. *A Is a Libertarian Directory.* Brighton, MI: Mega, 1972.

Hayes, M. Horace, and Francis Donald Tutt John. *Veterinary Notes for Horse Owners: A Manual of Horse Medicine and Surgery.* 16th rev. ed. London: S. Paul, 1968.

Hazell, Lester Dessez. *Commonsense Childbirth.* New York: Putnam, 1969.

Heard, Gerald. *The Five Ages of Man: The Psychology of Human History.* New York: Julian Press, 1964.

Heckel, Alice, and Ehrenfried Pfeiffer. *The Pfeiffer Garden Book: Bio-dynamics in the Home Garden.* Stroudsburg, PA: Bio-dynamic Farming and Gardening Association, 1967.

Heloise. *Heloise's Kitchen Hints.* Englewood Cliffs, NJ: Prentice-Hall, 1963.

Heloise. *Heloise's Work and Money Savers.* Englewood Cliffs, NJ: Prentice-Hall, 1967.

Henderson, John. *Emergency Medical Guide.* New York: McGraw-Hill, 1963.

Henry, Mary E., and Vasser Bishop. *Guide to Correspondence Study in Colleges and Universities 1966–1967.* Minneapolis: National University Extension Association, 1967.

Herbert, Frank. *Dune.* Philadelphia: Chilton Books, 1965.

Herkimer, Herbert. *Engineers' Illustrated Thesaurus.* New York: Chemical Publishing, 1952.

Herndon, James. *How to Survive in Your Native Land.* New York: Simon & Schuster, 1971.

Hertel, Heinrich. *Structure, Form, Movement.* New York: Reinhold, 1966.

Herter, George Leonard, Jacques Pierre Herther, and the North Star Guide Association. *Professional Guide's Manual.* 6th ed. Waseca, MN: Herter's, 1964.

Hertzberg, Hendrik. *One Million.* A Gemini Smith Book. New York: Simon & Schuster, 1970.

Hillcourt, William, and Ernest Kurt Barth. *The Golden Book of Camping: Tents and Tarpaulins, Packs and Sleeping Bags; Building a Camp; Firemaking and Outdoor Cooking; Canoe Trips, Hikes, and Indian Camping.* Rev. ed. New York: Golden Press, 1971.

Hiscock, Eric C. *Cruising under Sail.* 2nd ed. New York: Oxford University Press, 1965.

Hiscox, Gardner Dexter, T. O'Conor Sloane, and Harry E. Eisenson. *Henley's Twentieth Century Book of Formulas, Processes and Trade Secrets: A Valuable Reference Book for the Home, Factory, Office, Laboratory and the Workshop.* New rev. and enl. ed. New York: Books, Inc., 1970.

Hoffer, Abram, and Humphry Osmond. *The Hallucinogens.* New York: Academic Press, 1967.

Hoffer, Abram, and Humphry Osmond. *How to Live with Schizophrenia.* New Hyde Park, NY: University Books, 1966.

Hoffer, Abram, and Humphry Osmond, and the Canadian Schizophrenia Foundation. *Megavitamin Therapy: In Reply to the American Psychiatric Association Task Force Report on Megavitamins and Orthomolecular Psychiatry.* Regina, Sask.: Canadian Schizophrenia Foundation, 1971.

Holden, Alan, and Phylis Morrison. *Crystals and Crystal Growing. Science Study Series, S7.* Garden City, NY: Anchor Books, 1960.

Hollis, Nesta. *Successful Sewing: A Modern Guide.* New York: Taplinger, 1969.

Holm, Don. *The Old-Fashioned Dutch Oven Cookbook: Complete with Authentic Sourdough Baking, Smoking Fish and Game, Making Jerky, Pemmican, and Other Lost Campfire Arts.* Caldwell, ID: Caxton Printers, 1969.

Holt, John Caldwell. *How Children Learn.* New York: Pitman, 1967.

Howard, Albert. *An Agricultural Testament.* New York: Oxford University Press, 1940.

How to Improve Your Cycling. Chicago: Athletic Institute, n.d.

How to Make and Fly Paper Airplanes. New York: Bantam Books, 1970.

Huang, Eugene Yuching. *Manual of Current Practice for the Design, Construction, and Maintenance of Soil-Aggregate Roads.* University of Illinois Engineering Experiment Station Circular. Urbana, IL, 1959.

Hubbard, Frank. *Three Centuries of Harpsichord Making.* Cambridge, MA: Harvard University Press, 1965.

Hungry Wolf, Adolf. *Good Medicine: Life in Harmony with Nature.* Golden, BC: Good Medicine Books, 1970.

Hunt, W. Ben. *The Golden Book of Indian Crafts and Lore.* New York: Simon & Schuster, 1954.

Hunter, Beatrice Trum. *The Natural Foods Cookbook.* New York: Simon & Schuster, 1961.

Hydroponics as a Hobby. Urbana: College of Agriculture, University of Illinois, 1983.

Hyman, Harold Thomas. *The Complete Home Medical Encyclopedia.* Rev. and enl. ed. New York: Avon Books, 1965.

Ickis, Marguerite. *The Standard Book of Quilt Making and Collecting.* New York: Greystone Press, 1949.

Indian Agricultural Research Institute. *A New Technology for Dry Land Farming.* New Delhi: 1970.

Information. San Francisco: W. H. Freeman, 1966.

Innes, William T. *Exotic Aquarium Fishes: A Work of General Reference.* Norristown, PA: Aquarium, 1935.

Inter-American Housing and Planning Center and United Nations Department of Economic and Social Affairs. *Soil-Cement: Its Use in Building.* New York: United Nations, 1964.

Isaacs, Ken. *Culture Breakers: Alternatives and Other Numbers.* New York: MSS Educational Publishing, 1970.

Itten, Johannes. *The Art of Color: The Subjective Experience and Objective Rationale of Color.* New York: Reinhold, 1974.

Iyengar, B. K. S. *Light on Yoga: Yoga Dīpikā.* New York: Schocken Books, 1966.

Jackson, Gainor W., and W. Morley Sutherland. *Concrete Boatbuilding: Its Technique and Its Future.* London: Allen & Unwin, 1969.

Jackson, John Brinckerhoff, and Ervin H. Zube. *Landscapes: Selected Writings of J. B. Jackson.* Amherst, MA: University of Massachusetts Press, 1970.

Jacobs, Harold R. *Mathematics, a Human Endeavor: A Textbook for Those Who Think They Don't Like the Subject.* A Series of Books in Mathematics. San Francisco: W. H. Freeman, 1970.

Jaeger, Ellsworth. *Wildwood Wisdom.* New York: Macmillan, 1945.

Jameson, Norma. *Batik for Beginners.* New York: Studio Vista/Watson-Guptill, 1970.

Janov, Arthur. *The Primal Scream.* London: Garnstone Press, 1970.

The Japan Woodworker Catalog. Berkeley, CA: Japan Woodworker, n.d.

Jarvis, D. C. *Folk Medicine: A Vermont Doctor's Guide to Good Health.* New York: Holt, 1958.

Jenkins, J. Geraint. *Traditional Country Craftsmen.* London: Routledge & Kegan Paul, 1965.

J. K. Lasser Tax Institute. *Accounting for Everyday Profit.* Larchmont, NY: American Research Council, 1961.

Johnson, Stanley. *The Green Revolution.* New York: Harper & Row, 1972.

Johnston, Edward. *Writing and Illuminating and Lettering: With Diagrams and Illustrations by the Author and Noel Rooke.* London: Pitman, 1969.

Johnston, Randolph Wardell. *The Book of Country Crafts: On Working with Wood, Clay, Metals, Stone, and Color, with Many New Recipes and Secrets of the Crafts.* 2nd ed. New York: Barnes, 1964.

Jolley, John Lionel. *Data Study.* World University Library. London: Weidenfeld & Nicolson, 1968.

Jones, Dave. *Practical Western Training.* Princeton, NJ: Van Nostrand, 1968.

Jones, Ita. *The Grubbag: An Underground Cookbook.* Westminster, MD: Random House, 1971.

Jones, J. Christopher. *Design Methods: Seeds of Human Futures.* New York: Wiley Interscience, 1970.

Jung, C. G. *Psychological Reflections: An Anthology of the Writings of C. G. Jung.* New York: Harper, 1961.

Jung, C. G., and Marie-Luise von Franz. *Man and His Symbols*. Garden City, NY: Doubleday, 1964.

Jungk, Robert, Johan Galtung, Institut für Zukunftsfragen, Mankind 2000, and International Peace Research Institute. *Mankind 2000*. London: Allen & Unwin, 1969.

Kahn, Herman. *The Emerging Japanese Superstate: Challenge and Response*. Englewood Cliffs, NJ: Prentice-Hall, 1970.

Kahn, Herman, Anthony J. Wiener, and Hudson Institute. *The Year 2000: A Framework for Speculation on the Next Thirty-three Years*. New York: Macmillan, 1967.

Kahn, Lloyd. *Domebook Two*. Bolinas, CA: Pacific Domes, 1971.

Kaysing, William. *Intelligent Motorcycling*. Long Beach, CA: Parkhurst, 1966.

Kaysing, William. *Thermal Springs of the Western United States*. Santa Barbara, CA: Paradise Publishers, 1970.

Kazantzakis, Nikos, Kimon Friar, and Homer. *The Odyssey: A Modern Sequel*. New York: Simon & Schuster, 1958.

Keep Your Wig On. New York: Ryko, 2004. Sound recording, RCD 10666 Ryko.

Keesee, Allen. *The Sitar Book: A Beginning Instruction Method for Playing the Indian Sitar, with Musical Exercises, Detailed Explanation of Indian Musical Notation System, Ten Ragas, and a Brief History of Indian Music*. New York: Oak Publications, 1968.

Kellogg, Rhoda, and Scott O'Dell. *The Psychology of Children's Art*. A Psychology Today Book. San Diego, CA: CRM, distributed by Random House, 1967.

Kelso, Louis O., and Patricia Hetter Kelso. *Two-Factor Theory: The Economics of Reality; How to Turn Eighty Million Workers into Capitalists on Borrowed Money, and Other Proposals*. New York: Vintage Books, 1968.

Kepes, Gyorgy. *Education of Vision*. Vision + Value Series. New York: G. Braziller, 1965.

Kepes, Gyorgy. *The Man-Made Object*. Vision + Value Series. New York: G. Braziller, 1966.

Kepes, Gyorgy. *Module, Proportion, Symmetry, Rhythm*. Vision + Value Series. New York: G. Braziller, 1966.

Kepes, Gyorgy. *The Nature and Art of Motion*. Vision + Value Series. New York: G. Braziller, 1965.

Kepes, Gyorgy. *Sign, Image, Symbol*. Vision + Value Series. New York: G. Braziller, 1966.

Kepes, Gyorgy. *Structure in Art and in Science*. Vision + Value Series. New York: G. Braziller, 1965.

Kephart, Horace. *Camping and Woodcraft: A Handbook for Vacation Campers and for Travelers in the Wilderness*. New ed. 2 vols. New York: Macmillan, 1967. (Originally published 1921.)

Kern, Ken. *The Owner-Built Home*. Oakhurst, CA: Owner-Built Publications, 1961.

Kesey, Ken. *Sometimes a Great Notion: A Novel*. New York: Viking, 1964.

King, Franklin Hiram. *Farmers of Forty Centuries; or, Permanent Agriculture in China, Korea and Japan*. Madison, WI: Mrs. F. H. King, 1911.

Kirk, Donald R. *Wild Edible Plants of the Western United States, Including Also Most of Southwestern Canada and Northwestern Mexico*. Healdsburg, CA: Naturegraph Publishers, 1970.

Kirsch, Dietrich, and Jutta Kirsch-Korn. *Make Your Own Rug*. New York: Watson-Guptill, 1969.

Klinger, Herb, and Judy Klinger. *Knapsacking Abroad*. Harrisburg, PA: Stackpole Books, 1967.

Kneese, Allen V., Robert U. Ayres, and Ralph C. D'Arge. *Economics and the Environment: A Materials Balance Approach*. Washington, DC: Resources for the Future, 1970.

Knight, Austin Melvin, John Vavasour Noel, and Frank E. Bassett. *Knight's Modern Seamanship*. 16th ed. New York: Van Nostrand Reinhold, 1966.

Kodet, E. Russel, and Bradford Angier. *Being Your Own Wilderness Doctor: The Outdoorsman's Emergency Manual*. Harrisburg, PA: Stackpole Books, 1968.

Kodet, E. Russel, and Bradford Angier. *The Home Medical Handbook*. New York: Association Press, 1970.

Koehler, William R. *The Koehler Method of Dog Training*. New York: Howell Book House, 1962.

Koestler, Arthur. *The Act of Creation*. New York: Macmillan, 1964.

Koestler, Arthur. *The Ghost in the Machine*. London: Hutchinson, 1967.

Kohl, Herbert R. *The Open Classroom: A Practical Guide to a New Way of Teaching*. New York: New York Review, 1969.

Kormondy, Edward John. *Concepts of Ecology*. Concepts of Modern Biology Series. Englewood Cliffs, NJ: Prentice-Hall, 1969.

Korns, Michael F. *The Modern War in Miniature: A Statistical Analysis of the Period 1939 to 1945*. Lawrence, KS: M&J Independent Research, 1966.

Korzybski, Alfred. *Science and Sanity: An Introduction to Non-Aristotelian Systems and General Semantics*. International Non-Aristotelian Library. 4th ed. Lakeville, CT: International Non-Aristotelian Library, 1958.

Krampen, Martin, and Peter Seitz. *Design and Planning 2: Computers in Design and Communication*. Visual Communication Books. New York: Hastings House, 1967.

Krauss, Ruth, and Phyllis Rowand. *Bears*. New York: Harper, 1948.

Kübler-Ross, Elisabeth. *On Death and Dying*. New York: Macmillan, 1969.

Kuhn, Thomas S. *The Structure of Scientific Revolutions*. 2nd ed. Chicago: University of Chicago Press, 1970.

Lager, Mildred M., and Van Gundy Jones Dorothea. *The Soybean Cookbook*. New York: Devin-Adair, 1963.

La Leche League International. *The Womanly Art of Breastfeeding*. Rev. and enl. ed. Franklin Park, IL, 1963.

Lane, Frank W. *The Elements Rage*. Philadelphia: Chilton Books, 1965.

Langewiesche, Wolfgang, and Leighton Holden Collins. *Stick and Rudder: An Explanation of the Art of Flying*. London/New York: Whittlesey House/McGraw-Hill, 1944.

Lannon, Maurice. *Polyester and Fiberglass: And Information on Some Other Plastics*. 8th ed. North Hollywood, CA,1969.

Lanoue, Fred R. *Drownproofing: A New Technique for Water Safety*. Englewood Cliffs, NJ: Prentice-Hall, 1963.

Laozi and Archie J. Bahm. *Tao Teh King*. New York: F. Ungar, 1958.

Lapp, Ralph Eugene, and Time-Life Books. *Matter*. Life Science Library. New York: Time-Life Books, 1974.

Laubin, Reginald, and Gladys Laubin. *The Indian Tipi: Its History, Construction, and Use*. Norman: University of Oklahoma Press, 1957.

Living on the Earth: Celebrations, Storm Warnings, Formulas, Recipes, Rumors, and Country Dances. New York: Vintage Books, 1970.

Leach, Bernard. *A Potter's Book*. London: Faber & Faber, 1955.

LeBlance, Jerry. *Three Hundred Ways to Moonlight*. New York: Paperback Library, 1969.

LeCron, Leslie M. *Self Hypnotism: The Technique and Its Use in Daily Living*. Englewood Cliffs, NJ: Prentice-Hall, 1964.

Lederer, William J., and Joe Pete Wilson. *Complete Cross-Country Skiing and Ski Touring*. New York: Norton, 1970.

Lee, Gary. *The Wok: A Chinese Cook Book*. Concord, CA: Nitty Gritty Productions, 1970.

Lee, Marshall. *Bookmaking: The Illustrated Guide to Design and Production*. A Balance House Book. New York: Bowker, 1965.

Lehr, Paul E. *Weather: Air Masses, Clouds, Rainfall, Storms, Weather Maps, Climate*. A Golden Nature Guide. New York: Golden Press, 1957.

Leisure Camping. Saint Paul, MN: Camping Enterprises, 1969.

Leopold, Aldo. *A Sand County Almanac, and Sketches Here and There*. New York: Oxford University Press, 1949.

Lesch, Alma. *Vegetable Dyeing: 151 Color Recipes for Dyeing Yarns and Fabrics with Natural Materials*. New York: Watson-Guptill, 1970.

LeShan, Lawrence L., Douglas Dean, Aristide H. Esser, Stanley Krippner, Henry Margenau, Gardner Murphy, and Gertrude Raffel Schmeidler. *Toward a General Theory of the Paranormal*. n.p.: Big Sur Recordings, 1969. Sound recording, 2 cassettes.

Lessing, Doris. *The Four-Gated City*. Vol. 5 of *Her Children of Violence*. London: MacGibbon & Kee, 1969.

Lessing, Doris. *The Golden Notebook*. New York: Simon & Schuster, 1962.

Lévi-Strauss, Claude. *The Savage Mind*. The Nature of Human Society Series. Chicago: University of Chicago Press, 1966.

Lévi-Strauss, Claude. *Tristes Tropiques*. Terre Humaine. Paris: Plon, 1955.

Lewis, Jack. *Bow and Arrow Archer's Digest: The Encyclopedia for All Archers*. Northfield, IL: Digest Books, 1971.

Lieberman, Jay Benjamin. *Printing as a Hobby*. London: Oak Tree Press, 1963.

Lilly, John Cunningham. *The Mind of the Dolphin: A Nonhuman Intelligence*. Garden City, NY: Doubleday, 1967.

Lilly, John Cunningham. *Programming and Metaprogramming in the Human Biocomputer: Theory and Experiments*. Miami, FL: Communication Research Institute, 1968.

Limbourg, Pol de, Jean Colombe, Jean Longnon, Raymond Cazelles, Berry Jean de France, Musée Condé Bibliotheque Chantilly, and Catholic Church. *The Très Riches Heures of Jean, Duke of Berry: Musée Condé, Chantilly*. New York: G. Braziller, 1969.

Lobeck, A. K. *Things Maps Don't Tell Us: An Adventure into Map Interpretation*. New York: Macmillan, 1956.

Lockwood, Arthur. *Diagrams: A Visual Survey of Graphs, Maps, Charts and Diagrams for the Graphic Designer*. New York: Studio Vista/Watson-Guptill, 1969.

Lorenz, Konrad. *King Solomon's Ring: New Light on Animal Ways*. New York: Crowell, 1952.

Lowndes, Douglas. *Film Making in Schools*. New York: Batsford/Watson-Guptill, 1968.

Lungwitz, A., and John William Adams. *A Textbook of Horseshoeing for Horseshoers and Veterinarians*. Corvallis: Oregon State University Press, 1966.

Lurie, Ellen. *How to Change the Schools: A Parents' Action Handbook on How to Fight the System*. New York: Random House, 1970.

Lurie, Jesse Zel, and Samuel Segev. *The Israel Army Physical Fitness Book*. New York: Grosset & Dunlap, 1969.

Luzadder, Warren Jacob. *Basic Graphics for Design, Analysis, Communications, and the Computer*. 2nd ed. Englewood Cliffs, NJ: Prentice-Hall, 1968.

Mackenzie, David. *Goat Husbandry*. 2nd ed. London: Faber, 1967.

Maile, Anne. *Tie and Dye*. New York: Ballantine Books, 1969.

Making Building Blocks with the Cinva-Ram Block Press. Mount Rainier, MD: Volunteers in Technical Aassistance, 1966.

Maltz, Maxwell. *Psycho-cybernetics: A New Way to Get More Living Out of Life*. Englewood Cliffs, NJ: Prentice-Hall, 1960.

Mander, Jerry, George Dippel, and Howard Luck Gossage. *The Great International Paper Airplane Book*. New York: Simon & Schuster, 1967.

Marcus, Abraham, and William Marcus. *Elements of Radio*. 5th ed. Prentice-Hall Industrial Arts Series. Englewood Cliffs, NJ: Prentice-Hall, 1965.

Marks, Robert W. *The Dymaxion World of Buckminster Fuller*. New York: Reinhold, 1960.

Marks, Robert W. *The New Mathematics Dictionary and Handbook*. Bantam Matrix Editions. New York: Bantam Books, 1964.

Marshall, Robert. *Alaska Wilderness: Exploring the Central Brooks Range*. 2nd ed. Berkeley: University of California Press, 1970.

Marvel, M. E. *Hydroponic Culture of Vegetable Crops*. Gainesville: Florida Agricultural Extension Service, 1966.

Matematicheskiĭ Institut im V. A. Steklova and A. D. Aleksandrov. *Mathematics: Its Content, Methods, and Meaning*. 3 vols. Cambridge, MA: MIT Press, 1964.

Matterson, Elizabeth Mary. *Play and Playthings for the Preschool Child*. Rev. ed. Penguin Handbook 115. Baltimore: Penguin, 1967.

Mattingly, E. Grayson, and Welby A. Smith. *Introducing the Single-Camera VTR System: A Layman's Guide to Videotape Recording*. New York: Scribner, 1973.

May, Rollo. *Love and Will*. New York: Norton, 1969.

Mazurkiewicz, Albert J., and Harold J. Tanyzer. *The ITA Handbook for Writing and Spelling*. New York: Initial Teaching Alphabet, 1964.

McCulloch, Warren S. *Embodiments of Mind*. Cambridge, MA: MIT Press, 1965.

McDonnell, Leo P. *Hand Woodworking Tools*. Albany: Delmar, 1962.

McHarg, Ian L., and American Museum of Natural History. *Design with Nature*. Garden City, NY: Natural History Press, 1969.

McKenny, Margaret, and Daniel E. Stuntz. *The Savory Wild Mushroom*. Seattle: University of Washington Press, 1971.

McLuhan, Marshall. *Culture Is Our Business*. New York: McGraw-Hill, 1970.

McLuhan, Marshall. *Understanding Media: The Extensions of Man*. New York: McGraw-Hill, 1964.

McNeill, William Hardy. *The Rise of the West: A History of the Human Community*. Chicago: University of Chicago Press, 1963.

Mead, Margaret, and James Baldwin. *A Rap on Race*. Philadelphia: Lippincott, 1971.

Meilach, Dona Z. *Macramé: Creative Design in Knotting*. New York: Crown Publishers, 1971.

Menninger, Karl. *Number Words and Number Symbols: A Cultural History of Numbers*. Cambridge, MA: MIT Press, 1969.

The Merck Manual of Diagnosis and Therapy. Rahway, NJ: Merck Sharp & Dohme Research Laboratories, 1966.

Mesirow, Kit. *The Care and Use of Japanese Tools*. Berkeley, CA: Japanese Woodworker, 1975.

Michell, E. B. *The Art and Practice of Hawking*. Boston: C. T. Branford, 1960. (Originally published 1900.)

Milam, Lorenzo W. *Sex and Broadcasting: A Handbook on Starting a Radio Station for the Community*. 3rd ed. Los Gatos, CA: Dildo Press, 1972.

Miller, Arthur C., and Walter Strenge, and the American Society of Cinematographers. *American Cinematographer Manual*. 3rd ed. Hollywood: American Society of Cinematographers, 1969.

Miller, Carey D., Katherine Bazore, and Mary Bartow. *Fruits of Hawaii: Description, Nutritive Value, and Recipes*. 4th ed. Honolulu: University of Hawaii Press, 1965.

Miller, George A. *Plans and the Structure of Behavior*. New York: Holt, 1960.

Millerson, Gerald. *The Technique of Television Production*. 6th rev. and enl. ed. Communication Arts Books. New York: Hastings House, 1968.

Milne, Lorus Johnson, and Joan Greene Milne Margery. *Patterns of Survival*. Prentice-Hall Series in Nature and Natural History. Englewood Cliffs, NJ: Prentice-Hall, 1967.

Mitchell, Howie. *The Mountain Dulcimer: How to Make It and Play It (after a Fashion)*. Sharon, CT: Folk-Legacy Records, 1966.

Mitchell, John G., and Sierra Club. *Ecotactics: The Sierra Club Handbook for Environmental Activists*. New York: Pocket Books, 1970.

Modern Plastics Encyclopedia. New York: McGraw-Hill, 1965.

Momentary Visions. Sound recording. Eighth Sea 8 Songs, 2006.

Moneysworth: The Cheap Book: The Moneysworth Consumer Encyclopedia. New York: Moneysworth, 1973.

Monroe, Robert A. *Journeys Out of the Body*. Garden City, NY: Doubleday, 1971.

Mookerjee, Ajit. *Tantra Asana: A Way to Self-Realization*. New York: George Wittenborn, 1971.

Moolman, Valerie. *How to Buy Food*. New York: Cornerstone Library, 1970.

Moore, Patrick. *The Atlas of the Universe*. New York: Rand McNally, 1970.

Morgan, Ernest. *A Manual of Simple Burial*. Burnsville, NC: Celo Press, 1968.

Morgan, Robin. *Sisterhood Is Powerful: An Anthology of Writings from the Women's Liberation Movement*. New York: Random House, 1970.

Morse, Edward Sylvester. *Japanese Homes and Their Surroundings*. New York: Dover, 1961.

Morse, Grant W. *The Concise Guide to Library Research*. New York: Washington Square Press, 1966.

Motor (New York). *Motor's Truck Repair Manual*. Hackensack, NJ: Wehman Brothers, 1969.

Moulds, George Henry. *Thinking Straighter*. Dubuque, IA: William C. Brown, 1965.

Mountaineers (Society) and Harvey Manning. *Mountaineering: The Freedom of the Hills*. 2nd ed. Seattle, 1967.

Mountain Safety Research Newsletter. Seattle, WA: Mountain Safety Research, n.d.

Mowat, Farley. *Never Cry Wolf*. Boston: Little, Brown, 1963.

Muir, John. *How to Keep Your Volkswagen Alive! A Manual of Step by Step Procedures for the Compleat Idiot; For 1950–1969 Sedans, Ghias and Transporters, Types I and II*. Santa Fe, NM: J. Muir Publications, 1969.

Mumford, Lewis. *The Pentagon of Power*. Vol. 2 of *The Myth of the Machine*. New York: Harcourt Brace Jovanovich, 1970.

Mumford, Lewis. *Technics and Civilization*. New York: Harcourt, 1934.

Murray, Max. *The King and the Corpse*. New York: Farrar, 1948.

Naranjo, Claudio, and Robert E. Ornstein. *On the Psychology of Meditation*. An Esalen Book. New York: Viking, 1971.

National Federation of Young Farmers Clubs. *Farm Horses*. Kenilworth, Warwickshire, England, n.d.

National Fire Protection Association. "Electrical Code for One- and Two-Family Dwellings: Excerpted from the National Electrical Code." In *NFPA 70A*. Quincy, MA: National Fire Protection Association, n.d.

National Fire Protection Association and National Electrical Contractors Association Codes and Standards Committee. *National Electrical Code, 1968: Tables*. Washington, DC: National Electrical Contractors Association, 1968.

Natural Life Styles: On the Road: Touring Organic America: Restaurants, Farms, Herb Gardens, Stores, Hostelries. New York: Gordon & Breach Science Publications, 1973.

Nearing, Helen, and Scott Nearing. *Living the Good Life: How to Live Sanely and Simply in a Troubled World*. New York: Schocken Books, 1970. Originally published 1954.

Nearing, Helen, and Scott Nearing. *The Maple Sugar Book, Together with Remarks on Pioneering as a Way of Living in the Twentieth Century*. New York: Schocken Books, 1970.

Needham, Joseph. *Physics and Physical Technology: Part I, Physics. Science and Civilisation in China*. Cambridge: Cambridge University Press, 1962.

APPENDIX

263

Needham, Joseph, and Ling Wang. *History of Scientific Thought. Science and Civilisation in China.* Cambridge: Cambridge University Press, 1956.

Needham, Joseph, and Ling Wang. *Introductory Orientations. Science and Civilisation in China.* Cambridge: Cambridge University Press, 1954.

Needham, Joseph, and Ling Wang. *Mathematics and the Sciences of the Heavens and the Earth. Science and Civilisation in China.* Cambridge: Cambridge University Press, 1959.

Needleman, Jacob. *The New Religions.* Garden City, NY: Doubleday, 1970.

Negroponte, Nicholas. *The Architecture Machine.* Cambridge, MA: MIT Press, 1970.

Nelms, Henning. *Thinking with a Pencil.* New York: Barnes & Noble, 1964.

Nelson, Glenn C. *Ceramics: A Potter's Handbook.* 3rd ed. New York: Holt, 1971.

Nelson, Richard K. *Hunters of the Northern Ice.* Chicago: University of Chicago Press, 1969.

Nesbitt, Paul H. *The Survival Book.* Princeton, NJ: Van Nostrand, 1959.

Newell, Peter. *Topsys and Turvys.* New York: Dover, 1964. (Originally published 1894.)

Newman, James R. *The World of Mathematics: A Small Library of the Literature of Mathematics from A'h-Mosé the Scribe to Albert Einstein,* 4 vols. New York: Simon & Schuster, 1956.

Newman, Thelma R. *Plastics as an Art Form.* 1964. Rev. ed. Philadelphia: Chilton, 1969.

New York (State) Department of Environmental Conservation and New York (State) Conservation Department. *The Conservationist.* 36 vols. Albany, NY: New York State Department of Environmental Conservation, 1945–.

Nichols, Herbert L. *Modern Techniques of Excavation.* Greenwich, CT: North Castle Books, 1956.

Nichols, Herbert L. *Moving the Earth: The Workbook of Excavation.* 2nd ed. Greenwich, CT: North Castle Books, 1962.

Nicholson, Barbara, S. G. Harrison, G. B. Masefield, and Michael Wallis. *The Oxford Book of Food Plants.* London: Oxford University Press, 1969.

Nicholson, John Bernard. *Modern Motorcycle Maintenance.* Saskatoon: Nicholson Brothers Motorcycles, 1974.

Nicks, Oran W. *This Island Earth.* NASA SP-250. Washington, DC: Scientific and Technical Information Division, Office of Technology Utilization, National Aeronautics and Space Administration; for sale by the Supt. of Docs., 1970.

Nicolaïdes, Kimon. *The Natural Way to Draw: A Working Plan for Art Study.* Boston: Houghton Mifflin, 1941.

Norman, Gurney. *Divine Right's Trip: A Folk Tale.* New York: Dial Press, 1972.

Nourse, Alan Edward, and Time-Life Books. *The Body.* Life Science Library. New York: Time-Life Books, 1971.

Oberg, Erik, and Franklin Day Jones. *Machinery's Handbook.* New York: Industrial Press, 1959.

Odell, Peter R. *Oil and World Power: A Geographical Interpretation.* Pelican Geography Series. Harmondsworth: Penguin, 1970.

Odier, Daniel, and William S. Burroughs. *The Job: Interview with William Burroughs.* London: Cape, 1970.

Odum, Eugene P. *Fundamentals of Ecology.* 2nd ed. Philadelphia: Saunders, 1959.

Odum, Howard T. *Environment, Power, and Society.* New York: Wiley Interscience, 1970.

Okakura, Kakuzō. *The Book of Tea.* New York: Dover, 1964. (Originally published in 1906.)

Oliver, Paul. *Shelter and Society.* New York: F. A. Praeger, 1969.

Olsen, Don, and Ray Olsen. *Modern Art of Candle Creating.* New York: Barnes, 1971.

Olsen, Larry Dean. *Outdoor Survival Skills.* Provo, UT: Extension Publications, Division of Continuing Education, 1967.

OM Collective. *The Organizer's Manual.* New York: Bantam Books, 1971.

One Hundred Embroidery Stitches. Coats & Clark's Book No. 130. New York: Coats & Clark, 1964.

O'Neill, Dan. *Hear the Sound of My Feet Walking, Drown the Sound of My Voice Talking: An Odd Bodkins Book.* San Francisco: Glide Urban Center Publications, 1969.

Opie, Iona Archibald, and Peter Opie. *Children's Games in Street and Playground: Chasing, Catching, Seeking, Hunting, Racing, Duelling, Exerting, Daring, Guessing, Acting, Pretending.* Oxford: Clarendon Press, 1969.

Orage, A. R. *Psychological Exercises and Essays: The Active Mind.* New York: S. Weiser, 1972.

Oravetz, Jules A. *Building Maintenance.* 3rd ed. Indianapolis: T. Audel, 1977.

Organic Gardening Staff. *The Organic Way to Plant Production.* Emmaus, PA: Rodale Books, 1966.

Orton, Vrest. *Observations on the Forgotten Art of Building a Good Fireplace: The Story of Sir Benjamin Thompson, Count Rumford, an American Genius, and His Principles of Fireplace Design Which Have Remained Unchanged for 174 Years.* Dublin, NH: Yankee, 1969.

Ostrander, Sheila, and Lynn Schroeder. *Psychic Discoveries behind the Iron Curtain.* Englewood Cliffs, NJ: Prentice-Hall, 1970.

Otto, Frei, Rudolf Trostel, and Friedrich Karl Schleyer. *Tensile Structures: Design, Structure, and Calculation of Buildings of Cables, Nets, and Membranes,* 2 vols. Cambridge, MA: MIT Press, 1967.

Ozenfant, Amédée. *Foundations of Modern Art.* New York: Dover, 1952.

Papanek, Victor J. *Design for the Real World: Human Ecology and Social Change.* New York: Pantheon Books, 1972.

Parkin, N., and C. R. Flood. *Welding Craft Practice, Part I.* New York: Pergamon Press, 1969.

Pauling, Linus. *Vitamin C and the Common Cold.* San Francisco: W. H. Freeman, 1970.

Pawlak, Vic. *Conscientious Guide to Drug Abuse.* Hollywood, CA: Do It Now Foundation, 1971.

Pearce, William Edward, and Aaron E. Klein. *Transistors and Circuits: Electronics for Young Experimenters.* Garden City, NY: Doubleday, 1971.

Peattie, Lisa Redfield. *The View from the Barrio.* Ann Arbor: University of Michigan Press, 1968.

Pei, Mario. *The Story of Language.* Rev. ed. Philadelphia: Lippincott, 1965.

Perls, Frederick S., and John O. Stevens. *Gestalt Therapy Verbatim*. Moab, UT: Real People Press, 1969.

Peter, Laurence J., and Raymond Hull. *The Peter Principle*. New York: W. Morrow, 1969.

Petersen, Grete, and Elsie Svennås. *Handbook of Stitches: 200 Embroidery Stitches, Old and New, with Descriptions, Diagrams and Samplers*. New York: Van Nostrand Reinhold, 1970.

Petersen Publishing Company. *Basic Auto Repair Manual*. 5th rev. ed. Los Angeles, CA: Petersen, 1973.

Peterson, Harold. *The Last of the Mountain Men*. New York: Scribner, 1969.

Peterson, Roger Tory. *A Field Guide to the Birds, Giving Field Marks of All Species Found East of the Rockies*. 2nd rev. and enl. ed. Boston: Houghton Mifflin, 1947.

Peterson, Roger Tory. *A Field Guide to Western Birds: Field Marks of All Species Found in North America West of the 100th Meridian, with a Section on the Birds of the Hawaiian Islands*. 2nd ed. The Peterson Field Guide Series, 2. Boston: Houghton Mifflin, 1961.

Peterson, Roger Tory, and the Cornell University Laboratory of Ornithology. *A Field Guide to Bird Songs of Eastern and Central North America*. Boston: Houghton, Mifflin, 1959. Sound recording.

Pfeiffer, John. *New Look at Education: Systems Analysis in Our Schools and Colleges*. New York: Odyssey Press, 1968.

Pfeiffer, John E., and Time-Life Books. *The Cell. Life Science Library*. New York: Time-Life Books, 1972.

Philbrick, Helen Louise, and Richard Bartlett Gregg. *Companion Plants and How to Use Them*. New York: Devin-Adair, 1966.

Philbrick, John H., and Helen Louise Philbrick. *Gardening for Health and Nutrition: An Introduction to the Method of Bio-dynamic Gardening Inaugurated by Rudolf Steiner*. Blauvelt, NY: R. Steiner, 1971.

Phillips, Mary Walker. *Step-by-Step Macramé: A Complete Introduction to the Craft of Creative Knotting*. The Golden Press Step-by-Step Craft Series. New York: Golden Press, 1970.

Pieper, Josef. *Leisure: The Basis of Culture*. With an introduction by T. S. Eliot. London: Faber & Faber, 1952.

Pincus, Edward, and Jairus Lincoln. *Guide to Filmmaking*. A Signet Book. New York: New American Library, 1969.

Platt, John Rader. *Perception and Change: Projections for Survival*. Ann Arbor: University of Michigan Press, 1970.

Platt, John Rader. *The Step to Man*. New York: Wiley, 1966.

Pólya, George. *How to Solve It: A New Aspect of Mathematical Method*. Princeton, NJ: Princeton University Press, 1945.

Popko, Edward. *Geodesics*. University of Detroit, 1968.

Popper, Karl R. *The Logic of Scientific Discovery*. 3rd ed. London: Hutchinson, 1968.

Portland Cement Association. *Design and Control of Concrete Mixtures*. Skokie, IL: Portland Cement Association, 1948.

Postman, Neil, and Charles Weingartner. *Teaching as a Subversive Activity*. New York: Delacorte Press, 1969.

Pound, Ezra. *ABC of Reading*. A New Directions Paperbook, No. 89. Philadelphia: New Directions, 1934.

Pound, Ezra, and T. S. Eliot. *Literary Essays of Ezra Pound*. Philadelphia: New Directions, 1968.

Preparation for Childbirth. New York: Maternity Center Association, 1969.

Private Independent Schools. Wallingford, CT: J. E. Bunting, 1947.

Procedures for Processing and Storing Black and White Photographs for Maximum Possible Permanence. Grinnell, IA: East Street Gallery, 1970.

Propst, Robert. *The Office: A Facility Based on Change*. Elmhurst, IL: Business Press, 1968.

Pruning Handbook. A Sunset Book. New ed. Menlo Park, CA: Lane Books, 1972.

Psychology Today: An Introduction. Del Mar, CA: CRM Books, 1970.

Pye, David William. *The Nature and Art of Workmanship*. Cambridge: University Press, 1968.

Rabbit, Peter. *Drop City*. New York: The Olympia Press, 1971.

Ram Dass, Baba. *Be Here Now*. San Cristobal, NM: Lama Foundation, 1971.

Ramsey, Charles George, and Harold Reeve Sleeper, and the American Institute of Architects. *Architectural Graphic Standards*. 6th ed. New York: J. Wiley, 1970.

Rand, Ayn. *Atlas Shrugged*. New York: Random House, 1957.

Rasberry, Salli, and Robert Greenway. *How to Start Your Own School and Make a Book*. Sebastopol, CA: Freestone, 1970.

Reich, Hanns, and Oto Bihalji-Merin. *The World from Above*. A Terra Magica Book. New York: Hill and Wang, 1968.

Reichardt, Jasia, and the Institute of Contemporary Arts (London). *Cybernetic Serendipity: The Computer and the Arts*. New York: Praeger, 1969.

Repo, Satu. *This Book Is about Schools*. New York: Pantheon Books, 1970.

Reynolds, John. *Windmills and Watermills*. Excursions into Architecture. New York: Praeger, 1970.

Reyntiens, Patrick. *The Technique of Stained Glass*. New York: Batsford/Watson-Guptill, 1967.

Rhodes, Daniel. *Clay and Glazes for the Potter*. Arts and Crafts Series. New York: Greenberg, 1957.

Rhodes, Daniel. *Kilns: Design, Construction, and Operation*. Philadelphia: Chilton, 1968.

Richards, Mary Caroline. *Centering in Pottery, Poetry, and the Person*. Middletown, CT: Wesleyan University Press, 1964.

Richter, H. P. *Wiring Simplified*. 30th ed. Minneapolis: Park Publishing, 1971.

Rife Hydraulic Engine Manufacturing Co. *Rife Hydraulic Ram for the Water Supply of Towns, Villages, Mansions, Estates, Cottages, Farm Houses, Railroad Tanks, Irrigation*. New York: Rife Hydraulic Engine Mfg. Co., 1915.

Riviere, Bill. *Pole, Paddle and Portage*. New York: Van Nostrand Reinhold, 1969.

Robbins, Maurice, and Mary B. Irving. *The Amateur Archaeologist's Handbook*. 2nd ed. New York: Crowell, 1973.

Robert, Henry M., and Henry M. Robert. *Robert's Rules of Order, Newly Revised: A New and Enl. Ed.* Glenview, IL: Scott, 1970.

Roberts, Rex. *Your Engineered House.* New York: Evans, 1964.

Roberts, Ronald. *Musical Instruments Made to Be Played.* Leicester: Dryad Press, 1965.

Rodale, Jerome I. *How to Grow Vegetables and Fruit by the Organic Method.* Emmaus, PA: Rodale Press, 1961.

Rodale, Jerome I. *The Encyclopedia of Organic Gardening: By the Staff of Organic Gardening and Farming Magazine.* Emmaus, PA: Rodale Books, 1959.

Rodale, Robert, and Glenn F. Johns. *The Basic Book of Organic Gardening.* New York: Ballantine Books, 1971.

Rogers, Carl R., and Barry Stevens. *Person to Person: The Problem of Being Human; A New Trend in Psychology.* Walnut Creek, CA: Real People Press, 1967.

Roland, Conrad. *Frei Otto: Tension Structures.* New York: Praeger, 1970.

Root, Amos Ives. *The ABC and XYZ of Bee Culture.* 33rd ed. Medina, OH: A. I. Root, 1966.

Rose, Augustus F., and Antonio Cirino. *Jewelry Making and Design: An Illustrated Textbook for Teachers, Students of Design and Craft Workers.* 4th rev. ed. New York: Dover, 1967. (Originally published 1918.)

Rosenberg, Jay F. *The Impoverished Students' Book of Cookery, Drinkery, and Housekeepery.* 4th ed. Portland, OR: Reed College Alumni Association, 1967.

Rosenberg, Sharon, and Joan Wiener. *The Illustrated Hassle-Free Make Your Own Clothes Book.* San Francisco: Straight Arrow Books, New York, 1971.

Rosenfeld, Albert. *The Second Genesis: The Coming Control of Life.* Englewood Cliffs, NJ: Prentice-Hall, 1969.

Rothenberg, Jerome. *Technicians of the Sacred: A Range of Poetries from Africa, America, Asia and Oceania.* Garden City, NY: Doubleday, 1968.

Rothenberg, Jerome, Dennis Tedlock, Stony Brook Poetics Foundation, and Boston University. *Alcheringa.* New York: J. Rothenberg & D. Tedlock, 1970.

Röttger, Ernst, and Dieter Klante. *Creative Drawing: Point and Line.* Creative Play Series. New York: Reinhold, 1964.

Roukes, Nicholas. *Sculpture in Plastics.* New York: Watson-Guptill, 1968.

Rowlands, John J. *Cache Lake Country: Life in the North Woods.* New York: Norton, 1959.

Royal Astronomical Society of Canada. *The Observer's Handbook.* Toronto: Royal Astronomical Society of Canada, 1966.

Royce, Joseph. *Surface Anatomy.* Philadelphia: F. A. Davis, 1965.

Royce, Patrick M. *Sailing Illustrated.* 4th ed. Newport Beach, CA: Royce Publications, 1968.

Rudofsky, Bernard. *Architecture without Architects: An Introduction to Nonpedigreed Architecture.* New York: Museum of Modern Art, distributed by Doubleday, Garden City, NY, 1964.

Russ, Stephen. *Practical Screen Printing. A Studio Handbook.* New York: Studio Vista/Watson-Guptill, 1969.

Rutstrum, Calvin. *The Wilderness Cabin. Illustrated by Les Kouba.* New York: Macmillan, 1961.

Safdie, Moshe. *Beyond Habitat.* Cambridge, MA: MIT Press, 1970.

Salisbury, Gordon. *Catalog of Free Teaching Materials.* Ventura, CA: Catalog of Free Teaching Materials, 1973.

Salvadori, Mario George, and Matthys Levy. *Structural Design in Architecture.* Englewood Cliffs, NJ: Prentice-Hall, 1967.

Samuelson, Paul A. *Economics.* 8th ed. New York: McGraw-Hill, 1970.

Sandage, Allan, and Edwin Hubble. *The Hubble Atlas of Galaxies.* Carnegie Institution of Washington Publication 618. Washington, DC: Carnegie Institution of Washington, 1961.

Saunby, T. *Soilless Culture.* Levittown, NY: Transatlantic Arts, 1953.

Sax, Joseph L. *Defending the Environment: A Strategy for Citizen Action.* New York: Knopf, 1971.

Schachner, Erwin. *Step-by-Step Printmaking: A Complete Introduction to the Craft of Relief Printing.* The Golden Press Step-by-Step Craft Series. New York: Golden Press, 1970.

Schäfer, Georg, and Cuz Nan. *In the Kingdom of Mescal: A Fairy-Tale for Adults.* London: Macdonald & Co, 1969.

Schoenfeld, Eugene. *Dear Doctor Hippocrates.* New York: Gove Press, 1968.

Schön, Donald A. *Technology and Change: The New Heraclitus.* New York: Delacorte Press, 1967.

Scorer, R. S., and Harry Wexler. *Cloud Studies in Colour.* The Commonwealth and International Library Meteorology Division. New York: Pergamon Press, 1967.

Scott, R. W. *Handy Medical Guide for Seafarers.* West Byfleet: Fishing News, 1969.

Sebrell, W. H., J. James. Haggerty, and Time-Life Books. *Food and Nutrition.* Rev ed. Life Science Library. Alexandria, VA: Time-Life Books, 1970.

Self, Margaret Cabell. *Horses: Their Selection, Care and Handling.* New York: A. S. Barnes, 1943.

Selye, Hans. *The Stress of Life.* New York: McGraw-Hill, 1956.

Shaw, Steve. *Surfboard Builder's Yearbook.* La Mesa, CA: Transmedia, 1974.

Shelton, John S. *Geology Illustrated.* San Francisco: W. H. Freeman, 1966.

Shepard, Paul, and Daniel McKinley. *The Subversive Science: Essays toward an Ecology of Man.* Boston: Houghton Mifflin, 1969.

Sheppard, Muriel Earley. *Cabins in the Laurel.* Chapel Hill: University of North Carolina Press, 1935.

Shklovskiĭ, I. S., and Carl Sagan. *Intelligent Life in the Universe.* San Francisco: Holden-Day, 1966.

Shortney, Joan Ranson. *How to Live on Nothing.* Garden City, NY: Doubleday, 1961.

Shumway, George, and Howard C. Frey. *Conestoga Wagon, 1750–1850: Freight Carrier for 100 Years of America's Westward Expansion.* 3rd ed. York, PA: G. Shumway, 1968.

Simeons, A. T. W. *Man's Presumptuous Brain: An Evolutionary Interpretation of Psychosomatic Disease.* New York: Dutton, 1961.

Simon, Bernard. *Simon's Directory of Theatrical Materials, Services and Information.* New York: Bernard Simon, 1955.

Simon, Herbert. *The Science of the Artificial.* Cambridge, MA: MIT Press, 1969.

Simonson, Leroy. *Private Pilot Study Guide.* Glendale, CA: L. Simonson, 1970.

Siu, Ralph Gun Hoy. *The Tao of Science: An Essay on Western Knowledge and Eastern Wisdom.* Cambridge: Technology Press, 1957.

Skeist, Irving. *Plastics in Building.* New York: Reinhold, 1966.

Skolnick, Jerome H., and the University of California–Berkeley Center for the Study of Law and Society. *Justice without Trial: Law Enforcement in Democratic Society.* New York: Wiley, 1966.

Sloane, Eric. *A Museum of Early American Tools.* New York: Wilfred Funk, 1964.

Sloane, Eugene A. *The Complete Book of Bicycling.* New York: Trident Press, 1970.

Sloane, Irving. *Classic Guitar Construction: Diagrams, Photographs, and Step-by-Step Instructions.* New York: E. P. Dutton, 1966.

Small Engine Service Manual. Overland Park, KS: Intertec, 1992.

Smith, Adam. *The Money Game.* New York: Random House, 1968.

Smith, Alexander H. *The Mushroom Hunter's Field Guide.* Ann Arbor: University of Michigan Press, 1958.

Smoking Fish in a Cardboard Smokehouse. Mount Rainier, MD: Volunteers in Technical Assistance, 1966.

Snyder, Gary. *The Back Country.* New York: New Directions, 1968.

Snyder, Gary. *Earth House Hold: Technical Notes and Queries to Fellow Dharma Revolutionaries.* New York: New Directions, 1969.

Snyder, Gary. *Regarding Wave.* New York: New Directions, 1970.

Snyder, Robert M. *What Everyone Should Know about Oceanography.* Jupiter, FL: Snyder Oceanography Services, 1970.

Soleri, Paolo. *Arcology: The City in the Image of Man.* Cambridge, MA: MIT Press, 1969.

Soleri, Paolo. *The Sketchbooks of Paolo Soleri.* Cambridge, MA: MIT Press, 1971.

Sourdough Jack. *Sourdough Jack's Cookery: Authentic Sourdough Cookery from His Country Kitchen.* Nevada City, CA: Osborn/ Woods, 1965.

Sparks, Laurance. *Self-Hypnosis: A Conditioned-Response Technique.* New York: Grune & Stratton, 1962.

Spencer, Charles Louis, and Percy William Blandford. *Knots, Splices and Fancy Work, Revised by P. W. Blandford.* 5th ed. Glasgow: Brown Son & Ferguson, 1956.

Spencer-Brown, G. *Laws of Form.* London: Allen & Unwin, 1969.

Spiro, Melford E. *Kibbutz: Venture in Utopia.* New York: Schocken Books, 1963.

Spock, Benjamin. *Baby and Child Care.* New York: Meredith Press, 1968.

Stamm, G. W. *Veterinary Guide for Farmers.* New and rev. ed. New York: Hawthorn Books, 1963.

Stanley Works Inc. Stanley Tools Division, Robert William Campbell, and N. H. Mager. *How to Work with Tools and Wood.* Newly rev. and enl. ed. New York: Pocket Books, 1965.

Stapledon, Olaf. *Star Maker.* London: Methuen, 1937.

Starrett, L. S. *The Starrett Book for Student Machinists.* Athol, MA: L. S. Starrett, 1941.

Steinhaus, Hugo. *Mathematical Snapshots.* 3rd ed. New York: Oxford University Press, 1969.

Stevens, L. Clark. *Est: The Steersman Handbook: Charts of the Coming Decade of Conflict.* Santa Barbara, CA: Capricorn Press, 1970.

Steward, Robert M. *Boatbuilding Manual.* Lakeville, CT: Poseidon, 1969.

Stewart, George Rippey, and Robert Chesley Osborn. *Not So Rich as You Think.* Boston: Houghton Mifflin, 1968.

Stone, Robert. *A Hall of Mirrors.* Boston: Houghton Mifflin, 1967.

Storer, John H. *The Web of Life: A First Book of Ecology.* New York: Devin-Adair, 1953.

Stout, Ruth. *How to Have a Green Thumb without an Aching Back: A New Method of Mulch Gardening.* A Banner Book. New York: Exposition Press, 1955.

Strauss, Rick. *How to Win Games and Influence Destiny I and II.* Tolucca Station, North Hollywood, CA: Gryphon House, 1969.

Stromberg, Loy. *Stromberg's Beginner's Falconry Bulletin.* Fort Dodge, IA: Stromberg's Chicks and Pets, n.d.

Strouse, Jean. *Up against the Law: The Legal Rights of People under Twenty-One.* A Signet Book. New York: New American Library, 1970.

Strunk, William. *The Elements of Style.* New York: Macmillan, 1959.

Study of Critical Environmental Problems and Massachusetts Institute of Technology. *Man's Impact on the Global Environment: Assessment and Recommendations for Action; Report.* Cambridge, MA: MIT Press, 1970.

Sturt, George. *The Wheelwright's Shop.* New York: Cambridge University Press, 1923.

Sun Bear. *At Home in the Wilderness.* Healdsburg, CA: Naturegraph Publishers, 1968.

Survival, Evasion and Escape. Washington, DC: US Government Printing Office, 1969.

Suzuki, Shunryåu, and Trudy Dixon. *Zen Mind, Beginner's Mind.* New York: Weatherhill, 1970.

Swatek, Paul. *The User's Guide to the Protection of the Environment.* Walden Edition. New York: Friends of the Earth/ Ballantine Books, 1970.

Sweeney, James Johnson, and José Luis Sert. *Antoni Gaudí.* Rev. ed. New York: Praeger, 1970.

Swezey, Kenneth M. *Formulas, Methods, Tips, and Data for Home and Workshop.* New York: Popular Science, 1969.

Sze, Mai-mai, and Gai Wang. *The Way of Chinese Painting: Its Ideas and Technique; With Selections from the Seventeenth-Century Mustard Seed Garden Manual of Painting.* Modern Library Paperbacks 57. New York: Random House, 1959.

Taft, J. Richard. *Understanding Foundations: Dimensions in Fund Raising.* New York: McGraw-Hill, 1967.

Tamplin, Arthur R., and John W. Gofman. *Population Control through Nuclear Pollution.* Chicago: Nelson-Hall, 1970.

Tangerman, E. J. *Whittling and Woodcarving.* New York: Whittlesey House/ McGraw-Hill, 1936.

Tanner, J. M., Gordon Rattray Taylor, and Time-Life Books. *Growth*. Rev. ed. Life Science Library. Alexandria, VA: Time-Life Books, 1970.

Tart, Charles T. *Altered States of Consciousness*. 3rd ed. San Francisco: Harper, 1990.

Taussig, Harry. *Folk Style Autoharp: An Instruction Method for Playing the Autoharp and Accompanying Folk Songs*. New York: Oak Publications, 1967.

Teilhard de Chardin, Pierre. *The Phenomenon of Man*. New York: Harper, 1959.

Templeton, George S. *Domestic Rabbit Production*. 4th ed. Danville, IL: Interstate Printers & Publishers, 1968.

Templeton, George S., and Charles Edward Kellogg. *Rabbit Production*. Rev. ed. Washington, DC: US Department of Agriculture, 1950.

Theobald, Robert. *An Alternative Future for America II: Essays and Speeches*. 2nd ed. Chicago: Swallow Press, 1970.

Theobald, Robert, and Jean M. Scott. *Teg's 1994: An Anticipation of the Near Future*. 2nd ed. Chicago: Swallow Press, 1972.

Thomas Register of American Manufacturers. New York: Thomas, 1905.

Thomas, William Leroy. *Man's Role in Changing the Face of the Earth*. Chicago: Published for the Wenner-Gren Foundation for Anthropological Research and the National Science Foundation by the University of Chicago Press, 1956.

Thompson, D'Arcy Wentworth. *On Growth and Form*. Cambridge: Cambridge University Press, 1917; 1943; abridged ed., 1961.

Thompson, Stith. *Tales of the North American Indians*. Bloomington: Indiana University Press, 1966.

Thoreau, Henry David, and Thomas Carew. *Walden; or, Life in the Woods*. Boston: Ticknor & Fields, 1854.

Time-Life Books. *The Camera*. New York: Life Library of Photography, 1970.

Time-Life Books. *Light and Film*. New York: Life Library of Photography, 1970.

Time-Life Books. *Photojournalism*. New York: Life Library of Photography, 1971.

Time-Life Books. *The Print*. New York: Life Library of Photography, 1970.

Tovmasian, G. M., and Byowrakani Astghaditaran. *Extraterrestrial Civilizations: Proceedings*. NASA TT F-438. Jerusalem: Israel Program for Scientific Translations; available from the US Department of Commerce, Clearinghouse for Federal Scientific and Technical Information, Springfield, 1967.

Townsend, Robert. *Up the Organization*. New York: Knopf, 1970.

Tremaine, Howard M. *Audio Cyclopedia*. 2nd ed. Indianapolis: H. W. Sams, 1969.

Trungpa, Chögyam. *Meditation in Action*. London: Stuart & Watkins, 1969.

Tunis, Edwin. *Colonial Craftsmen and the Beginnings of American Industry*. Cleveland: World, 1965.

Tunis, Edwin. *Frontier Living*. Cleveland: World, 1961.

Tyler, Hamilton A. *Organic Gardening without Poisons*. New York: Van Nostrand Reinhold, 1970.

Ulrey, Harry F. *Carpentry and Building*. Indianapolis: Audel, 1966.

Underhill, Evelyn. *Mysticism: A Study in the Nature and Development of Man's Spiritual Consciousness*. London: Methuen, 1911.

UNESCO. *Catalogue of Colour Reproductions of Paintings Prior to 1860*. UNESCO publication no. 629. Paris: UNESCO, 1950.

UNESCO. *Catalogue of Colour Reproductions of Paintings Prior to 1860*. 8th ed. Paris: UNESCO, 1968.

UNESCO. *Seven Hundred Science Experiments for Everyone*. Garden City, NY: Doubleday, 1958.

United Nations Secretary-General 1961– (Thant) and United Nations Department of Economic and Social Affairs. *New Sources of Energy and Energy Development: Report on the United Nations Conference on New Sources of Energy: Solar Energy, Wind Power, Geothermal Energy, Rome, 21 to 31 August, 1961*. United Nations Document E/3577/Rev 1 ST/ECA/72. New York: United Nations, 1962.

University of Michigan Architectural Research Laboratory and Stephen C. A. Paraskevopoulos. *Architectural Research on Structural Potential of Foam Plastics for Housing in Underdeveloped Areas*. 2nd ed. Ann Arbor: 1966.

Untracht, Oppi. *Metal Techniques for Craftsmen: A Basic Manual for Craftsmen on the Methods of Forming and Decorating Metals*. Garden City, NY: Doubleday, 1968.

Urban, John T. *A White Water Handbook for Canoe and Kayak*. Boston: Appalachian Mountain Club, 1969.

US Agricultural Research Service. *Home Freezing of Fruits and Vegetables*. Slightly rev. ed. Home and Garden Bulletin No. 10. Washington, DC: US Government Printing Office, 1969.

US Agricultural Research Service, Human Nutrition Research Division. *Home Canning of Fruits and Vegetables*. Slightly rev. ed. Home and Garden Bulletin No. 8. Washington, DC: US Government Printing Office, 1969.

US Agricultural Research Service, Market Quality Research Division, and James H. Beattie. *Storing Vegetables and Fruits in Basements, Cellars, Outbuildings, and Pits*. Slightly rev. ed. Home and Garden Bulletin No. 119. Washington, DC: US Government Printing Office, 1970.

US Bureau of Medicine and Surgery. *Handbook of the Hospital Corps, United States Navy, 1949*. Washington, DC: US Government Printing Office, 1949.

US Bureau of Naval Personnel. *Basic Hand Tools*. ITS Navy Training Course. Rev. ed. Washington, DC: Washington, 1963.

US Department of the Army. *Concrete and Masonry*. ITS Technical Manual 5–742. Washington, DC: Washington, 1970.

US Military Academy, Department of Earth Space and Graphic Sciences, and James L. Scovel. *Atlas of Landforms*. New York: J. Wiley, 1966.

US National Aeronautics and Space Administration. *Earth Photographs from Gemini VI through XII*. NASA SP-171. Washington, DC: for sale by the Supt. of Docs., 1968.

US National Aeronautics and Space Administration, Scientific and Technical Information Division. *Earth Photographs from Gemini III, IV, and V*. NASA SP-129. Washington, DC: for sale by the Supt. of Docs., 1967.

US Office of Civil Defense and US Public Health Service. *Family Guide to Emergency Health Care*. Washington, DC: US Government Printing Office, 1965.

US Public Health Service, Division of Environmental, Engineering, and Branch Food Protection, Special Engineering Services. *Manual of Individual Water Supply Systems*. Washington, DC: US Government Printing Office, 1962.

Valens, Evans G., and Berenice Abbott. *The Attractive Universe: Gravity and the Shape of Space*. Cleveland: World, 1969.

Vallentine, H. R. *Water in the Service of Man*. Pelican Books, A852. Baltimore: Penguin, 1967.

Van Amerongen, C., and Bibliographisches Institut (Mannheim, Germany). *The Way Things Work: An Illustrated Encyclopedia of Technology*. 2 vols. New York: Simon & Schuster, 1967.

Van der Ryn, Sim, and Jim Campe. *Farallones Scrapbook*. Point Reyes Station, CA: Farallones Star Route, 1971.

Van Valkenburgh Nooger and Neville Inc. *Basic Electricity*. 5 vols. A Rider Publication, No. 469-1-5. New York: J. F. Rider, 1954.

Van Valkenburgh Nooger and Neville Inc. *Basic Electronics: A Course of Training Developed for the United States Navy*, 6 vols. London: Technical Press, 1959.

Vogel, Virgil J. *American Indian Medicine*. The Civilization of the American Indian Series. 1st ed. Norman: University of Oklahoma Press, 1970.

Volunteers in Technical Assistance. *Village Technology Handbook*. Rev. ed. Mount Rainier, MD: Volunteers in Technical Assistance, 1970.

Von Bertalanffy, Ludwig, and Ervin Laszlo. *The Relevance of General Systems Theory: Papers Presented to Ludwig Von Bertalanffy on His Seventieth Birthday*. International Library of Systems Theory and Philosophy. New York: G. Braziller, 1972.

von Brandt, Andres. *Fish Catching Methods of the World*. London: Fishing News, 1964.

Von Foerster, Heinz, and James W. Beauchamp. *Music by Computers*. New York: J. Wiley, 1969.

Von Hilsheimer, George. *How to Live with Your Special Child: A Handbook for Behavior Change*. Washington, DC: Acropolis Books, 1970.

Vosburgh, John. *Living with Your Land: A Guide to Conservation for the City's Fringe*. Cranbrook Institute of Science Bulletin.

Bloomfield Hills, MI: Cranbrook Institute of Science, 1968.

Wagner, Edmund G., and J. N. Lanoix. *Excreta Disposal for Rural Areas and Small Communities*. World Health Organization Monograph Series. Geneva: World Health Organization, 1958.

Wagner, Edmund G., and J. N. Lanoix. *Water Supply for Rural Areas and Small Communities*. World Health Organization Monograph Series. Geneva: World Health Organization, 1959.

Wagner, Philip L. *The Human Use of the Earth*. Glencoe, IL: Free Press, 1960.

Wagner, Richard H. *Environment and Man*. New York: Norton, 1971.

Walton, Harry. *Home and Workshop Guide to Sharpening*. A Popular Science Skill Book. New York: Popular Science, 1967.

Wambold, Homer R. *Bowhunting for Deer*. Harrisburg, PA: Stackpole, 1964.

Wang, J. Y., and Raymond Balter. *A Survey of Environmental Science Organizations in the USA*. San Francisco: Ecology Center Press, 1970.

Waters, Frank, and White Bear Fredericks Oswald. *Book of the Hopi*. New York: Viking, 1963.

Watson, Aldren Auld. *Hand Bookbinding: A Manual of Instruction*. New York: Reinhold, 1963.

Watt, Bernice K., and Annabel L. Merrill. *Composition of Foods: Raw, Processed, Prepared*. US Department of Agriculture Agriculture Handbook. Rev. December 1963. Washington, DC: Consumer and Food Economics Research Division, Agricultural Research Service, US Department of Agriculture, 1964.

Watt, Kenneth E. F. *Ecology and Resource Management: A Quantitative Approach*. McGraw-Hill Publications in the Biological Sciences Series in Population Biology. New York: McGraw-Hill, 1968.

Watts, Nay Theilgaard. *Master Tree Finder: A Manual for the Identification of Trees by Their Leaves*. Berkeley: Nature Study Guild, 1963.

Watts, Tom. *Pacific Coast Tree Finder: A Pocket Manual for Identifying Pacific Coast Trees*. 2nd ed. Rochester, NY: Nature Study Guild Publishers, 2004.

Watts, Tom, and Bridget Watts. *Rocky Mountain Tree Finder: A Pocket Manual for Identifying Rocky Mountain Trees*. 2nd ed. Rochester, NY: Nature Study Guild Publishers, 2008.

Weast, Robert C. *CRC Handbook of Chemistry and Physics*. 1st student ed. Boca Raton, FL: CRC Press, 1988.

Well Drilling Operations. Army and Air Force Technical Manual. Washington, DC: US Government Printing Office, 1965.

Wells, Malcolm. *What You Can Do*. Washington, DC: World Wildlife Fund, n.d.

Wesley, Silas M. *Short-Cut Shorthand*. New York: Cowles Education, 1967.

Westbrook, Adele, and Oscar Ratti. *Aikido and the Dynamic Sphere: An Illustrated Introduction*. Rutland, VT: C. E. Tuttle, 1970.

Wheat, Margaret M. *Survival Arts of the Primitive Paiutes*. Reno: University of Nevada Press, 1967.

White, Lynn. *Dynamo and Virgin Reconsidered*. Cambridge, MA: MIT Press, 1968.

White, Minor. *Zone System Manual: Previsualization, Exposure, Development, Printing: The Ansel Adams Zone System as a Basis of Intuitive Photography*. New rev. ed. Hastings-on-Hudson, NY: Morgan & Morgan, 1968.

White, T. H. *The Goshawk*. London: Cape, 1951.

Whiteford, Andrew Hunter. *North American Indian Arts*. A Golden Science Guide. New York: Golden Press, 1970.

Whitener, Jack R. *Ferro-cement Boat Construction*. Cambridge, MD: Cornell Maritime Press, 1971.

Whyte, Lancelot Law. *Aspects of Form: A Symposium on Form in Nature and Art*. New York: Pellegrini & Cudahy, 1951.

Whyte, Lancelot Law. *The Next Development in Man*. New York: H. Holt, 1948.

Wiener, Norbert. *Cybernetics; or, Control and Communication in the Animal and the Machine*. 2nd ed. Cambridge, MA: MIT Press, 1961.

Wiener, Norbert. *The Human Use of Human Beings: Cybernetics and Society*. Boston: Houghton Mifflin, 1950.

Wilhelm, Richard, and Cary F. Baynes. *The I Ching; or, Book of Changes.* 3rd ed. Bollingen Series. Princeton, NJ: Princeton University Press, 1967.

Wilhelm, Richard, and C. G. Jung. *The Secret of the Golden Flower: A Chinese Book of Life.* 4th rev. ed. New York: Harcourt Brice &World, 1962.

Williamson, Charles Owen. *Breaking and Training the Stock Horse.* New Plymouth, ID, 1950.

Wilson, E. Bright. *An Introduction to Scientific Research.* New York: McGraw-Hill, 1952.

Wilson, Forrest. *Architecture: A Book of Projects for Young Adults.* New York: Reinhold, 1968.

Wilson, J. Douglas, and Clell M. Rogers. *Simplified Carpentry Estimating.* 4th ed. New York: Simmons-Boardman, 1948.

Wilson, John Rowan. *The Mind.* Life Science Library. New York: Time-Life Books, 1969.

Win. Rifton, NY: Win, Box 547.

Winkler, A. J. *General Viticulture.* Berkeley: University of California Press, 1962.

Wolff, Ernest N., Richard H. Byrns, and the University of Alaska Mineral Industry Research Laboratory. *Handbook for the Alaskan Prospector.* 2nd ed. College: Mineral Industry Research Laboratory, 1969.

Wolfskill, Lyle A., Wayne A. Dunlap, Bob M. Callaway, US Agency for International Development, and US Department of Housing and Urban Development, Office of International Affairs. *Handbook for Building Homes of Earth.* Washington, DC: Department of Housing and Urban Development, Office of International Affairs, 1963.

Wood, Donald G. *Space Enclosure Systems: The Variables of Packing Cell Design.* Ohio State University Engineering Experiment Station Bulletin. Columbus, OH: Engineering Experiment Station, 1968.

Wood, Marion A., and Katharine W. Harris. *Quantity Recipes.* Ithaca, NY: Cornell Home Economics Extension, 1966.

Woodman, Jim. *Air Travel Bargains.* New ed. Miami, FL: Air Travel Bargains Worldwide Guidebook, 1969.

Woodson, Thomas T. *Introduction to Engineering Design.* New York: McGraw-Hill, 1966.

Woodson, Wesley E. *Human Engineering Guide for Equipment Designers.* Berkeley: University of California Press, 1954.

Wooldridge, Dean E. *The Machinery of the Brain.* New York: McGraw-Hill, 1963.

Wright, Frank Lloyd, and Edgar Kauffman. *Writings and Buildings.* New York: Meridian Books, 1960.

Yanes, Samuel, and Cia Holdorf. *Big Rock Candy Mountain: Resources for Our Education.* New York: Delacorte Press, 1972.

Yarsley, Victor E., and Edward Gordon Couzens. *Plastics in the Modern World.* Rev. ed. Pelican Books, A1016. Harmondsworth: Penguin, 1968.

Yoshimura, Yūji, and Giovanna M. Halford. *The Japanese Art of Miniature Trees and Landscapes: Their Creation, Care, and Enjoyment.* Rutland, VT: C. E. Tuttle, 1957.

Young, J. Z. *A Model of the Brain.* The William Withering Lectures, 1960. Oxford: Clarendon Press, 1964.

Youngblood, Gene. *Expanded Cinema.* New York: Dutton, 1970.

Zablocki, Benjamin David. *The Joyful Community: An Account of the Bruderhof, a Communal Movement Now in Its Third Generation.* Pelican Books. Baltimore: Penguin, 1971.

Zheng, Manqing, and Robert W. Smith. *T'ai-Chi: The "Supreme Ultimate" Exercise for Health, Sport, and Self-Defense.* Rutland, VT: Tuttle, 1967.

Zhuangzi, and Burton Watson. *Chuang Tzu: Basic Writings.* Trans. B. Watson. New York: Columbia University Press, 1964.

Zipf, George Kingsley. *Human Behaviour and the Principle of Least Effort.* New York: Hafner, 1965.

Znamierowski, Nell. *Step-by-Step Weaving: A Complete Introduction to the Craft of Weaving, Including Photographs in Full Color.* New York: Golden Press, 1967.

Zuckerman, David W., Robert E. Horn, and Paul A. Twelker. *The Guide to Simulation Games for Education and Training.* Cambridge, MA: Information Resources, 1970.

Zwicky, F. *Discovery, Invention, Research through the Morphological Approach.* New York: Macmillan, 1969.

Index